Lecture Notes in Computer Science 579

Edited by G. Goos and J. Hartma

Advisory Board: W. Brauer   D.

S. Toueg   P. G. Spirakis   L. Kirousis (Eds.)

Distributed Algorithms

5th International Workshop, WDAG '91
Delphi, Greece, October 7–9, 1991
Proceedings

Springer-Verlag

Berlin Heidelberg New York
London Paris Tokyo
Hong Kong Barcelona
Budapest

Series Editors

Gerhard Goos
Universität Karlsruhe
Postfach 69 80
Vincenz-Priessnitz-Straße 1
W-7500 Karlsruhe, FRG

Juris Hartmanis
Department of Computer Science
Cornell University
5148 Upson Hall
Ithaca, NY 14853, USA

Volume Editors

Sam Toueg
Dept. of Computer Science, Cornell University
4106 Upson Hall, Ithaca, NY 14853, USA

Paul G. Spirakis
Dept. of Computer Science and Engineering
and Computer Technology Institute, Patras University
P. O. Box 1122, 26110 Patras, Greece

Lefteris Kirousis
Dept. of Computer Engineering and Informatics
Patras University
P. O. Box 1045, 26110 Patras, Greece

CR Subject Classification (1991): F.1, D.1.3, F.2.2, C.2.2, C.2.4, D.4.4–5

ISBN 3-540-55236-7 Springer-Verlag Berlin Heidelberg New York
ISBN 0-387-55236-7 Springer-Verlag New York Berlin Heidelberg

Typesetting: Camera ready by author
Printing and binding: Druckhaus Beltz, Hemsbach/Bergstr.
45/3140-543210 - Printed on acid-free paper

Preface

The fifth International Workshop on Distributed Algorithms (WDAG 91) took place October 7–9, 1991, in Delphi (where the oracle comes from). The workshop covers the area of distributed algorithms and is intended to provide a forum for researchers and other parties interested in distributed algorithms, communication networks and decentralized systems. The aim is to present recent research results, explore directions for future research and identify common fundamental techniques that serve as building blocks in many distributed algorithms.

The workshop followed four successful workshops in Ottawa (1985), Amsterdam (1987), Nice (1989) and Bari (1990). Since 1987 The WDAG proceedings have been published by Springer-Verlag in the Lecture Notes in Computer Science series.

The 23 papers were selected by the Program Committee from about fifty extended abstracts submitted within the announced deadline in response to a call for papers. The selection was based—first and foremost—on perceived originality and quality, but also on thematic appropriateness and topical balance. The careful screening by the Program Committee may be compared to the standard refereeing process.

It is expected that the authors will prepare extended versions of their papers to be submitted for refereed publication in one of the scientific journals. The Program Committee wishes to thank all who submitted papers for consideration.

The Program Committee consisted of:

Y. Afek (Hebrew U.)
V. Hadzilacos (U. Toronto)
J. Van Leeuwen (Utrecht U.)
N. Santoro (Cartelon U.)
P. M. B. Vitányi (CWI and U. Amsterdam)
D. Dolev (Hebrew U. and IBM Almaden)
A. Itai (Technion U.)
F. Mattern (U. Kaiserslautern)

P. Spirakis (co-chairman) (Patras U.)
M. Yung (IBM Yorktown)
C. Dwork (IBM Almaden)
S. Kutten (IBM Yorktown)
B. Sanders (ETH Zürich)
S. Toueg (co-chairman) (Cornell U.)
S. Zaks (Technion U.)

WDAG 91 was organized by the Computer Technology Institute (CTI) of Patras University, Greece and the local Arrangements Chair was Prof. L. Kirousis (Patras U.)

WDAG 91 was followed by an ALCOM activity on Distributed Computing. ALCOM stands for the project Algorithms and Complexity of ESPRIT Basic Research Action 3075 of the European Communities. The ALCOM activity took place in Delphi during the afternoon of October 9 and S. Kutten (invited speaker, IBM Yorktown) lectured on broadcasting in dynamic networks.

We wish to express our gratitude to all the members of the Program Committee for their cooperation. We would also like to thank all the referees who assisted them. The list of referees is as complete as we can and we apologise for any omissions or errors.

We gratefully acknowledge the institutions which financially supported this Conference: CTI, Techical Chamber of Greece, ALCOM Project of ESPRIT.

Last but not least, we wish to express our gratitude to the members of the Organizing Committee and to the CTI researchers who helped in many things: L. Kirousis, P. Spirakis, L. Gourdoupi, Y. Garofalakis, C. Bouras, Ph. Tsigas, V. Kapoulas. Through their dedication and effort they made the conference possible. Many thanks are due to the people of the Cultural Center of Delphi who hosted the workshop. We also thank Ms Fotini Anastasopoulou of Albatros Travel for her excellent and highly professional job in handling the organization of the conference.

Patras, December 1991   Sam Toueg
                                                          Paul Spirakis
                                                       Lefteris Kirousis

List of Referees

Afek, Y.
Attiya, H.
Ben-David, S.
Biran, O.
Burns, J.
Chandra, T.
Cohen, R.
Dolev, D.
Dwork. C.
Francet, N.
Gopal, A.
Hadzilacos, V.
Heuberger, P.

Israeli, A.
Itai, A.
Katz, S.
Kranakis, E.
Kutten, S.
Lalis, S.
Leeuwen, J. Van
Makowski, J.
Malka, Y.
Mattern, F.
Merritt, M.
Moran, S.
Papatriantafillou, M.

Reiter, M.
Sanders, B.
Santoro, N.
Spirakis, P.
Tampakas, B.
Taubenfeld, G.
Toueg, S.
Tsigas, Ph.
Vitányi, P. M. B.
Waarts, O.
Yadin I.
Yung, M.
Zaks, S.

Table Of Contents

On the limitation of the Global Time Assumption
in Distributed Systems

(Extended Abstract)

Uri Abraham*
Dept. of Mathematics and Computer Science
Ben-Gurion University
Beer-Sheva, Israel

Shai Ben-David [†]
Dept. of Computer Science
Technion
Haifa, Israel

Shlomo Moran [‡]
Dept. of Computer Science
Technion
Haifa, Israel

Abstract

An ongoing debate among theoreticians of distributed systems concerns the global time issue. The basic question seems to be to what extent does a model with global time reflect the 'real' behavior of a distributed system. The assumption of the existence of global time simplifies the analysis of distributed algorithms to an extent that makes it almost irresistible. On the other hand it should be clear that when the operations discussed are of time duration that is comparable to that of the time needed for a signal to pass between different components of the system, then such an assumption is unrealistic. The debate is of a somewhat philosophical nature mainly because, so far, there were no known examples of faulty conclusions caused by excessive use of the global time model.

In this note we demonstrate, for the first time, a protocol that is guaranteed to perform well as long as it is run in a system that enjoys the existence of global time, yet it may fail in some other conceivable circumstances. The protocol is very simple and is often used as a basic step in protocols for mutual exclusion.

By the results of [5], such a phenomenon could not have occurred had we been running our protocol on one of the three common types of shared registers (safe, regular or atomic). Another contribution of this work is the introduction of a new class of shared memory registers - 'weakly regular registers'. Our weak regularity is a natural variant on Lamport's definition of regularity of registers.

1 Introduction

The global time axiom is very appealing as it simplifies the analysis of distributed protocols. The use of the global time assumption (or some equivalent assumption) for the analysis of distributed

---
*email: abraham@bungus.ac.il
[†]email: shai@cs.technion.ac.il
[‡]email: moran@cs.technion.ac.il

protocols is so common that one can hardly find any example of work in this area that does not take this route either explicitly or implicitly (the outstanding counter example being the work of Leslie Lamport). The problem is that researchers seem to neglect the need to justify the use of this convenient simplification.

As shown in [5], the use of this axiom is equivalent to viewing all the events in a system as being executed during intervals on a mutual (time) axis. Somewhat intuitively speaking, allowing the use of a global time in the analysis of a system amounts to assuming that an interleaving model can faithfully reflect all that we care about in a run of that system. We wish to focus on the possible costs of relying on this axiom.

The difficulty with the global time assumption is that it clearly fails in some distributed environments. This failure becomes noticeable as one considers operation executions that are quick relative to the communication lags in the system. To demonstrate such a failure Lamport [10] evokes the theory of Special Relativity. Let us offer a different perspective.

When one discusses 'time' in the context of shared memory in a distributed system, the real issue is usually not so much minutes and seconds as it is the causality relations between events. We apply time mostly as a tool, or even just a metaphor, to help analyze causality. By saying " a read execution R precedes a write W" we usually wish to say "R cannot be affected by the value written by W" rather than "If R terminates at 07:00 p.m. then W should not commence before 07:01 p.m.".

Under such an interpretation the arrow relations '$A \longrightarrow B$' and '$A - - \to B$' (see section 2) should be read as "The outcome of A is available to the process executing B when the execution begins" and "A can causally affect B" (respectively). The global time axiom should now be interpreted as stating the existence of a causal relation, at least in one direction, between any pair of operations in the system.

Viewing the arrow relations this way it is not hard to imagine a distributed system in which the global time axiom fails, especially when one bears in mind that in reality any shared memory register is implemented with the aid of message passing devices (such an example is explicitly described in section 5).

It should therefore come as a surprise that such an unrealistically permissive framework has not led to faulty conclusions when applied to truly distributed systems. A partial explanation to this phenomenon is given by Theorem 10 of [5]. The theorem states that, for a wide class of protocols and specifications, if the protocols meet their specifications in every system that enjoys the existence of a global time then they are guaranteed to do so in any system complying with Lamport's general axioms (as listed in section 2 below).

In other words, although the global time axiom itself may be false, certain types of conclusions derived from it are valid. The class of protocols to which that theorem applies includes all protocols for implementing one type of Lamport's registers by another, as well as all protocols for implementing mutual exclusion that are based on registers in one of these classes.

At this point one could wonder whether all the fuss about the use of the global time assumption is of a purely philosophical nature. Maybe for all questions that we may care about, the conclusions implied by adding this axiom are already consequences of the basic properties of distributed systems.

This paper provides the first clear indication that this is not the case. We present a very simple and natural algorithm that , on one hand, is guaranteed to meet its specification whenever it is run in a system with a global time, and on the other hand, fails in some other setting. It follows, using the results of [5], that a correctness proof for that algorithm exists when the global time axiom is available but not from the set of Lamport axioms without it.

We begin our exposition by a brief presentation of Lamport's axiomatic framework. In section 3 we present a new class of registers which is the basic tool for our main example. Section 4 describes our protocols and provides the proofs of its claimed properties. We conclude by a 'concrete example' of a communication scenario where our protocol fails to achieve what it would succeed in any interleaving run.

2 A short review of Lamport's formal theory

This is a minimal outline of Lamport's theory, the reader is encouraged to consult [9, 10], for elaborate presentation and discussion.

Lamport bases his formal theory on two abstract relations over operation executions. For operation executions A, B, "$A \longrightarrow B$" stands for "A precedes B" and "$A \; -- \to B$" for "A can causally affect B".

A *System Execution* is a triple $< E, \longrightarrow, \; -- \to >$ where E is a set of *operation executions* and $\longrightarrow, \; -- \to$, are binary relations over E. Lamport offers the following axioms:

A_1: $\longrightarrow$ is an irreflexive transitive relation.

A_2: If $A \longrightarrow B$ then $A \; -- \to B$ and $\neg(B \; -- \to A)$.

A_3: $(A \longrightarrow B$ and $B \; -- \to C)$ or $(A \; -- \to B$ and $B \longrightarrow C)$ implies $A \; -- \to C$.

A_4: If $A \longrightarrow B \; -- \to C \longrightarrow D$ then $A \longrightarrow D$.

A_5: For every A, "$\{B : \neg(A \longrightarrow B)\}$" is finite. (This is assuming all events are terminating.)

An intuition for these axioms can be gained by considering the following model for it:

Let S be a partially ordered set and let E be a collection of non-empty subsets of S.
For $A, B \epsilon E$, define

- $A \longrightarrow B$ iff $\forall x \epsilon A \; \forall y \epsilon B \; (x < y)$ (in the sense of S).

- $A \; -- \to B$ iff $\exists x \epsilon A \; \exists y \epsilon B \; (x \leq y)$.

A straightforward checking shows that such models satisfy axioms A_1, A_2, A_3, A_4, and also the following axiom:

A_4^*: If $A \; -- \to B \longrightarrow C \; -- \to D$ then $A \; -- \to D$.

This last axiom was suggested by Anger[4] and by Abraham and Ben-David [2] where a completeness theorem is proved for the above mentioned class of models with respect to $\{A_1, A_2, A_3, A_4, A_4^*\}$.

An important class of models is obtained when S is a linear (=total) ordering. In such a case the system satisfies an additional axiom:

Global Time Axiom : For all A, B, $(A \; -- \to B$ or $B \longrightarrow A)$.

2.1 The register axioms

On top of the above axioms there are the register axioms that specify the communication mechanism of a system.

Any register is assumed to satisfy the following basic demands:

$B0$: The set of all write operations to a given register is linearly ordered by the $\longrightarrow$ relation.

$B1$: For any read R and any write W to the same register, $W \dashrightarrow R$ or $R \dashrightarrow W$ (or both).

$B2$: A read of a register always obtains one of the values that may be written in the register.

Lamport [10] defines three classes of registers. The classes vary in the behavior they can guarantee when a read operation from a register is executed concurrently with a write operation to the same register.

We shall quote here Lamport's informal description of the types of registers. As this work focuses on one of these types - the regular register - we shall present (in section 3) a detailed formal definition for that class.

Following is Lamport's description of the register classes [10]:

- "The weakest possibility is a *safe* register. In which it is assumed that only a read not concurrent with a write obtains the correct value - that is, the most recently written one. No assumption is made about the value obtained by a read that overlaps a write, except that it must obtain one of the possible values of the register."

- "The next stronger possibility is a *regular* register, which is safe (a read not concurrent with a write obtains the correct value) and in which a read that overlaps a write obtains either the old or the new value."

- "The final possibility is an *atomic* register, which is safe and in which reads and writes behave as if they occur in some definite order."

3 A new class of registers

We define our notion of *a weakly regular register* in terms of the two arrow relations.

Definition: *A register is* weakly regular *if any read operation from it, R, obtains a value written to it by a write operation, W, satisfying the following two conditions:*

1. $W \dashrightarrow R$.

2. *There exist no W' such that $W \longrightarrow W' \longrightarrow R$.*

Intuitively, this is just what one could understand from Lamport's informal definition of regularity. In fact, if one checks the formal definition of such registers in [10], it is easy to see that his definition amounts to demanding that the write W whose value is obtained by R satisfies:

1. $W \dashrightarrow R$.

2. There exist no W' such that $W \longrightarrow W'$ and $\neg(R \dashrightarrow W')$.

Note that the global time axiom implies the equivalence of '$A \longrightarrow B$' with '$\neg(B \dashrightarrow A)$' and therefore, assuming the existence of global time, weak regularity coincides with regularity.

The focal point of Lamport's definition is the notion of concurrency. If a write operation is concurrent with a given read then there may be a difficulty in obtaining the value of either that write or the write preceding it.

The point where we vary from his idea is the definition of such concurrency. Lamport considers '$(A \dashrightarrow B) and (B \dashrightarrow A)$' while our definition is justified by claiming that '$(A \dashrightarrow B) and \neg(A \longrightarrow B)$' suffices to indicate some concurrency. Consequently, it is conceivable to consider a register for which, as long as there is no full precedence of a read by a write, this read may still obtain the value that was in the register prior to that last write operation.

The above weaker form of concurrency has similar implications to the definitions of safe and atomic registers as well. We postpone a fuller discussion of this issue to the full version of this paper.

4   The protocol

The following protocol is for a system with two processors, each *owning* a register into which only the owner can write, we allow both processors reading access to both of the registers.

We assume both processors hold an initial value 0. The protocol for each process is as follows:

1. *Write 1 to your register.*

2. *Read the other process register.*

Consider the following statement:

* "If both the above registers are weakly regular then, upon termination of the protocol by both processes, at least one process have read the value 1"

This statement is a simple correctness claim. We shall see that it is provable on the basis of the global time axiom and yet it may fail in some system execution (in the sense of section 2).

Lemma 1 : *If the global time axiom holds then the claim * holds.*

Proof: : Let us denote by W_i^0, W_i, R_i the operations of processor i (i.e. the initial 'write 0' the 'write 1' and the 'read' respectively). For both $i = 0$ and $i = 1$ we have $W_i^0 \longrightarrow W_i \longrightarrow R_i$. Let us consider two cases:

If $R_0 \dashrightarrow W_1$ then by axiom A_4 we have $W_0 \longrightarrow R_1$ and then by the weak regularity R_1 is bound to obtain the value written by W_0 - namely 1.

On the other hand, if $\neg(R_0 \dashrightarrow W_1)$ then we apply to the global time axiom. Assuming the global time axiom weak regularity is as strong as regularity (or, if one wishes, $W_1 \longrightarrow R_0$), and therefore R_0 is bound to obtain the value written by W_1 - namely 1.

Note that we have actually shown that our protocol is correct as long as it is run on regular registers. The only role of the global time was to turn the weakly regular registers into regular ones.

Lemma 2 : *There exists an execution system for which * fails.*

Proof: : We shall use the notation of the previous lemma. Let us define the arrow relations among our operations. We assume as before $W_i^0 \longrightarrow W_i \longrightarrow R_i$ for both $i = 0$ and $i = 1$. We shall also impose a $--\rightarrow$ from every write to every read and whenever it is implied by a $\longrightarrow$ relation. No further arrow relations hold in our system. It is straightforward to check that what we've just defined complies with the axioms $\{A_1, A_2, A_3, A_4, A_4^*, A_5, B0, B1\}$ and therefore is a legitimate system execution. Still, as we have for both i's $\neg(W_i \longrightarrow R_{1-i})$, both read operations are allowed to return the values written by the W^0's without violating the weak regularity of the registers.

5 Example

We describe below a simple model in which the protocol discussed in the previous section indeed fails to satisfy statement *. Rather than settle for a formal definition of a 'System Execution' as above, we wish to demonstrate how such a model can result from some conceivable implementation of shared-memory registers by a message-passing design. This design seems to reflect realistic physical constraints, and hence our example implies that the limitation of the global time assumption demonstrated in Lemma 2 cannot be ignored in practical considerations.

We start by describing the message-passing net, in this level of granularity we do assume the existence of a global time. It may be thought of as the real physical time (we assume that differences in the physical clocks of the different processes are negligible). It is only when we pass to the shared-memory representation of the system that the global time axiom fails. As noted in the introduction, at that level of abstraction, the global time assumption should be viewed as a statement about causality relations rather than about physical time.

The system consists of two processes, P_0 and P_1, and two registers, G_0 and G_1. Register G_i is owned by process P_i. Process P_i and register G_{1-i} are *located* at location L_i. We assume that there is a global time in the system, and that each read or write operation is subject to a non-negative delay, that depends on the locations of the corresponding process and register: an operation of process at location L_i at register at location L_j is subject to delay $\delta_{i,j}$, where $\delta_{i,i} = 0$ (this model reflects a situation where a process can read from its read-only register with no delay, but operations to its read/write register requires communication time).

For simplicity, we assume that a write operation of process P_i (at location L_i) to register G_i (at location L_{1-i}) takes zero time at the location of P_i. Thus, such an operation which occurs at time t in P_1 (meaning that at this time P_1 transmit the value to be written at G_1), occurs at time $t + \delta_{1,0}$ in G_1 (meaning that at this time the value is actually written at G_1).
A read operation of P_1 from G_i ($i = 0, 1$) is subject to similar rules. Such an operation consists of transmitting a request signal from P_1 to G_i, and then transmitting the value in G_i back to P_1. When such a read operations is initiated at time t at P_1, it occurs at G_i and terminates at P_1 at the same time, $t + \delta_{1,i}$. The value read by P_1 is the one written in G_i at time $t + \delta_{1,i}$. The write and read operations of process P_0 are defined in a similar way.
We also assume, as in [10], that the registers hold some initial value, by assuming that the first write operation to each register precedes all read operations to that register. For simplicity, we assume that this initial write operation is subject to no delay.

We now turn to the shared-memory representation of the system. The operations at this level are the reads and the writes of the two participating processes. When we define the arrow relations in the system we wish to reflect the possible causality relationships that are implied by the implementation model described above.

These guidelines imply the following arrow relations: For operations A and B on the same register and which are subject to delays δ_A, δ_B respectively, let us denote by $[t_A, t_A']$ and $[t_B, t_B']$ the time intervals along which A and B occur (in the global clock of the massage-passing implementation model). Then $A \longrightarrow B$ if $t'_A + \delta_A < t_B$, and $A \; \text{-} \text{-} \rightarrow B$ if $t_A + \delta_A \leq t'_B$. As for operation executions of the same processor, we impose the demand that the $\longrightarrow$ relation induces a total order among the events of any process.

We leave to the reader the straight forward, yet crucial for our claims, checking that the system thus defined is exactly the system described in the proof of lemma 2 of the previous section (and therefore, it satisfies the axioms $\{A_1, A_2, A_3, A_4, A_4^*, A_5, B0, B1\}$ as well as the weak regularity axiom for both registers).

Figure 1 below depicts an execution of the protocol of the previous section in the system described above, in which both processes read 0, and hence statement $*$ does not hold. We assume in this example that $\delta_{1,2} = \delta_{2,1} = 2$. (i.e., a write of P_i that occurs at time t in P_i occurs at time $t + 2$ in G_i).

time	P_0	G_0	P_1	G_1
0	W_0^0	W_0^0	W_1^0	W_1^0
1	W_0		W_1	
2	R_0	R_1	R_1	R_0
3		W_0		W_1

Figure 1

For $i = 1, 2$, W_i^0 is the initial write operations that write 0 to G_i. W_i, R_i stands for write (read) operation of P_i. Each entry indicates the time and location of the operation (thus, W_0 occurs in P_0 at time 1, and in G_0 at time 3). It is easy to see that both read operations return the value 0.

We conclude this section by noting that a protocol which satisfies statement $*$ for the above specific system is obtained by forcing each process P_i to read its own register G_i after it writes to it, and only then to read the other process' register. This additional read operation prevents P_i from reading R_{1-i} before the value of R_i is 1 (in global time!), which implies that $*$ must hold. The protocol for P_i is, therefore:

1. *Write 1 to your register.*

2. *Read your own register*

3. *Read the other process register.*

References

[1] Abraham, U., "On interprocess communication and the problem of common atomic registers", manuscript, 1989.

[2] Abraham, U., and Ben-David, S., "Informal and Formal Correctness Proofs for Programs", manuscript, November 1987.

[3] Avraham, U., Ben-David, S., and Magidor, M., "On Global Time and Inter-Process Communication", Semantics for Concurrency, M.Z. Kwiatkowska, M.W. Shield, and R.M. Thomas (Eds.), Springer - Verlag, July 1990, 311-323.

[4] Anger, F. D., "On Lamport's interprocess communication model", ACM Transactions on Programming Languages and Systems, Vol. 11 No. 3, July 1989, 404-417.

[5] Ben-David, S., "The global-time assumption and semantics for concurrent systems", Proceedings of the 7th Annual ACM Symposium on Principles of Distributed Computing, ACM Press, 1988, 223-232.

[6] Ben-David, S., "On Lamport's Shared Memory Registers", in preparation.

[7] Van Benthem, J. F. A. K., " Time, Logic and computation", in Bakker, Roever and Rozenberg (Eds), Linear Time, Branching Time and partial Order in Logics and Models for Concurrency, pp.1-49, Springer, Berlin, 1989.

[8] Fishburn, P.C., "Interval orders and interval graphs", Wiley, New-York (Wiley-Interscience series in discrete mathematics), 1987.

[9] Lamport, L., " The mutual Exclusion Problem: Part I–A Theory of interprocess Communication; Part II–Statements and Solutions", J. of the A.C.M., Vol 33, No.2(1986), pp. 313-326.

[10] Lamport, L., "On Interprocess Communication, Part I: Basic formalism, Part II: Algorithms", Distributed Computing, Vol. 1(1986), pp. 77 - 101.

[11] Wiener, N., "A contribution to the theory of relative position", Proc. Camb. Philos. Soc. 17(1914), pp.441-449.

Causal Memory*

Mustaque Ahamad
James E. Burns
Phillip W. Hutto
Gil Neiger

College of Computing, Georgia Institute of Technology
Atlanta, Georgia 30332-0280 U.S.A.

Abstract

We define and implement one member of a family of *causal memories* that we have developed. Causal memories are *weakly consistent* and admit more executions, hence allowing more concurrency, than either atomic or sequentially consistent memories. Informally, causal memories require the apparent order of writes to respect a shared memory analog of Lamport's potential causality. In this paper we:

- motivate weakly consistent memories, particularly their use as distributed shared memories;

- precisely and formally characterize one causal memory;

- demonstrate that our causal memory can be effectively programmed;

- present a simple implementation using reliable, "FIFO" message passing, and prove it correct;

- present a fault-tolerant causal memory implementation that tolerates lossy channels and stopping failures and prove it correct.

1 Introduction

Lamport [21] introduced and axiomatically defined three related classes of memory called safe, regular and atomic registers. He has also defined *sequential consistency* [19], which can be viewed as another class of memory. In an earlier work [16], we showed that sequentially consistent memory is weaker than atomic memory and incomparable to safe and regular memories[1] and suggested an approach for developing new weakly consistent memories. We advocate the study and use of weakly consistent memories and believe this area represents a rich source of research that may yield insight into the nature of concurrency. Moreover, weakly consistent memories are the subject of active research in the hardware community and serve as a promising bridge between theoretical and practical work. In this paper we present the first formal study, to our knowledge, of a memory strictly weaker than sequential consistency. In this paper, we precisely define, implement and prove correct one member of a family of weakly consistent *causal memories* that we have developed.

---

*This work was supported in part by the National Science Foundation under grants CCR-8619886, CCR-8806358, CCR-8909663, and CCR-9106627.

[1]Sequentially consistent memory is weaker than multi-writer atomic memory and incomparable to both single-writer and multi-writer safe and regular memories.

Causal memories are based on Lamport's concept [18] of *potential causality*. Potential causality provides a natural ordering on events in a distributed system where processes communicate via message passing. We introduce a similar notion of causality based on reads and writes in a shared memory environment. Causal memories require that reads return values consistent with causally related reads and writes and we say that "reads respect the order of causally related writes." Since causality only partially orders events, readers may disagree on the relative ordering of concurrent writes. This provides independence between concurrent writers and greatly reduces consistency maintenance (synchronization) costs. Various interpretations of this informal definition give rise to a family of related memories; stronger definitions simplify programming while weaker definitions allow more concurrency. In this paper we present one particularly natural definition.

Weakly consistent memories are of more than theoretical interest. Many researchers [4,6,8,13,22, 27] seek to provide a shared memory abstraction across distributed hardware as an alternative to the message-passing primitives traditionally available. Such *virtual* or *distributed shared memories* (DSM) are attractive because they allow processes to access local and remote information uniformly, thereby simplifying programming. But traditional consistent shared memory can be costly to implement in a distributed environment. Most existing or proposed DSM implementations resemble multiprocessor cache coherency protocols, but the cost of communicating between processors in a distributed system can be orders of magnitude greater than in a tightly-coupled multiprocessor. Moreover, implementing consistent memory involves a classical tradeoff between memory access latency and scalability. A simple argument [23,24] can be used to show that no consistent memory can both retain low latency and scalability. This tradeoff between latency and scalability represents a significant efficiency problem since it forces applications to pay the costs of consistency even if they are highly parallel and involve little synchronization. A number of techniques [5,9,17] have been suggested to improve the efficiency of DSM implementations, but all provide only partial remedies to the fundamental problem of latency and scale for consistent memory. Designers of large multiprocessors [11,14] that strive to develop scalable, low-latency parallel computers face this same problem. Recent research [1,3,14,16,23] suggests that a weakening of memory consistency can reduce the costs of consistency maintenance while maintaining a viable "target" model for programmers.

Causal memory is closely related to the ISIS causal broadcast and, thereby, to the notion of causally ordered messages [7]. But causal memory is more than a collection of "locations" updated by causal broadcasts, and there are significant differences in the two models. First, memory has overwrite semantics and messages have queuing semantics. A message recipient can be assured that it will eventually receive all messages that have been sent, but repeated reads cannot guarantee that all values written will be read. "Hidden writes," values overwritten before they are read, are always possible. A message recipient is forced to receive all earlier messages in order to receive a later, interesting message. Since a process may read memory locations in any order it chooses, it may read a value v_1 from location x much later that a value v_2 from location y, even when the write operation that stores v_1 in x is causally before the write of v_2 to y. Such a behavior would violate causal ordering in a message system. However, causal broadcast is similar to the definition of causal memory given here in that any two causally related messages must be received in the same order at all processes.

It is far from clear that there is a "best" kind of shared memory model for use with distributed systems. By exploring the various possibilities, we hope to get a better understanding of what is feasible and what is desirable. This paper carefully examines one such possibility.

2 Shared Memory Systems

In a *distributed system with shared memory S*, processes in $\mathcal{P}$ read and write shared memory locations in $\mathcal{M}$. We assume both $\mathcal{P}$ and $\mathcal{M}$ to be finite. Each process can execute any of a number of *operations* (and the execution of every operation of S is associated with some process in $\mathcal{P}$). Two operations are externally visible: *read* and *write*. Each read or write operation by process p acts on some named location x and has an associated value v. A write operation by process p, $w_p(x)v$, associates value v with location x, while a read operation, $r_p(x)v$, reports that v is in this location. A special value, $\perp$, which cannot be written, is used to denote the value of a location before any writes have acted on it. We may omit the parameters of a read or write operation when they are not important. For example, $w(x)$ denotes a write by some process acting on location x with an unspecified value.

2.1 Partially Ordered Sets

In a distributed system, processes are usually assumed to execute asynchronously and without reference to a global clock. The operations of an execution are thus not generally ordered sequentially, but are related by some partial order. We use partially ordered sets to represent executions histories. In this subsection, we define terms and notation that will be used throughout the remainder of the paper.

We will need to define various irreflexive[2] partially ordered sets (posets) on operations. A *poset* is a pair $(S, \rightarrow)$ where S is a set and $\rightarrow$ is a relation on S (i.e., $\rightarrow \subseteq S \times S$) that is irreflexive, transitive, and antisymmetric.[3] We normally use infix notation for expressing the relation $\rightarrow$, writing $s \rightarrow s'$ rather than say, $(s, s') \in \rightarrow$.

Sometimes, we need to define a poset that uses a restricted version of a relation defined on another poset. If $\vec{S} = (S, \underset{S}{\rightarrow})$ is a poset and $T \subseteq S$, we write $\vec{T} = (T, \underset{S}{\rightarrow})$ to mean $(T, \underset{T}{\rightarrow})$, where $\underset{T}{\rightarrow}$ is the restriction of $\underset{S}{\rightarrow}$ to $T \times T$. This economy of notation allows us to avoid introducing excess relation names such as "$\underset{T}{\rightarrow}$."

For any poset $\vec{S}$ and element $s \in S$, the *prefix of s in S* is the set $\vec{S}(s) = (S(s), \underset{S}{\rightarrow}) = \{s\} \cup \{s' \in S : s' \underset{S}{\rightarrow} s\}$. Poset $\vec{S}$ has the *finite prefix property* if, for every operation $s \in S$, $\vec{S}(s)$ is finite.

A *sequential* poset $\vec{S} = (S, \underset{S}{\rightarrow})$ has the additional property ("relatedness") that for all $s, s' \in S$, if $s \neq s'$, then either $s \underset{S}{\rightarrow} s'$ or $s' \underset{S}{\rightarrow} s$. We will have occasion to concatenate sequential posets. The following lemma shows this makes sense.

Lemma 2.1: Let $\vec{S}$ and $\vec{T}$ be sequential posets such that $S \cap T = \emptyset$. Let $\underset{ST}{\rightarrow}$ be the union of $\underset{S}{\rightarrow}$, $\underset{T}{\rightarrow}$, and all pairs (s, t) such that $s \in S$ and $t \in T$. Then $(S \cup T, \underset{ST}{\rightarrow})$ is a sequential poset.

[The proof is easy, though a bit tedious, and is omitted.]

Note that the lemma does not hold, in general, if $S \cap T \neq \emptyset$. When the intersection is empty, however, the described poset does exist, and we are justified in giving it a name, $\vec{S} \cdot \vec{T} = (S \cup T, \underset{ST}{\rightarrow})$, where $\underset{ST}{\rightarrow}$ is defined as in the lemma. For convenience, we let $\vec{S} \cdot t$ where $t \notin S$ stand for $\vec{S} \cdot (\{t\}, \emptyset)$, so that we can concatenate single elements to a sequential poset.

A poset $\vec{T} = (T, \underset{T}{\rightarrow})$ *respects* poset $\vec{S} = (S, \underset{S}{\rightarrow})$ if for all $t, t' \in T \cap S$, $t \underset{S}{\rightarrow} t'$ implies $t \underset{T}{\rightarrow} t'$. If sequential poset $\vec{T}$ respects $\vec{S}$ and $S \subseteq T$, then $\vec{T}$ *sequences* $\vec{S}$, and we sometimes say $\vec{T}$ is a *sequencing* of $\vec{S}$. Note that T can have more elements than S in this case. When $T = S$ and $\vec{T}$ sequences $\vec{S}$, then $\vec{T}$ is a *topological sort* of $\vec{S}$. It is well-known that such a topological sort exists for every poset.

---

[2] A "partial order" is usually taken to be reflexive. Our definition is more convenient for our purposes. Note that a poset by our definition is equivalent to a transitive directed acyclic graph.

[3] We generally use the symbol "$\rightarrow$" for relations. Other symbols (such as $\rightsquigarrow, \Rightarrow$) are used to distinguish special relations that have specific meaning for this paper.

2.2 Execution Histories

Consider any finite or infinite execution of a distributed system with shared memory, S. We assume that every operation in the execution is distinct, so that, for example, two successive reads that happen to return the same value are distinguished. Let H be the set of operations that occur in the execution. In general, an *execution history* is just a poset $(H, \underset{H}{\rightarrow})$. We will consider several special types of execution histories.

We assume that processes are sequential (execute one operation after another). The operations of each process in H should thus be sequential. We define *process order*, $\underset{H}{\Rightarrow}$, to be the relation on operations of H defined so that $o \underset{H}{\Rightarrow} o'$ if and only if o and o' are operations of the same process and o' immediately follows o from the point of view of the process (i.e., if o is an operation of p, $o \underset{H}{\Rightarrow} o'$ represents a single step of p). Let H^p be the subset of H consisting of all operations of H. Then $(H^p, \underset{H}{\overset{+}{\Rightarrow}})$ is a sequential poset, where $\underset{H}{\overset{+}{\Rightarrow}}$ is the transitive closure of $\underset{H}{\Rightarrow}$. We refer to $\overrightarrow{H} = (H, \underset{H}{\overset{+}{\Rightarrow}})$ as the *process execution history*.

For simplicity of exposition, we assume that all writes to a location are uniquely valued (Misra [26] makes a similar assumption). For every read, $r(x)v$, there is at most one write, $w(x)v$, acting on x with value v. The *reads* relation, $\underset{H}{\mapsto}$, on H is defined so that $o \underset{H}{\mapsto} o'$ if and only if o' is a read, $o = r(x)v$, and o is the unique corresponding write, $o = w(x)v$. Note that $(H, \underset{H}{\mapsto})$ is a poset because $\underset{H}{\mapsto}$ is trivially an irreflexive partial order.

In a distributed system, operations are not instantaneous and can overlap in complex ways. A *timing* τ assigns a real time (generally not available to the processes) to the points at which an operation o is invoked, $inv_\tau(o)$, and the time at which it responds, $rsp_\tau(o)$. Timing τ is *proper* if, for all operations $o \in H$, we have $inv_\tau(o) \leq rsp_\tau(o)$. Given a proper timing, we can define the time ordering, $\underset{H^\tau}{\rightarrow}$, that relates operations in H that do not overlap in time to be $\{(s, t) : s, t \in H \text{ and } rsp_\tau(s) < inv_\tau(t)\}$. Thus, $s \underset{H^\tau}{\rightarrow} t$ means that operation s finishes before, and hence does not overlap, operation t. It is should be clear that $\underset{H^\tau}{\rightarrow}$ is an irreflexive partial order. A *timed execution history* of an implementation of a system S with timing τ is $\overrightarrow{H^\tau} = (H, \underset{H^\tau}{\rightarrow})$.

A timed execution history $\overrightarrow{H^\tau} = (H, \underset{H^\tau}{\rightarrow})$ is *well-formed* if τ is proper and

- $\overrightarrow{H^\tau}$ has the finite prefix property,

- $\overrightarrow{H^\tau}$ respects $(H, \underset{H}{\overset{+}{\Rightarrow}})$, and

- there are not operations $o, o' \in H$ such that $o \underset{H^\tau}{\rightarrow} o'$ and $o' \underset{H}{\mapsto} o$.

We assume that, for any system execution, there is a timing τ such that $\overrightarrow{H^\tau}$ is well-formed. (If not, the execution has some strange property; for example, the last part of the condition prohibits a read that finishes before the corresponding write has begun.) Note that well-formedness is related to how the set of possible executions is generated and is not something that is normally controllable by the algorithms written for the processes. We do not treat the generation of executions formally here.

Another important type of execution history relates the operations of H causally. A *single causal step* connects either a write to the corresponding read or two successive operations of the same process. Notationally, we write $\underset{H}{\overset{1}{\rightsquigarrow}}$ for the union of $\underset{H}{\mapsto}$ and $\underset{H}{\Rightarrow}$. Define the *causality relation* $\underset{H}{\rightsquigarrow}$ to be the transitive closure of $\underset{H}{\overset{1}{\rightsquigarrow}}$. (We will use the symbol "$\rightsquigarrow$" throughout to stand for causality type relations.) The *causal execution history* of an execution is $\widetilde{H} = (H, \underset{H}{\rightsquigarrow})$. It is immediate that $\widetilde{H}$ respects $(H, \underset{H}{\overset{+}{\Rightarrow}})$ and $(H, \underset{H}{\mapsto})$. Note that our assumption of the existence of a well-formed timed execution history for every execution implies that $\underset{H}{\leadsto}$ is an irreflexive partial order, as the notation indicates.

The notion of causality that we use is similar to that defined by Lamport [18]; we use write and read operations instead of send and receive actions. In Section 5, we will give a definition of "message causality," which is equivalent to Lamport's definition of potential causality.

We normally expect that a read should return the value of the "most recent" write. "Most recent" is not well-defined for an arbitrary poset, but for sequential posets we have the following definition.

> Let $\overrightarrow{H} = (H, \underset{H}{\rightarrow})$ be an execution history that is a sequential, and let o be any operation in H. If there is any write acting on x in $H(o)$, the *last write to location x before o in H* is the unique write acting on x, $w(x)$, such that there is no write $w'(x)$ in H for which $w(x) \underset{H}{\rightarrow} w'(x) \underset{H}{\rightarrow} o$. Define $last(\overrightarrow{H}(o), x)$ to be the value written by o, if o is a write to x, and otherwise to be the last value written to x before o, if defined, or $\perp$, if not.

A natural definition for a sequential execution history to be legal can now be given.

> A sequential execution history $(H, \underset{H}{\rightarrow})$ is *legal* if and only if for every read operation $r(x)v \in H$, $last(\overrightarrow{H}(r(x)v), x) = v$.

Note that the existence of a legal sequential execution history $(H, \underset{H}{\rightarrow})$ implies that for every read in H, there is a corresponding write (except for the special case of a read returning $\perp$).

2.3 Atomic Memory

An atomic memory should behave as if all operations are sequential. A timed execution history $\overrightarrow{H^t}$ satisfies *atomicity* if there exists a legal sequencing $\overrightarrow{S}$ of $\overrightarrow{H^t}$. A memory is *atomic* if for each execution there is a proper timing τ and a well-formed timed execution history $\overrightarrow{H^t}$ that satisfies atomicity.

Atomic memory requires that the order of nonoverlapping operations in H be preserved in the sequential poset $\overrightarrow{S}$. Single-writer atomic memory was formalized by Lamport [20] and the multi-writer case by Misra [26]. Herlihy and Wing's consistency condition [15] for general objects, linearizability, is also precisely equivalent to atomic memory when restricted to objects (registers) that support read and write operations.

2.4 Sequentially Consistent Memory

Sequential consistency, introduced by Lamport [19], is a weakening of atomic memory that does not require operations to "take effect" during their execution interval. Executions must "appear," however, to be the result of some sequential interleaving of the individual process histories. Note that a sequentially consistent memory is weaker than an atomic one because the former need not preserve, in the sequencing, the order of nonoverlapping operations at different processes.

A process execution history $\overrightarrow{\overrightarrow{H}}$ satisfies *sequential consistency* if there exists a legal poset $\overrightarrow{S}$ that sequences $\overrightarrow{\overrightarrow{H}}$. (Note that the legality of $\overrightarrow{S}$ also implies that $\overrightarrow{S}$ respects $\underset{H}{\mapsto}$.) A memory is *sequentially consistent* if each of its process execution histories satisfies sequential consistency. Note that the definition of sequential consistency does not refer to timed execution histories, but depends only on the definition of process order.

$$p: \quad w(x)1 \quad r(x)2$$
$$q: \quad w(x)2 \quad r(x)1$$

Figure 1: Causal but not Sequentially Consistent

2.5 Causal Memory

Atomic and sequentially consistent memories require all processes to agree on a single equivalent sequential history. Causal memory allows each process to "perceive" a different sequential history as long as these histories are consistent with causality.

In describing the other forms of memory, we implicitly assumed that processes did not fail. One of our implementations of causal memory allows processes to fail by stopping. The definition below thus refers to "correct" processes, those that do not fail in the execution. In the case that all processes are assumed to be non-failing, the definition below compares directly with the other forms of memory.

When processes may fail and messages can be dropped, some writes may be completed without any correct process knowing of them. Our definition will only require correct processes to schedule writes that are known. An operation o is *known* to process p if o is an operation of p or there is an operation o' of p such that $o \underset{H}{\leadsto} o'$. Let H^{known} be the set of all *write* operations in H that are known to some correct process. The we will require that process p schedule all of its own operations, H^p and all known writes, H^{known}. Let $H_p = H^p \cup H^{known}$.

We now define correctness for an implementation of causal memory. A causal execution history $\widetilde{H}$ *satisfies causality* if, for each process $p \in \mathcal{P}$, there is a legal poset $\overrightarrow{S_p} = (S_p, \underset{S_p}{\rightarrow})$ such that $S_p \subseteq H$ and $\overrightarrow{S_p}$ sequences $(H_p, \underset{H}{\leadsto})$ and respects $\widetilde{H}$. A memory is *causal* if each of its causal execution histories satisfies causality.

The definition of causal requires that all known writes be scheduled. But a write does not become known until its location is read (or a causally later write is read). In general, we might an algorithm to schedule writes as they are learned about (say from a message), rather than waiting until they are read. This "early scheduling" is allowed by the definition because $\overrightarrow{S_p}$ is not a topological sort of $(H_p, \underset{H}{\leadsto})$ and so can contain additional operations. Note that these additional operations must be from H, and they must still be ordered with respect to $\underset{H}{\leadsto}$.

Causal memory derives its flexibility by allowing each process a different sequential poset S_p and by eliminating reads of other processes from that history. In effect, concurrent writes may appear to occur in different orders to different processes. Figure 1 shows a correct execution on causal memory that is not sequentially consistent. The sequential posets demonstrating that the execution satisfies causality are $\overrightarrow{S_p}$ defined so that $w(x)1 \underset{S_p}{\rightarrow} w(x)2 \underset{S_p}{\rightarrow} r(x)2$ and $\overrightarrow{S_q}$ defined by $w(x)2 \underset{S_q}{\rightarrow} w(x)1 \underset{S_q}{\rightarrow} r(x)1$.

This definition of causal memory implies a liveness requirement. Since $\overrightarrow{S_p}$ must contain the writes of all correct processes in H (because each is trivially known to a correct process), each such write must appear in some finite prefix of $\overrightarrow{S_p}$. Thus, the "effects" of the writes of correct processes cannot be indefinitely delayed.

Other forms of weakly consistent memory can be defined using this formalism including *PRAM* (Pipelined RAM) defined by Lipton and Sandberg [23] and *slow memory* given by Hutto and Ahamad [16].

```
COORDINATOR                          WORKER_i
while (¬done)                        while (¬done)
   wait until (∀i complete_i)           t_i := (b_i − Σ_{j=1}^{i−1} a_{i,j} x_j − Σ_{j=i+1}^{n} a_{i,j} x_j) / a_{i,i}
   foreach i                            complete_i := true
      complete_i := false                  wait until (¬complete_i)
   wait until (∀i changed_i)           x_i := t_i
   if (converged())                    changed_i := true
      done := true                        wait until (¬changed_i)
   foreach i
      changed_i := false
```

Figure 2: Synchronous Iterative Linear Solver on Causal Memory

3 Programming with Causal Memory

Causal memory can be easily programmed. We show a synchronous iterative linear equation solver that works correctly on atomic memory and show it correct on causal memory as well.

Consider a parallel iterative algorithm that solves $Ax = b$, where A is a known matrix, b is a known vector, and x is the vector that is the solution. We use x_i^{k+1} to represent the value of the ith component of x in phase $k + 1$: $x_i^{k+1} = \left(b_i − \sum_{j=1}^{i-1} a_{i,j} x_j^k − \sum_{j=i+1}^{n} a_{i,j} x_j^k\right) / a_{i,i}$. Computing x_i^{k+1} requires access to all x_j^k ($i \neq j$) from the previous iteration.

Figure 2 shows the solver. All booleans are initially *false* and "wait until (B)" means "while $(\neg B)$ *skip*". Worker$_i$ computes x_i. The coordinator tests for convergence and synchronizes each Worker$_i$ twice per iteration using a barrier technique: before reading the various x_j from phase k and before writing x_i for phase $k + 1$. (By having workers read alternately from x_i and t_i, we could eliminate the first synchronization.)

The protocol in Figure 2 is easily shown to be correct on atomic memory. To show that the protocol is correct on causal memory, we argue that the reads of each x_i in phase k return the value written in phase $k − 1$, just as with atomic memory. A similar argument holds for the handshake bits (*complete_i* and *changed_i*). Let $w_i^k(x_i)v$ denote a write of x_i by p_i in phase k. Consider the causal relations established between p_j's write of x_j in phase $k − 1$ and p_i's read of x_j in phase k in some execution H of the code in Figure 2. (Subscript 'c' denotes the coordinator.)

$$(1) \quad w_j^{k-1}(x_j)v \underset{H}{\Rightarrow} w_j^{k-1}(changed_j)true \underset{H}{\mapsto} r_c^{k-1}(changed_j)true$$

$$(2) \quad r_c^{k-1}(changed_j)true \underset{H}{\overset{+}{\Rightarrow}} w_c^{k-1}(changed_i)false$$

$$(3) \quad w_c^{k-1}(changed_i)false \underset{H}{\mapsto} r_i^{k-1}(changed_i)false \underset{H}{\overset{+}{\Rightarrow}} r_i^k(x_j)v$$

Above, (1) holds because the two writes are consecutive operations of p_j and the coordinator eventually notices that p_j has copied t_j to x_j. (2) is the key to the argument and holds because the coordinator must read *changed* true from *all* workers (j in particular) before setting *changed* false for all workers (i in particular) and initiating the next phase. (3) holds because p_i must eventually see *changed_i* false and begin the next phase. At the beginning of phase k, p_i reads all components of x (except x_i) and, in particular, x_j. Taken together (1)–(3) imply that $w_j^{k-1}(x_j)v \underset{H}{\leadsto} r_i^k(x_j)v$. Since p_j is the only process to write to x_j, and since all prior writes to x_j by p_j causally precede $w_j^{k-1}(x_j)v$, and since all subsequent writes to x_j by p_j causally follow $w_j^{k-1}(x_j)v$, p_i's phase k read of x_j *must* return p_j's phase $k − 1$ write. Since we assumed an arbitrary i and j this argument shows that all reads of x in the computation return the value computed in the previous iteration, the same value returned when the computation is executed on atomic memory.

4   The Simple Algorithm

This section presents a simple and efficient implementation of causal memory using message passing primitives. We then prove its correctness with respect to the formal definition introduced in Section 2. This implementation uses an adaptation of *vector timestamps* [12,25]. It requires ordered ("FIFO") channels and is not fault-tolerant. That is, the implementation assumes no processes fail and that all messages sent are delivered exactly once and in the order sent. Since all process are correct, all writes are known, so for this section, H^{known} is the set of all writes. In the next section, we present a more complex implementation that tolerates stopping failures and lossy communication channels.

In our implementation, each process p maintains and accesses a local (private) copy M of the abstract shared causal memory $\mathcal{M}$. A process reads from memory simply by reading from this copy of memory. In addition to M, process p maintains a vector clock t, which it uses to timestamp outgoing messages. This is a vector of natural numbers, one for each process in the system. Informally, $t[q]$ is the number of writes by q of which p is aware. Two vectors can be compared by comparing their components. Vector t_1 is *less than or equal* to t_2 ($t_1 \leq t_2$) if each of t_1's components is less than or equal to t_2's corresponding component; t_1 is *less than* t_2 ($t_1 < t_2$) if it is less than or equal to t_2 and is not equal to t_2. Before writing a value v to location x, a process increments its own component of its vector clock t. It then writes to its local copy of memory and informs the other processes of its write by sending them $\langle x, v, t \rangle$. Such a message is called a *write message* and has *timestamp* t. We sometimes informally talk of a process receiving a write instead of a write message. Upon receiving a write message, a process updates the component of its vector clock that corresponds to the sending process. The vector timestamps on these messages are used to determine which writes can be applied to the process's memory and which must be left pending.

The implementation is shown in Figure 3. It consists of an initialization routine and three basic operations. The *read* operation is executed whenever a read of a location x by p is invoked. The value stored in $M[x]$ is returned to p. The *write* operation is executed whenever a write to a location x by p is invoked. Process p increments $t[p]$, writes to the value $M[x]$, and then sends an appropriate write message to all other processes. The *receive* operation is executed when a process p receives such a message. Process p first updates the appropriate component of its logical clock; recall that the component for q counts the number of q's writes of which p is aware. The message is then added to p's set of pending writes. A write message must remain pending as long as its vector timestamp reflects some write of which p is not aware. This can be detected by comparing p's vector clock to the timestamp of the write; the write is processed only if its timestamp is less than or equal to p's vector clock. Each time a write message is received, all pending messages are checked; those that can be processed are applied to memory in an order consistent with their timestamps. Note that the implementation admits the execution shown in Figure 1. Thus, our implementation is neither atomic nor sequentially consistent.

In order to prove the correctness of the implementation, we introduce the following notation: if o is an operation of a process q, let $ts(o)$, the *timestamp* of o, be the value of q's vector clock immediately after o completes. Note that for a write operation o, $ts(o)$ is the same as the timestamp included in the write message.

The following two lemmas are used in the proof of correctness. The first asserts that the causal relation $\underset{H}{\leadsto}$ is reflected in vector timestamps:

Lemma 4.1: *Suppose that $\overrightarrow{H} = (H, \underset{H}{\leadsto})$ is a causal execution history of the implementation and let o and o' be a two operations such that $o \underset{H}{\leadsto} o'$. Then $ts(o) \leq ts(o')$. Furthermore, if o' is a write operation, then $ts(o) < ts(o')$.*

Proof: Since $o \underset{H}{\leadsto} o'$, there is some shortest list of operations $o = o_0, o_1, ..., o_k = o'$, where $k > 0$ and the operations are connected by single causal steps, i.e., $o_i \underset{H}{\overset{1}{\to}} o_{i+1}$ for $0 \leq i < k$. We continue by induction on k.

```
/* Initialization: */

    foreach x ∈ M do
        M[x] := ⊥
    foreach q ∈ P do
        t[q] := 0
    pending := ∅

/* Read action: to read from x */

    return(M[x])

/* Write action: for p to write v to x */

    t[p] := t[p] + 1
    M[x] := v
    send ⟨x, v, t⟩ to all others

/* Receive action: upon receipt of ⟨x_q, v_q, t_q⟩ from q */

    t[q] := t_q[q]
    pending := pending ∪ {⟨x_q, v_q, t_q⟩}
    foreach ⟨x, v, s⟩ ∈ pending do
        if s ≤ t then
            pending := pending − {⟨x, v, s⟩}
            process := process ∪ {⟨x, v, s⟩}
    Let T⃗ be a topological sort of process by vector timestamp

    foreach ⟨x, v, s⟩ ∈ process do   /* In order by T⃗ */
        M[x] := v         /* This applies the write */
        remove ⟨x, v, s⟩ from process
```

Figure 3: Simple Implementation of Causal Memory

For the basis, assume $k = 1$. Then either $o \underset{H}{\Rightarrow} o'$ or $o \underset{H}{\mapsto} o'$. If $o \underset{H}{\Rightarrow} o'$, then o and o' are operations of the same process p. Since no process ever decrements any component of its vector clock, $ts(o)$ must be less than or equal to $ts(o')$. Furthermore, if o' is a write operation, then p increments its local component before executing the write, so $ts(o) < ts(o')$.

If $o \underset{H}{\mapsto} o'$, then o is a write operation, say $w_q(x)v$, and o' is a corresponding read, say $r_p(x)v$. Note that the write message associated with o includes the timestamp $ts(o)$. It is clear that p cannot read v from x until p has applied the write to its memory. Process p will not apply the write until its timestamp is greater than or equal to $ts(o)$. Since no component of p's timestamp is ever decremented, it is still greater than or equal to $ts(o)$ when it reads v, so $ts(o) \leq ts(o')$.

Now assume $k > 1$ and the lemma holds for lists shorter than k in length. Then $ts(o) \leq ts(o_{k-1})$ and $ts(o_{k-1}) \leq ts(o')$ by the induction hypothesis, so $ts(o) \leq ts(o')$. If o' is a write operation, then o_{k-1} is an operation of p, so $ts(o_{k-1}) < ts(o')$ by the induction hypothesis and hence $ts(o) < ts(o')$, as required.

□

The next lemma is used to show the liveness of the implementation:

Lemma 4.2: *Let $\overrightarrow{H}$ be a causal execution history of the implementation and suppose that o is a write operation of process p. Then each process q eventually applies o to its memory.*

Proof: Suppose that $ts(o) = t$. First, for arbitrary processes p and q, write o by p will eventually be received (via a message) by q, since message passing is reliable. Further, since timestamps are finite vectors of natural numbers and every write is assigned a distinct timestamp, there can only be a finite number of writes that precede o, that is, that have timestamps less than t. By the argument above, all of these writes will eventually be received (in FIFO order from the senders) by q. Once all writes preceding o are received by q, these writes, including o, are applied to q's memory in some topological order. □

We can now assert the correctness of the implementation:

Theorem 4.3: *Let $\widetilde{H}$ be a causal execution history of the implementation. Then $\widetilde{H}$ satisfies causality.*

Proof: To prove this theorem, it is necessary to show that for each process p there is a legal sequential poset $\overrightarrow{S_p} = (S_p, \underset{S_p}{\rightarrow})$ with $H_p \subseteq S_p \subseteq H$ such that $\overrightarrow{S_p}$ respects $\widetilde{H}$.

The sequential poset $\overrightarrow{S_p}$ for p is generated by simply concatenating all writes as they are applied to p's memory and all reads as they occur. $\overrightarrow{S_p}$ is a validly defined by Lemma 2.1 because each write operation has exactly one message and hence is concatenated just once. By Lemma 4.2, $\overrightarrow{S_p}$ includes all write operations in H, and thus all of H_p. $\overrightarrow{S_p}$ is clearly legal because all reads and writes apply directly to p's copy of memory and each read thus reads the value most recently written. It remains to be seen that $\overrightarrow{S_p}$ respects $\widetilde{H}$.

Let o_1 and o_2 be two operations in S_p such that $o_1 \underset{H}{\leadsto} o_2$. One of the following five cases must hold:

- Both o_1 and o_2 are operations by p. Note $o_1 \underset{H}{\leadsto} o_2$ implies that o_1 is applied by p before o_2. Since p's operations are always scheduled immediately when they are applied, $o_1 \underset{S_p}{\rightarrow} o_2$.

- o_1 is a write by another process (call it q) and o_2 is a read operation by p. By Lemma 4.1, $ts(o_1) \leq ts(o_2)$. In particular, $ts(o_1)[q] \leq ts(o_2)[q]$, so since messages are received from q in FIFO order, p receives o_1 before o_2 takes place. Consider the time at which p sets its vector timestamp to $ts(o_2)$ (this happens during some receive occurring *before* o_2 executes). The write o_1 is applied no later than this time and thus is sequenced before o_2, and hence $o_1 \underset{S_p}{\rightarrow} o_2$.

- o_1 is a write by another process and o_2 is a write operation by p. By the definition of $\underset{H}{\leadsto}$, this can only be the case if there is a read operation o' by p such that $o_1 \underset{H}{\leadsto} o' \underset{H}{\leadsto} o_2$. By applying the two cases above, we conclude that $o_1 \underset{S_p}{\rightarrow} o' \underset{S_p}{\rightarrow} s_2$, so $o_1 \underset{S_p}{\rightarrow} o_2$.

- o_1 is an operation by p and o_2 is a write by another process. By Lemma 4.1, $ts(o_1) < ts(o_2)$. Since the value of p's vector clock is nondecreasing, when o_1 occurs, o_2 cannot have been applied by p because o_2's timestamp is too high. Thus, o_2 is sequenced after o_1, so $o_1 \underset{S_p}{\rightarrow} o_2$.

- o_1 and o_2 are both writes by processes other than p. By Lemma 4.1, $ts(o_1) < ts(o_2)$. Consider the point at which p moves o_2 from *pending* to *process*. Let r be the process writing o_1. Since $ts(o_1)[r] \leq ts(o_2)[r] \leq t[r]$, where t is p's vector clock value at this point, the message containing

o_1 has been received. If o_1 has already been applied at this point, then $o_1 \underset{S_p}{\rightarrow} o_2$. If o_1 has not been applied at this point, then, because $ts(o_1) < ts(o_2)$, it must also be in *process*. Since $ts(o_1) < ts(o_2)$, p applied o_1 before o_2 (the topological sort preserves timestamp order). Thus, $o_1 \underset{S_p}{\rightarrow} o_2$.

In all cases, $o_1 \underset{S_p}{\rightarrow} o_2$, so the proof is complete. $\qquad\qquad\qquad\qquad\qquad\qquad\qquad\square$

5 The Robust Algorithm

In this section we present a fault-tolerant implementation of causal memory. We call this implementation the Robust algorithm to distinguish it from the Simple (but fragile) algorithm presented in the last section. The Robust algorithm is very fault-tolerant but requires large messages (the message size is $O(mn^2)$ bits, where $m = |\mathcal{M}|$, $n = |\mathcal{P}|$, and we assume memory values and the components of vector timestamp need only a constant number of bits each). The Robust algorithm tolerates unreliable channels (lost, duplicate or out-of-order messages) and correct processes may continue execution despite up to $n - 1$ stopping failures. Processes never block on reads or writes, so the Robust algorithm is wait-free. While the algorithm potentially requires large messages, processes may send and receive according to any live schedule. Thus, sends and receives may be scheduled to the advantage of the process and of the system.

5.1 Causal Memory with Failures

We consider computations in which process might fail by stopping (without announcement). A process *fails* in a computation if it takes only a finite number of steps (we do not consider any other form of process failure).

In addition to process failures, we tolerate failures in the message passing system. Messages may be reordered, duplicated or dropped (but not corrupted). We do require the following liveness condition on channels:

> If process p sends an infinite number of distinct messages to q, then an infinite number of distinct messages are delivered to q, assuming q attempts to receive an infinite number of times.

Note that this does not prohibit the channel from dropping an infinite number of messages. The restriction on distinct messages prohibits the channel from repeating a fixed message an infinite number of times to satisfy the infinite delivery requirement. We assume that the send and receive operations of our algorithms to continue to execute forever, even when no more reads or writes are required.

5.2 Description of the Algorithm

As in the Simple algorithm, each process maintains a complete local copy of the shared causal memory M and a vector timestamp t. No *process* or *pending* sets are used, but the Robust algorithm keeps a set of location name, value, vector timestamp triples: $Z = \{\langle x_1, v_1, t_1 \rangle, \langle x_2, v_2, t_2 \rangle, \ldots, \langle x_k, v_k, t_k \rangle\}$. The restriction $Z[x]$ of Z contains all the triples in Z with first element x. (Essentially, $Z[x]$ holds all the values written to x by the most recent concurrent writes the process has learned about.) We will show that $Z[x]$ contains at most n triples, where n is the number of processes. When a process receives a message (containing the Z of some other process), it updates its copy of Z and modifies M appropriately. We will show that this implements causal memory correctly.

Besides initialization, there are four basic operations: read, write, send and receive. A read operation, as for the Simple algorithm, simply returns the value of the appropriate location from

/* Initialization: */

$Z := \emptyset$
foreach $x \in \mathcal{M}$ do
   $M[x] := \perp$
foreach $q \in \mathcal{P}$ do
   $t[q] := 0$

/* Read action: to read from x: */

return($M[x]$)

/* Write action: for p to write v to x: */

$t[p] := t[p] + 1$
$M[x] := v$
/* Delete all triples in Z covered by $\langle x, v, t \rangle$ */
$Z := (Z - Z[x]) \cup \{\langle x, v, t \rangle\}$

/* Send action: by any live schedule */

send $\langle Z, t \rangle$ to all others

/* Receive action: by any live schedule, on receipt of $\langle Z_q, t_q \rangle$ from q */

foreach $r \in \mathcal{P}$ do
   $t[r] := \max(t[r], t_q[r])$
/* Delete all triples in Z_q dominated by Z */
$Z'_q := Z_q - \{\langle x_q, v_q, s_q \rangle \in Z_q : \exists \langle x, v, s \rangle \in Z$ such that $s_q \leq s\}$
/* Choose some fixed topological sort, $\overrightarrow{T}$, of Z'_q by timestamps */
foreach $x \in \mathcal{M}$ do
   if $Z'_q[x] \neq \emptyset$ then
      foreach $\langle x, v, s \rangle \in Z[x]$do   /* In order by $\overrightarrow{T}$ */
         $M[x] := v$ /* This applies the write */
/* Delete triples in Z covered by Z'_q */
$Z' := Z - \{\langle x, v, s \rangle \in Z : \exists \langle x, v_q, s_q \rangle \in Z'_q$ such that $s < s_q\}$
$Z := Z'_q \cup Z'$

Figure 4: Robust Implementation of Causal Memory

M, the local copy of memory. The write operation is almost as simple, incrementing the process's own component of t (the local copy of the vector timestamp) and storing the value in M. Then Z is updated, which will be explained below. Notice that a write does not execute a send (as in the Simple algorithm). Intuitively, this is because message delivery is not guaranteed; a message sent by a single send operation associated with a write might never be received. Instead, there is a separate send operation that sends the local Z value and timestamp to all other processes whenever it is invoked. We make no requirements on when sends should be invoked except that a process must continue to send (i.e., a correct process will send an infinite number of times).

Before describing the purpose of the Z set, we need some terminology. Each triple is originally created by a write operation that writes the given value to the given location and has the stated timestamp. Thus, we can speak of the write operation that corresponds to each triple in Z. Informally, we use Z to refer to the set of operations associated with Z and sometimes refer to triples as if they were the corresponding write.

As in the Simple algorithm, each operation o has a timestamp $ts(o)$, which is the local value of t of the process executing o when the operation is completed. For any operations o and o', we say o *dominates* o' if $ts(o') \leq ts(o)$. We also apply the term "dominates" to timestamps themselves and say a timestamp t dominates an operation o when $ts(o) \leq t$. A set of operations O dominates set O' if every element of O' is dominated by some element of O. If o and o' are writes to the same location and o dominates o', then we say that write o *covers* write o'. We extend the cover relation to sets of operations as follows: a set of operations O covers set O' if for every write operation $o' \in O'$ there is a write operation $o \in O$ such that o covers o'.

In the Simple algorithm, writes were accumulated locally until they could be applied. Effectively, the Robust algorithm applies writes as soon as it learns of them from a message. Informally, we say a process *learns about a write* w when it first receives a message that contains an operation dominating w. We say a write has been *scheduled* when it has been effectively applied. By "effectively applied" we mean the action of the write has been taken into account, even though the write itself might not be applied (because a covering write is actually applied). The important thing is that a write will never be applied after it has been effectively applied. Informally, the invariant we want to maintain is that, just after operation o by process p is executed, the following holds:

- local timestamp t dominates all operations p has learned about,

- all the writes that p has learned about have been scheduled, and

- the Z set covers all writes that p has learned about.

We now return to the discussion of how a write updates Z. Consider a write operation by p acting on location x. All triples corresponding to writes to x are deleted from p's Z set and then the current write triple is added. Since the current timestamp is strictly greater than all operations that p has learned about, it is clear that the new write covers all the operations that were deleted and, by transitivity, all writes to x that p has learned about. Writes to other locations in Z are not affected. Thus, Z still covers all writes that p has learned about.

We now consider the most complicated operation of the algorithm, the receive. A receive message has the form $\langle Z_q, t_q \rangle$. The first part of the receive operation updates t to the componentwise maximum of the previous value of t and t_q. Since the previous value of t dominates all operations that p learned about previously and t_q dominates all those covered by Z_q, the new value of t clearly dominates all the writes that p has learned about now (including those in Z_q itself). Thus, the first condition of the invariant is obtained.

The next action of the receive operation sets Z'_q to Z_q, minus those operations dominated by Z. Thus, Z'_q dominates all the new writes about which p is to learn but does not contain any writes that p has already scheduled. As will be shown in the proof, the topological sort $\overrightarrow{T}$ (treated as a poset on the writes corresponding to the triples of Z'_q) respects $\overleftrightarrow{H}$, so it gives a valid order in which to apply the new writes. The proof will show that $\overrightarrow{T}$ can be extended to include new writes that are dominated by Z'_q but are not explicitly represented in Z'_q. Thus, in applying $\overrightarrow{T}$, we achieve the second condition of the invariant and effectively apply all writes p has now learned about. Before Z'_q is merged with the current value of the Z set, triples in Z that are dominated by Z'_q are deleted. This is purely an efficiency step and is not required for correctness. Since the reduced set still dominates all writes p has now learned about, the final condition of the invariant is maintained.

5.3 Some New Notation

To prove the correctness of the Robust algorithm we require some new notation. In addition to the read and write operations described previously, execution histories may also contain *send* and *receive* operations. A send operation by process p, $send_{p,q}\langle Z,t\rangle$, initiates a transfer of the message $\langle Z,t\rangle$ from process p to process q while a receive operation by process q, $receive_{p,q}\langle Z,t\rangle$, completes transfer of the message $\langle Z,t\rangle$ from process q to process p. We omit the message contents and the secondary process where appropriate. For example, $send_{p,q}\langle Z,t\rangle$, $send_{p,q}$, and $send_p$ may all denote the same send operation but with decreasing specificity.

We have previously defined the causality relation $\underset{H}{\leadsto}$ on the operations in an execution history H as the transitive closure of the process order relation $\underset{H}{\Rightarrow}$ and the reads $\underset{H}{\mapsto}$ relation. In analogy to the reads relation, we define the *receives* relation on sends and receives in H. As with writes in H, we assume all send operations are unique. The receives relation pairs each receive operation $o_R = receive_{p,q}$ in H with the unique send operation $o_S = send_{q,p}$, also in H, from which it receives, and we write $o_S \underset{H}{\hookrightarrow} o_R$. Similarly to the definition of single step causality ($\underset{H}{\overset{1}{\leadsto}}$), we define single step message causality to be $\underset{H^m}{\overset{1}{\leadsto}}$, the union of $\underset{H}{\hookrightarrow}$ and $\underset{H}{\Rightarrow}$. *Message causality* is then, $\underset{H^m}{\leadsto}$, the transitive closure of $\underset{H^m}{\overset{1}{\leadsto}}$. The *message-causal execution history* is $\widetilde{H^m} = (H, \underset{H^m}{\leadsto})$.

According to the definition of prefix, for any operation o of process p, $\widetilde{H^m}(o) = (H^m(o), \underset{H^m}{\leadsto})$. Note that the writes in $H^m(o)$ are exactly the writes that p has learned about after executing o. Define $H^m_p(o)$ to be all operations of p and all writes in $H^m(o)$. Then let $\widetilde{H^m_p}(o) = (H^m_p(o), \underset{H^m}{\leadsto})$. For any operation o of process p in H, we will define a legal poset $\overrightarrow{S^m_p}(o)$, which is a topological sort of $\widetilde{H^m_p}(o)$. Taking the limit of $\overrightarrow{S^m_p}(o)$ over all the operations of p will give a sequential poset $\overrightarrow{S^m_p}$, which we will show sequences $(H_p, \underset{H^m}{\leadsto})$. From this, we can show $\overrightarrow{S^m_p}$ sequences $(H_p, \underset{H}{\leadsto})$.

The *memory projection* $M(\overrightarrow{S^m_p}(o))$ of $\overrightarrow{S^m_p}(o)$ is the memory defined by $M(\overrightarrow{S^m_p}(o))[x] = last(\overrightarrow{S^m_p}(o), x)$ for each x. This relates the sequential poset $\overrightarrow{S^m_p}$ to a state of memory. We refer to process p's copy of memory generally as M_p. We use the notation $M_\perp$ to refer to the initial state of memory. Initially, $M_p = M_\perp$ for all processes p. For any operation o of p, $M_p(o)$ is the state of p's copy of memory immediately after operation o is executed. We will show $M(\overrightarrow{S^m_p}(o)) = M_p(o)$.

Define $O(Z)$ to the the set of write operations corresponding to the triples in Z. For each process p and operation o of p, let $Z_p(o)$ be $O(Z)$, where Z is the value of p's Z set immediately after operation o. Initially, $Z_p = \emptyset$. We will show that for all operations o, $S^m_p(o)$ covers $Z_p(o)$.

5.4 Correctness

To prove that the Robust algorithm correctly implements a causal memory we must show that it only admits causal execution histories that satisfy causality. In particular, for each execution history $\widetilde{H}$ allowed by the Robust algorithm and for every correct process p there exists a legal poset $\overrightarrow{S_p}$ that sequences $(H_p, \underset{H}{\leadsto})$ and that respects $\overset{\leadsto}{H}$. Recall that H_p includes all operations of p and all writes known to correct processes.

In the proof of the critical lemma (5.3), we will need a slightly stronger fact about timestamps than we had in Lemma 4.1 for the Simple algorithm. In particular, we will need to know that the partial order on writes based on timestamps is identical to the partial order given by $\underset{H^m}{\leadsto}$. Recall that the timestamp of any operation is the value of the timestamp of the executing process immediately after the operation completes.

Lemma 5.1: *Suppose that $\overset{\leadsto}{H^m}$ is a message-causal execution history of the Robust algorithm and let o and o' be any two operations.*

- *If $o \underset{H^m}{\leadsto} o'$, then $ts(o) \leq ts(o')$. Furthermore, if o' is a write operation, then $o \underset{H^m}{\leadsto} o'$ implies $ts(o) < ts(o')$.*

- *If o is a write operation and $o \neq o'$, then $ts(o) \leq ts(o')$ implies $o \underset{H^m}{\leadsto} o'$.*

Proof:

- Assume $o \underset{H^m}{\leadsto} o'$. Since $o \underset{H^m}{\leadsto} o'$, there is some shortest list of operations $o = o_0, o_1, ..., o_k = o'$ for $k > 0$ such that $o_i \underset{H^m}{\overset{1}{\leadsto}} o_{i+1}$ for $0 \leq i < k$. We continue by induction on k.

  For the basis, assume $k = 1$. Then either $o \underset{H}{\Rightarrow} o'$ or $o \underset{H}{\hookrightarrow} o'$. If $o \underset{H}{\Rightarrow} o'$, then o and o' are operations of the same process p. Since no process ever decrements any component of its vector clock, $ts(o)$ must be less than or equal to $ts(o')$. Furthermore, if o' is a write operation, then p increments its local component before executing the write, so $ts(o) < ts(o')$.

  If $o \underset{H}{\hookrightarrow} o'$, then o is a send operation by some process q, and o' is receive operation by some process p, $p \neq q$. Since the algorithm sets p's value of t to the componentwise maximum of its current value and the value of $t_q = ts(o)$ of q (in the received message), clearly $ts(o) \leq ts(o')$.

  Now assume $k > 1$ and the lemma holds for lists shorter than k in length. Then $ts(o) \leq ts(o_{k-1})$ and $ts(o_{k-1}) \leq ts(o')$ by the induction hypothesis, so $ts(o) \leq ts(o')$. If o' is a write operation, then o_{k-1} is an operation of p, so $ts(o_{k-1}) < ts(o')$ by the induction hypothesis and hence $ts(o) < ts(o')$, as required.

- Assume o is a write operation, $o \neq o'$, and that $ts(o) \leq ts(o')$. Let p be the process executing write o. Let O be the set of operations including o' and all $o_1 \in H$ such that $ts(o)[p] \leq ts(o_1)[p]$ and $o_1 \underset{H^m}{\leadsto} o'$. (The notation $ts(o)[p]$ refers to p's component of $ts(o)$.) Since O is finite (by the finite prefix property) and nonempty, there is some $o_1 \in O$ such that, for all $o_2 \in H$, $o_2 \underset{H^m}{\overset{1}{\leadsto}} o_1$ implies $ts(o_2)[p] < ts(o_1)[p]$. Let o_1 be such an operation, and let the process executing o_1 be q.

  Suppose $q \neq p$. Note that o_1 is not the first operation of q because then $ts(o_1)[p] = 0$, while $ts(o)[p]$ is at least 1. Thus, there is an operation o_2 of process q such that $o_2 \underset{H}{\Rightarrow} o_1$. By the choice of o_1, $ts(o_2)[p] < ts(o)[p] \leq ts(o_1)[p]$. Since $q \neq p$, o_1 must be a receive (only writes and receives change the local timestamp, and a write would only change q's component). Let o_3 be the unique send operation such that $o_3 \underset{H}{\hookrightarrow} o_1$. Again by the choice of o_1, $ts(o_3)[p] < ts(o)[p] \leq ts(o_1)[p]$. But the receive operation sets $ts(o_1)[p]$ to the maximum of $ts(o_2)[p]$ and $ts(o_3)[p]$, both of which are less than $ts(o)[p]$. This contradicts the assumption that $ts(o)[p] \leq ts(o_1)[p]$, so it must be that $q = p$. Since $ts(o)[p] \leq ts(o_1)[p]$ and timestamp component values are nondecreasing, either $o = o_1$, or $o \underset{H}{\overset{+}{\Rightarrow}} o_1$ and hence $o \underset{H^m}{\leadsto} o_1$. In either case by the transitivity of $\underset{H^m}{\leadsto}$ and the assumption that $o \neq o'$, $o \underset{H^m}{\leadsto} o'$, as required.

□

In the critical lemma (5.3), we will need to show that every receive operation preserves causal correctness of $\widetilde{H}_p^{m}(o)$ by properly maintaining a sequencing $\overrightarrow{S_p^{m}}(o)$ of $\widetilde{H}_p^{m}(o)$ at each step. We will construct $\overrightarrow{S_p^{m}}(o)$ inductively, concatenating not only the new writes in Z_q for each receive operation, but also new writes that are covered, but not explicitly represented in Z_q. For this, the following easy lemma will be useful.

Lemma 5.2: *Let $\overrightarrow{S} = (S, \underset{S}{\rightarrow})$ be a poset and T be a subset of S. For any sequential poset $\overrightarrow{T} = (T, \underset{T}{\rightarrow})$ that is a topological sort of $(T, \underset{S}{\rightarrow})$, there exists a sequential poset that is a topological sort of $\overrightarrow{S}$ and that respects $\overrightarrow{T}$.*

> [The omitted proof is an easy consequence of a graph theory result, treating irreflexive partial orders as digraphs.]

We are now ready to present and prove the critical lemma. For convenience, for every process p, let o_0^p be a fictitious operation of p that represents the setting of its initial state. This allows us to address all reachable states of p in a consistent way. In particular, we can write $\widetilde{H}_p^{m}(o_0^p) = (\{o_0^p\}, \emptyset)$, $\overrightarrow{S_p^{m}}(o_0^p) = (\{o_0^p\}, \emptyset)$, $M_p(o_0^p) = M_\perp$, and $Z_p(o_0^p) = \emptyset$. (Note that since o_0^p is fictitious, $\{o_0^p\}$ is actually the empty set.)

Lemma 5.3: *Let $\widetilde{H}^{m}$ be a message-causal execution history of the Robust algorithm and p be a process. For every operation o of p, there is a poset $\overrightarrow{S_p^{m}}(o)$ that is a legal topological sort of $\widetilde{H}_p^{m}(o)$. Furthermore, the memory projection, $M(\overrightarrow{S_p^{m}}(o))$ is equal to $M_p(o)$, and $Z_p(o)$ covers $S_p^{m}(o)$.*

Proof: We tag every operation in H with a pair for purposes of induction. For any operation o of process p, let $tag(o) = (R, C)$, where R is the number of receive operations in $H_p^{m}(o)$ and C is the number of operations of p since the most recent receive operation of p, or the number of operations of p since the beginning of the computation if there is no receive by p in $H_p^{m}(o)$. By our convention $tag(o_0^p) = (0,0)$ for all p. Tags are ordered lexicographically: $(R_1, C_1) \leq (R_2, C_2)$ if and only if $R_1 < R_2$ or $R_1 = R_2$ and $C_1 \leq C_2$. We can now do induction over operations by tag order.

The basis holds trivially because for every process p, $\widetilde{H}_p^{m}(o_0^p) = (\{o_0^p\}, \emptyset)$, $\overrightarrow{S_p^{m}}(o_0^p) = (\{o_0^p\}, \emptyset)$, $M_p(o_0^p) = M_\perp$, and $Z_p(o_0^p) = \emptyset$.

Let p be any process and o be any operation of p in H such that $(0,0) < tag(o)$ (that is, o is any non-fictitious operation). Assume the lemma holds for all operations with tags less than o. There are two cases depending on whether o is a receive or not. Let o' be the operation of p just before o (if o is the first operation of p, then $o' = o_0^p$).

1. o is a not a receive.
   We let $\overrightarrow{S_p^{m}}(o) = \overrightarrow{S_p^{m}}(o') \cdot o$ and show that the conditions of the lemma are satisfied.

   - $[\overrightarrow{S_p^{m}}(o)$ is a legal topological sort of $\widetilde{H}_p^{m}(o)]$.

     Since $o \notin S_p^{m}(o')$, $\overrightarrow{S_p^{m}}(o)$ is a sequential poset by Lemma 2.1. Since o is not a receive, o' is the only operation such that $o' \underset{H^{m}}{\overset{1}{\rightarrow}} o$. Thus, $H_p^{m}(o) = \{o\} \cup H_p^{m}(o')$. Since, by the induction hypothesis, $S_p^{m}(o')$ contains exactly the operations of $H_p^{m}(o')$, $S_p^{m}(o)$ contains exactly the operations of $H_p^{m}(o)$. (Note that, even if o' is fictitious, this reasoning still works.) Since $\overrightarrow{S_p^{m}}(o')$ is legal, $\overrightarrow{S_p^{m}}(o)$ is trivially legal if o is a send or write operation. If o is a read of location x, it returns $M_p(o')[x] = M(\overrightarrow{S_p^{m}}(o'))[x] = last(\overrightarrow{S_p^{m}}(o'), x)$. Since this is the value of the most recent write of x in $S_p^{m}(o)$, $\overrightarrow{S_p^{m}}(o)$ is legal. Finally, $\overrightarrow{S_p^{m}}(o)$ respects

$\widetilde{H^m}$ because $\overrightarrow{S_p^m}(o')$ respects $\widetilde{H^m}$ and for all operations $o'' \in H_p^m(o')$, $o'' \underset{H^m}{\leadsto} o$. Thus, $\overrightarrow{S_p^m}(o)$ is a legal topological sort of $\widetilde{H_p^m}(o)$.

- $[M(\overrightarrow{S_p^m}(o)) = M_p(o)]$.

  Sends and reads do not affect memory, so in these cases $M_p(o) = M_p(o') = M(\overrightarrow{S_p^m}(o')) = M(\overrightarrow{S_p^m}(o') \cdot o) = M(\overrightarrow{S_p^m}(o))$. If o is a write of value v to location x, $M_p(o)[x] = v$ and for all $y \neq x$, $M_p(o)[y] = M_p(o')[y]$. Since the last write to x in $\overrightarrow{S_p^m}(o)$ is o, $last(\overrightarrow{S_p^m}(o), x) = v = M_p(o)[x]$. Also, for all $y \neq x$, $last(\overrightarrow{S_p^m}(o), y) = last(\overrightarrow{S_p^m}(o'), y) = M_p(o')[y] = M_p(o)[y]$. Thus, $M(\overrightarrow{S_p^m}(o)) = M_p(o)$, as required.

- $[Z_p(o) \text{ covers } S_p^m(o)]$.

  Sends and reads do not affect the Z set of a process, so if o is a send or a read, $Z_p(o) = Z_p(o')$. Since $Z_p(o')$ covers $S_p^m(o')$ and o is not a write, $Z_p(o)$ covers $S_p^m(o)$. If o is a write to location x, then the algorithm deletes all triples corresponding to writes to x from Z and inserts a triple corresponding to write o. By Lemma 5.1, $ts(o)$ is greater than the timestamp of every operation o'' such that $o'' \underset{H^m}{\leadsto} o$. Thus, o covers all other writes to x in $H_p^m(o)$ and hence in $S_p^m(o)$. Triples in Z for locations other than x are unaffected by operation o, so, since $Z_p(o')$ covers $S_p^m(o')$, $Z_p(o)$ covers $S_p^m(o)$.

2. o is a receive.

   Let Z_q, Z_q' and Z' be the sets assigned during o, and assume the message received, $\langle Z_q, t_q \rangle$, was sent by o_q of process q. Since $H_q^m(o_q)$ has fewer receives than $H_p^m(o)$, $tag(o_q) < tag(o)$ and the induction hypothesis applies to q as well as o'. Let Q be the set of write operations in $H_q^m(o_q)$ but not in $H_p^m(o')$.

   Let $\overrightarrow{T}$ be the topological sort of $O(Z_q')$. Note that, since $O(Z_q')$ consists only of write operations, $\overrightarrow{T}$ respects $\widetilde{H^m}$ by Lemma 5.1. Since $O(Z_q') \subseteq Q$, by Lemma 5.2, $\overrightarrow{T}$ can be extended to a sequential poset over the operations of Q, say $\overrightarrow{Q}$, that also respects $\widetilde{H^m}$. Now let $\overrightarrow{S_p^m}(o) = \overrightarrow{S_p^m}(o') \cdot \overrightarrow{Q} \cdot o$, which is a sequential poset by Lemma 2.1 because $S_p^m(o') \cap Q = \emptyset$ and o is in neither $S_p^m(o')$ nor Q.

   - $[\overrightarrow{S_p^m}(o) \text{ is a legal topological sort of } \widetilde{H_p^m}(o)]$.

     We have just noted that $\overrightarrow{S_p^m}(o)$ is a sequential poset. By construction, $S_p^m(o) = H_p^m(o)$. Since $\overrightarrow{S_p^m}(o')$ is legal and $\overrightarrow{Q} \cdot o$ contains no reads, $\overrightarrow{S_p^m}(o)$ is also legal. We claim that $\overrightarrow{S_p^m}(o') \cdot \overrightarrow{Q}$ respects $\widetilde{H^m}$. If not, because $\overrightarrow{S_p^m}(o')$ and $\overrightarrow{Q}$ both respect $\widetilde{H^m}$, there must be operations $o_1 \in Q$ and $o_2 \in S_p^m(o)$ such that $o_1 \underset{H^m}{\leadsto} o_2$. But by Lemma 5.1, $ts(o_1) \leq ts(o_2)$, and o_1 cannot be in Q, because it is dominated by an operation (o_2) in $S_p^m(o')$. Thus, $\overrightarrow{S_p^m}(o') \cdot \overrightarrow{Q}$ respects $\widetilde{H^m}$, and so does $\overrightarrow{S_p^m}(o) = \overrightarrow{S_p^m}(o') \cdot \overrightarrow{Q} \cdot o$, because for every operation $o'' \in H_p^m(o') \cup H_q^m(o_q)$, $o'' \underset{H^m}{\leadsto} o$. Thus, $\overrightarrow{S_p^m}(o)$ is a legal topological sort of $\widetilde{H_p^m}(o)$.

   - $[M(\overrightarrow{S_p^m}(o)) = M_p(o)]$.
     The only locations in M affected by a receive are those that occur in Z_q'. Let x be a location acted on by a write in Z_q'. Then, $M_p(o)[x]$ is set to the value of the last write to x in $\overrightarrow{T}$, which is exactly the same value as $last(\overrightarrow{Q}, x)$ (since $O(Z_q')$ covers Q). Since $\overrightarrow{Q}$ is concatenated to $\overrightarrow{S_p^m}(o')$ in forming $\overrightarrow{S_p^m}(o)$ and o is not a write, $last(\overrightarrow{S_p^m}(o), x) =$

$last(\vec{Q}, x) = M_p(o)[x]$. Furthermore, any location y not occurring in Q also does not occur in Z'_q, so $last(\overrightarrow{S^m_p}(o), y) = last(\overrightarrow{S^m_p}(o'), y)$. Since locations not in Z'_q are not changed, $M_p(o)[y] = M_p(o')[y] = last(\overrightarrow{S^m_p}(o'), y) = last(\overrightarrow{S^m_p}(o), y)$. Thus, $M(\overrightarrow{S^m_p}(o)) = M_p(o)$.

- $[Z_p(o)$ covers $S^m_p(o)]$.

  Let w be any write in $S^m_p(o) = H^m_p(o)$. Since o itself is not a write, w is either in $H^m_q(o_q)$ or $H^m_p(o')$, or equivalently by the definition of Q, in Q or $H^m_p(o')$. We will show there is a write in $Z_p(o) = O(Z'_q) \cup O(Z')$ that covers w in either case.

  First suppose w is in Q. Since $Q \subseteq H^m_q(o_q) = S^m_q(o_q)$, $Z_q(o_q) = O(Z_q)$ covers Q, so there is some write w' (possibly w itself) in $O(Z_q)$ that covers w. If w' is in $O(Z'_q)$ it is also in $Z_p(o)$, so suppose w' is not in $O(Z'_q)$. Then, since w' was deleted by the second step of the receive operation, some write w'' in $Z_p(o')$ dominates w' and, by transitivity of the timestamp ordering, also dominates w. But, since both w and w'' are both writes, by Lemma 5.1 we have $w \underset{H^m}{\rightsquigarrow} w''$, implying $w \in H^m_p(o')$. But this conflicts with the choice of Q, so $w' \in O(Z'_q)$. Thus, since $Z_p(o) = O(Z'_q) \cup O(Z')$, there is a write ($w'$) in $Z_p(o)$ that covers w.

  Now suppose w is in $H^m_p(o')$. Then w is covered by some write, $w' \in Z_p(o')$. If w' is in $O(Z')$ it is also in $Z_p(o)$. Otherwise, if w' is not $O(Z')$, then it was deleted because some write w'' in $O(Z'_q)$ covers w'. In this case, w'' is in $Z_p(o)$ and covers w. Thus in all cases, each write in $H^m_p(o) = S^m_p(o)$ is covered by a write in $Z_p(o)$, so $Z_p(o)$ covers $S^m_p(o)$.

$\square$

The theorem is now straightforward. Note that any execution of the Robust algoritm has both a message-causal and a causal execution history.

Theorem 5.4: *For every message-causal execution history $\widetilde{H^m}$ of the Robust algorithm, the corresponding causal execution history $\widetilde{H}$ satisfies causality.*

Proof: We will show $\overrightarrow{S^m_p} = (S^m_p, \underset{S^m_p}{\rightarrow})$ is a legal topological sort of $(H^m_p, \underset{H^m}{\rightsquigarrow})$, and that this implies $\overrightarrow{S^m_p}$ sequences $(H_p, \underset{H}{\rightsquigarrow})$. Let p be any process in $\mathcal{P}$, and let $o^p_1, o^p_2, \ldots$ be all the operations of p in H ordered by process order $(\underset{H}{\Rightarrow})$. Let $H^m_p = \bigcup_{i=1,2,\ldots} H^m_p(o^p_i)$, $S^m_p = \bigcup_{i=1,2,\ldots} S^m_p(o^p_i)$, and $\underset{S^m_p}{\rightarrow} = \bigcup_{i=1,2,\ldots} \overrightarrow{S^m_p(o^p_i)}$.

Every operation, $o \in S^m_p$, occurs in $S^m_p(o^p_i)$ for some i. Since $S^m_p(o^p_i) = H^m_p(o^p_i)$ by Lemma 5.3, $S^m_p \subseteq H^m_p$. Similarly, every operation o' in H^m_p occurs in $H^m_p(o^p_j) = S^m_p(o^p_j)$ for some j, implying $H^m_p \subseteq S^m_p$. So $S^m_p = H^m_p$.

Now showing that $\overrightarrow{S^m_p}$ is a legal sequential poset respecting $\widetilde{H^m}$ establishes that $\overrightarrow{S^m_p}$ is a legal topological sort of $(H^m_p, \underset{H^m}{\rightsquigarrow})$. It takes at most three operations to demonstrate a violation of one of the properties of a legal sequential poset or that $\widetilde{H^m}$ is not respected. Suppose o, o' and o'' are three distinct elements of S^m_p that can be used to demonstrate that $\overrightarrow{S^m_p}$ is not a sequential poset or does not respect $\widetilde{H^m}$. Since for all integers $i, j > 0$, $i \leq j$ implies $S^m_p(o_i) \subseteq S^m_p(o_j)$, there is some least i such that $o, o', o'' \in S^m_p(o_i)$. Since $\overrightarrow{S^m_p}(o_i)$ is a legal topological sort of $\widetilde{H^m_p}(o_i)$ by Lemma 5.3, it cannot be that o, o', o'' can be used in $\overrightarrow{S^m_p}(o_i)$ to refute legality, transitivity, non-relatedness or that $\widetilde{H^m}$ is not respected. Thus, there must be some $j > i$ such that $\overrightarrow{S^m_p}(o_j)$ adds pairs that violate one of the properties. But because $S^m_p(o_i) \subseteq S^m_p(o_j)$, $\overrightarrow{S^m_p}(o_j)$ also has this violation, contradicting Lemma 5.3.

All that remains to be shown is that $\widetilde{H^m}$ respects $\widetilde{H}$ and that $H_p \subseteq H^m_p$, since this implies $\overrightarrow{S^m_p}$ sequences $(H_p, \underset{H}{\rightsquigarrow})$. To see that $\widetilde{H^m}$ respects $\widetilde{H}$ note that for any two operations $o, o' \in H$, $o \underset{H}{\mapsto} o'$

implies that $o \underset{H^m}{\leadsto} o'$, because o is either a write by the same process execution o' or a write by another process that was received by a chain of messages.

We now show that $H_p \subseteq H_p^m$. Let o be any operation in H_p, so that o is either an operation of p or is a write known to some correct process. If o is an operation of p, it occurs in $H_p^m(o_i^p)$ for some i, and certainly occurs in H_p^m. Otherwise, o is a write known to some correct process, say q, with operations $o_1^q, o_2^q, \ldots$ in process order. Let o_i^q be the first operation at which q knows of o. After this point q knows of o, and every message sent by q contains a triple that covers o. Since the q-to-p channel eventually delivers one of these messages, o will be included in $H_p^m(o_j)$ for some j. Thus, $H_p \subseteq H_p^m$, and the theorem holds since $\overrightarrow{S_p^m}(o)$ is legal sequencing of $(H_p, \underset{H}{\leadsto})$. □

When introducing the Robust algorithm, we mentioned that messages required $O(mn^2)$ bits where $m = |\mathcal{M}|$, $n = |\mathcal{P}|$, and where we assume memory values and timestamp components need only a constant number of bits each (thus, a vector timestamp needs $O(n)$ bits). We now establish that claim by showing that at most n triples per location are needed in Z.

Lemma 5.5: *For any execution of the Robust algorithm, and any operation o by process p acting on location x, the set $Z[x]$ of p after executing o contains at most one triple written by each process.*

Proof: Formally, an induction on operations similar to that done in Lemma 5.3 is needed, but we continue informally since this result is easy to see. Read and send operations do not affect the Z set. A write operation to location x deletes all other writes to x, obviously maintaining the required condition. The only operation that could violate the condition is a receive.

Let o be a receive operation, and let Z refers to the set of triples at the beginning of the operation, Z_{new} be the set at the end, and Z_q, Z_q' and Z' be the sets referred to during the execution of o. By the inductive assumption, any process r has at most one write triple to location x in Z and at most one in Z_q. Suppose that r has two distinct writes, w_1 and w_2, to location x corresponding to triples in Z_{new}. Since both writes are by the same process, there timestamps are related by $\leadsto$, and so without loss of generality, we may assume $ts(w_1) < ts(w_2)$. Since $Z_q' \subseteq Z_q$ and $Z' \subseteq Z$, neither $O(Z_q')$ nor $O(Z')$ can contain both writes by the induction hypothesis. Thus, since $Z_{new} = Z' \cup Z_q'$, one of the writes is in $O(Z_q')$ and the other is in $O(Z')$.

Suppose w_1 is in $O(Z_q')$, implying w_2 is in $O(Z')$, and hence in $O(Z)$. But this is impossible because w_1 is dominated by w_2, and the triple of w_1 would be deleted when Z_q' was formed.

On the other hand, suppose w_1 is in $O(Z')$, so that w_2 is in $O(Z_q')$. This is also impossible because w_1 is covered by w_2, and the triple of w_1 would be deleted when Z' was formed. Thus, the Z_{new} cannot contain two writes by the same process. □

6   Concluding Remarks

We have introduced a type of memory called *causal memory*, which preserves the orderings of events defined by causality. Because different applications may require that different causal relationships be preserved, we can actually define a class of causal memories, one of which is presented in detail in this paper. We can define all of these memories using a simple formal model.

We show that greater concurrency is possible with causal memory than with atomic or sequentially consistent memories. This is evident from the simple implementation given in Section 4, which allows a process to complete its read or write operations without requiring communication with other processes (newly written values are sent in messages to other processes, but the sender can continue execution while the messages are in transit). In contrast, it can be easily shown that any implementation of an atomic or sequentially consistent memory must force a process to sometimes block execution until communication completes with one or more other processes. For example, in their implementation of sequentially consistent memory, Afek et al. [2] allow asynchrony by queuing updates, but they require an expensive atomic broadcast to complete when a process reads a variable with a queued update. By demonstrating that an application in which processes interact heavily can be programmed in an identical manner on both causal and atomic memories (other applications are discussed in a related paper [3]), we believe that causal memory provides a reasonable and viable target architecture for programming distributed and parallel applications.

Our implementations of causal memory can use a large amount of storage. (Note that the simple algorithm requires unbounded space, as can be seen by considering a computation in which messages on a single channel, say pq, are delayed, while writes continue and all other channels run normally. The *pending* set of process q will grow without bound.) Perhaps a large amount of space is the price paid for wait-free causal memory based on message passing, but we would like to have clear lower bound results. Some interesting open questions are

- Can bounded timestamps [10] (that are consistent with causality) be implemented in a wait-free manner using message passing?

- Can the upper bound on space established for the Robust algorithm be improved (ignoring the problem of unbounded timestamps)? (We have developed "hybrid" algorithms that blend the ideas of the Simple and Robust algorithm, but have not improved on the space bound of the Robust algorithm.)

In the absence of concrete results, it may be interesting to explore the performance of variants of the Robust algorithm under reasonable assumptions about delays in message passing.

In this paper, we have only considered one type of causal memory. However, there exist varieties of causal memory which are both stronger and weaker than the causal memory implemented by the Simple and Robust algorithms. For example, a weaker causal memory can be defined using *weak causality*. An execution history $\widetilde{H}$ satisfies weak causality if, for each process $p \in \mathcal{P}$ and each memory location $x \in \mathcal{M}$, there is a legal sequential poset $\overrightarrow{S_p^x}$ of the read and write operations of p to location x and the write operations of all other processes to x that respects $\widetilde{H}$. In fact, the linear solver example presented earlier still runs correctly even when the memory correctness is defined using weak causality. Our future work will address the relationships of various types of memories and characterizations of problems that can be solved with such memories. The goal of our practical work is to demonstrate that weak memories can be feasibly programmed and implemented to achieve overall performance gains.

References

[1] Sarita V. Adve and Mark D. Hill. Weak ordering — a new definition. In *Proceedings of the 17th Annual International Symposium on Computer Architecture*, pages 2–14, 1990.

[2] Yehuda Afek, Geoffrey Brown, and Michael Merritt. A lazy cache algorithm. In *Proceedings of the 1989 ACM Symposium on Parallel Algorithms and Architectures*, pages 209–223, June 1989.

[3] Mustaque Ahamad, Phillip W. Hutto, and Ranjit John. Implementing and programming causal distributed shared memory. In *11th International Conference on Dist. Comput.*, May 1991.

[4] Henri E. Bal and Andrew S. Tanenbaum. Distributed programming with shared data. In *Proceedings of the IEEE 1988 International Conference on Computer Languages*, pages 82–91, October 1988.

[5] J. K. Bennett, J. B. Carter, and W. Zwaenepoel. Adaptive software cache management for distributed shared memory architectures. In *Proceedings of the 17th Annual International Symposium on Computer Architecture*, May 1990.

[6] J. K. Bennett, J. B. Carter, and W. Zwaenepoel. Munin: Distributed shared memory based on type-specific memory coherence. In *Proceedings of the 2nd ACM Symposium on Principles and Practice of Parallel Programming*, pages 168–177, March 1990.

[7] Kenneth Birman, Andre Shiper, and Pat Stephenson. Lightweight causal and atomic group multicast. Technical Report 91-1192, Department of Computer Science, Cornell University, February 1991. To appear in *ACM Transactions on Computer Systems*.

[8] Roberto Bisiani, Andreas Nowatzyk, and Mosur Ravishankar. Coherent shared memory on a distributed memory machine. In *Proceedings of the 1989 International Conference on Parallel Processing*, volume I, pages 133–141, 1989.

[9] Patrick M. Clancey and Joan M. Francioni. Distribution of pages in distributed shared memory. In *Proceedings of the 1990 International Conference on Parallel Processing*, volume II, pages 258–265, August 1990.

[10] Danny Dolev and Nir Shavit. Bounded concurrent time-stamp systems are constructible. In *Proceedings of the Twenty-First ACM Symposium on Theory of Computing*, pages 454–466, May 1989.

[11] Michel Dubois, Christoph Scheurich, and Faye Briggs. Synchronization, coherence, and event ordering in multiprocessors. *IEEE Computer*, 21(2):9–22, February 1988.

[12] Colin Fidge. Timestamps in message-passing systems that preserve the partial ordering. In *Proceedings of the Eleventh Australian Computer Science Conference*, University of Queensland, Australia, 1988.

[13] Brett D. Fleisch and Gerald J. Popek. Mirage: A coherent distributed shared memory design. In *Proceedings of the Twelfth ACM Symposium on Operating Systems Principles*, pages 211–224, December 1989.

[14] Kourosh Gharachorloo, Daniel Lenoski, James Laudon, Phillip Gibbons, Anoop Gupta, and John Hennessy. Memory consistency and event ordering in scalable shared-memory multiprocessors. In *Proceedings of the 17th Annual International Symposium on Computer Architecture*, pages 15–26, 1990.

[15] Maurice P. Herlihy and Jeannette M. Wing. Linearizability: A correctness condition for concurrent objects. *ACM Transactions on Programming Languages and Systems*, 12(3):463–492, July 1990.

[16] Phillip W. Hutto and Mustaque Ahamad. Slow memory: Weakening consistency to enhance concurrency in distributed shared memories. In *Proceedings of the 10th International Conference on Distributed Computing Systems*, pages 302–311, 1990.

[17] R. E. Kessler and M. Livny. An analysis of distributed shared memory algorithms. In *Proceedings of the 9th International Conference on Distributed Computing*, pages 498–505, June 1989.

[18] Leslie Lamport. Time, clocks, and the ordering of events in a distributed system. *Communications of the ACM*, 21(7):558–565, July 1978.

[19] Leslie Lamport. How to make a multiprocessor computer that correct executes multiprocess programs. *IEEE Transactions on Computers*, C-28(9):690–691, September 1979.

[20] Leslie Lamport. On interprocess communication; part I: Basic formalism. *Distributed Computing*, 1(2):77–85, 1986.

[21] Leslie Lamport. On interprocess communication; part II: Algorithms. *Distributed Computing*, 1(2):86–101, 1986.

[22] Kai Li and Paul Hudak. Memory coherence in shared virtual memory systems. *ACM Transactions on Computer Systems*, 7(4):321–359, November 1989.

[23] Richard J. Lipton and Jonathan S. Sandberg. PRAM: A scalable shared memory. Technical Report 180-88, Princeton University, Department of Computer Science, September 1988.

[24] Richard J. Lipton and D. N. Serpanos. Uniform-cost communication in scalable multiprocessors. In *Proceedings of the 1990 International Conference on Parallel Processing*, pages I429–I432, August 1990.

[25] Friedemann Mattern. Time and global states of distributed systems. In Michel Cosnard, Patrice Quinton, Yves Robert, and Michel Raynal, editors, *Proceedings of the International Workshop on Parallel and Distributed Algorithms*. North-Holland, October 1988.

[26] J. Misra. Axioms for memory access in asynchronous hardware systems. *ACM Transactions on Programming Languages and Systems*, 8(1):142–153, January 1986.

[27] Umakishore Ramachandran, Mustaque Ahamad, and M. Yousef Khalidi. Coherence of distributed shared memory: Unifying synchronization and data transfer. In *Proceedings of the 18th International Conference on Parallel Processing*, pages 160–169, August 1989.

More on the Power of Random Walks:
Uniform Self-Stabilizing Randomized Algorithms
(Preliminary Report)

Efthymios Anagnostou   Ran El-Yaniv

Department of Computer Science
University of Toronto
Toronto M5S 1A4, Canada

September 6, 1991

Abstract

We present a self-stabilizing randomized protocol for the *Unique Naming* problem. In the Unique Naming problem an anonymous system assigns unique names to all the processors in the system. Let G be the underlying interconnection network. If N is a known bound on the network size then our protocol uses $O(C_G N log N)$ bits and stabilizes within $O(C_G)$ rounds where C_G is the *cover time* of G. The protocol is uniform, tolerates dynamic changes of the network topology, and works correctly under a very powerful adversary which at any stage has knowledge of a bounded number of future random choices of the processors and it can even bias all future random choices.

We then show that a small modification to our protocol provides a solution for another important problem; the *Topology* problem in which each node in an anonymous network computes an exact description of the network's topology. Moreover these two protocols yield uniform and bounded space solutions to many other important problems such as *Leader Election*, *Spanning Tree*, *Mutual Exclusion* (*Token Management*), etc.

1   Introduction

The concept of self-stabilization was known and used in mathematics for many years. Consider for example the Newton-Raphson method for finding roots of functions. For many functions no matter what the initial estimate for the root is, eventually the iteration will converge to the root. Self-Stabilization was introduced to computer science by Dijkstra [Dij74] and subsequently gained attention among the distributed computing community. Informally, a system is said to be *self-stabilizing* if starting from an arbitrary configuration, the system is guaranteed to reach a "legal" configuration within some finite period. Two important advantages that make self-stabilizing systems appealing are:

- A self stabilizing system need not be initialized to any particular configuration. When the system is started from an arbitrary configuration it then stabilizes to a legal configuration without any external intervention.

- As a consequence it follows that a self stabilizing system can tolerate transient faults. These are the type of faults that corrupt the state of one or more components of the system but keep those components "alive". Following such faults, the system will loose it's consistency but it will regain it within some finite period.

Like many other recent papers, we adopt here the anonymous, asynchronous, link-register distributed model that was formalized[1] by Dolev, Israeli, and Moran [DIM90]. Informally, the system is modeled as a set of processors, each residing on a node of some communication graph. Every two neighbours are connected by a bidirectional communication link. For each link there exists a shared register used by both neighbours for communication. The processors are anonymous. This means that they have no IDs. In order to achieve some computational task, every processor is provided with a program. To guarantee the correctness of our programs we must assume the worst scenario in every computation. Since the system is completely asynchronous, we must assume in particular that all processor activations are scheduled by some malicious scheduler. Given a problem, a desirable self-stabilizing solution for it should satisfy the following:

1. The solution must be *uniform*; i.e., all the processors are provided with the same program.

2. The space used by every processor should be *bounded*.

Note that we could add to this list many other desirable properties. For example, we would like our algorithms to work properly for a significant class of communication graphs, to stabilize to a legal configuration rapidly, to use a little space etc. The difference is that *uniformity* and *boundness* are a must, while the other properties mentioned will "just" contribute to the "goodness" of the algorithm; violating uniformity or boundness simply means compromising with our model. So far many of the attempts to solve self-stabilizing problems, compromised at least one of the desirable properties. Let us mention the most important solutions. In the original paper [Dij74], Dijkstra solved the *Token Management* (TM) problem on rings but he used a distinguished processor, giving up uniformity. At this point, it would be proper to mention that the TM problem was the most "attacked" problem in the area, and this comes as no surprise, since a solution for the TM provides a solution for another fundamental problem in distributed computing, the *Lockout Free Mutual Exclusion* problem. Kruijer [Kru79] solved a variation of the TM problem on trees but he also used a distinguished processor. A special processor was also by Brown, Gouda and Wu in [BGW89] to solve the TM problem on chains. Burns and Pachl [BP89] gave a sophisticated uniform solution for the TM problem on rings but only for the distinguished case where the ring is of prime size[2].

---

[1]This model is a variant on the model originally used by Dijkstra.

[2]Dijkstra observed that for rings of composite size no deterministic solution exists.

Dolev *et al.* [DIM90] gave solutions for the *Spanning Tree* problem and the *Mutual Exclusion* (ME) problem on trees and, by introducing a simple and very useful technique to superimpose two self-stabilizing algorithms (*Fair Protocol Composition*), they were able to solve the ME problem on general graphs. Unfortunately their solution suffered the same problem as most of the previous ones; it used a distinguished processor. Therefore, ME and Spanning Tree (with uniform solutions) were still left open for general graphs. Israeli and Jalfon [IJ90] added a new dimension to the area by introducing the tool of *random walks*. Since deterministic solutions were condemned to failure even for very basic problems (e.g. TM on rings of even size), they resorted to randomization. Their novel approach provided a solution for the TM problem on general graphs, but unlike the other solutions that gave up uniformity, they sacrificed boundedness. Their solution used an unbounded number of states even if some knowledge, such as the topology or the number of nodes in the graph, was known. In the same paper, a uniform and bounded solution for the case where the graph is a ring was presented ([IJ90] and also [Jal90]). Following this, the class of rings was the only class of graphs for which a desirable solution for the TM problem was known. Anagnostou *et al.* [AET91] observed that a simple reduction can extend this ring TM algorithm to include the classes of *trees* and *rooted DAGs*.

Afek, Kutten, and Yung [AKY90] presented a leader election and a spanning tree algorithm, which turns out to be sufficient to solve many other problems. Their solution assumes unique IDs. Katz and Perry [KP90] , showed how to "self-stabilize" distributed protocols by performing *snapshots*. Their protocols work in message passing systems but assumes the existence of a leader. Notice that these results can be combined together to establish a sort of a framework (or "compiler") for extending many distributed protocols to their self-stabilizing versions. More recently, Dolev, Israeli and Moran [DIM91] exhibited a leader election and a ranking algorithm for anonymous networks. In addition Afek, Kutten, and Yung [AKY91] established similar results to [DIM91]. Awerbuch, Patt-Shamir, and Varghese [APV91] provided two very useful self-stabilizing subroutines: a communication primitive *(end-to-end)*, and a *reset* protocol. Awerbuch and Varghese [AV91] used these protocols in their "compiler" that extends many deterministic, *non-interactive* distributed protocols to their self-stabilizing versions.

In this work we exploit the power of random walks and provide uniform and bounded solutions for many important problems. These include the Unique Naming, Topology, Ranking, Spanning Tree, TM (ME), *ℓ-Exclusion*, Leader Election, etc. Interestingly, all of these problems are solved with the same basic technique. If N is a bound on the network size and C_G is the *cover time* of the underlying interconnection network G, then a version of our basic protocol uses $O(C_G N \log N)$ bits and its expected stabilization time is $O(C_G)$. In addition, our protocols can accommodate dynamic changes of the network topology. As long as the number of nodes in the graph is bounded above by N, the protocol will stabilize after every change of the network topology. Furthermore, in this paper we introduce a new type of scheduler which is stronger than any other scheduler considered in the literature. This scheduler, besides controlling processor activations, knows at any stage the results of *any bounded number* of future random choices that will be made by the processors and in addition, it can *bias* by any constant *all* future random choices. We prove that our algorithms work properly even under such a powerful scheduler. Note that our complexities do not measure up with [DIM91] and [AKY91], but as we shall see our method is more general and provides a new paradigm for self-stabilizing protocols via the use of consistency tables.

The rest of this paper is organized as follows. In Section 2 we describe the model. Sections 3-4 are devoted for presenting the Unique Naming protocol and proving its correctness. In Section 5 we show how a small modification to the Unique Naming protocol yields the Topology protocol. Section 6 presents solutions to a variety of other problems. Every protocol in this section is either a reduction to Naming (or Topology), or a minor variation of them. In section 7 we derive an improved version of our Unique Naming protocol (and consequently, for all the other problems). Finally, in section 8, we summarize our conclusions.

2 The Model

Consider a system of n processors $p_1, p_2, \ldots, p_n$. Each processor is a (randomized) finite state machine. The processors are connected according to some undirected graph $G = (V, E)$ where $V = \{p_1, \ldots, p_n\}$. The processors are *anonymous*; that is, they have no identities. Each node of G represents a processor and each edge represents a bidirectional communication link. Each processor (node) of degree d has d ports through which it communicates with its neighbours. The links (edges) incident with each node are arbitrarily but uniquely numbered. Each processor associates a port with each one of the edges incident with the node. The communication between two neighbouring processors is facilitated by *link registers*. If p_i and p_j are neighbours, they communicate by using two registers: $R_{i,j}$ in which p_i writes and from which p_j reads, and $R_{j,i}$ in which p_j writes and from which p_i reads. We say that a processor *owns* all the registers to which it can write. We assume that two conflicting operations on the same register are serialized by the hardware.

We define a *uniform* protocol as a quadruple $\langle \mathcal{G}, S, \Sigma, \delta \rangle$ where $\mathcal{G}$ is a class of communication graphs for which the protocol is correct; S is a set of states, Σ is a finite alphabet which its symbols are read from and written to the link registers; $\delta = \langle \delta_1, \delta_2, \ldots, \delta_d \rangle$ where d is the maximal degree of any graph in $\mathcal{G}$, and δ_u, the (randomized) transition function of a processor with u neighbours, is a mapping from $S \times (\Sigma)^u$ to itself. The *execution state* of a processor p with d neighbours (which owns d registers) is defined to be a $d + 1$-tuple $\langle s, \alpha_1, \ldots, \alpha_d \rangle$ where $s \in S$, is p's state and $\alpha_i \in \Sigma$ are the contents of p's own registers. The *global* state of a uniform n-processor system is described by its *configuration*. A configuration of a system is an n-tuple $\langle e_1, \ldots, e_n \rangle$ where e_i is the execution state of the ith processor. A processor p_i is *enabled* in a configuration $C = \langle e_1, \ldots, e_n \rangle$ if $\delta_d(e_i) \neq e_i$ where the α_i are the values read by p_i from the registers of its neighbours in C (for randomized protocols we require that $Pr[\delta_d(e_i) = e_i] < 1$). If a processor is not enabled, it is said to be *disabled*. For two configurations C_1 and C_2, we say that C_2 is *reachable* from C_1 if there exists a non-empty set Q of processors all of which are enabled in C_1 such that one activation of every processor in Q results in the configuration C_2. A *computation* of a system is a sequence of configurations $C_1, C_2, \ldots$ where for each $i > 1$, C_i reachable from C_{i-1}. The sequence of subsets of activated processors is called a *schedule*.

To ensure correctness of an execution, we must assume the worst schedule; namely, the activity of the processors is controlled by a malicious *scheduler* (adversary). We let the scheduler choose its activated processors on-line using the current configuration as an input. Whenever the scheduler activates a processor, the processor executes a single atomic step. The three most common schedulers used in the literature are: (1) The *Central Demon* which activates processors in a sequence, one after another. The atomic step of a processor consists of reading all its neighbours' registers, changing its state, and writing all of its own registers. (2) The *Distributed Demon* can activate any set of processors together. The atomic step is exactly as in the central demon but now it is guaranteed that all activated processors simultaneously read the content of all the registers of their neighbours. (3) The *Read/Write Demon* can activate only one processor at a time but now, an atomic step consists of either reading the content of a single register, changing state, or writing on a single register it owns.

A scheduler is *fair* if in every infinite computation, every processor which is enabled infinitely often, is activated infinitely often. We call a configuration C, a *deadlock* configuration if no processor is enabled in C. A scheduler is *proper* if in every configuration which is not a deadlock configuration it activates at least one enabled processor. We say that a protocol satisfies the *no livelock* property if starting from any configuration and following any schedule, every processor will be activated infinitely often. Clearly, no livelock implies no deadlock.

We now define the requirements from a *self-stabilizing* protocol, P. Let P be a uniform protocol for a link-register distributed system. Let C denote the set of all configurations. The first requirement from P is the *no-livelock* requirement. To explain the other requirements, from a randomized self-stabilizing protocol, consider the following Markov process. C is the set of states and let $M_P = (p_{ij})$ be the transition probability matrix, where p_{ij} is the probability of reaching C_j from C_i. Since P satisfies the no-deadlock requirement, $p_{ii} < 1$ for every i. Let $p_{jk}^{(l)}$ be the probability of a transition from C_j to C_k in exactly l steps. Let $\mathcal{L} \subseteq \mathcal{C}$ be called the set of *legitimate* configurations (under P). The protocol P is *self-stabilizing with respect to $\mathcal{L}$ under A* if, whenever it is scheduled by an adversary A , the following hold:

(i) **Closure** the closure of $\mathcal{L}$ is $\mathcal{L}$.

(ii) **Stabilization** for every non-legitimate states C_i and C_j, $\lim_{l \to \infty} p_{ij}^{(l)} = 0$.

Informally, the first requirement simply says that once the system is in a legitimate configuration, it will remain in a legitimate configuration forever. The second requirement ensures that no matter what the initial configuration is, the system will eventually *stabilize*; namely it will enter a legitimate configuration.

To measure the efficiency of a self-stabilizing protocol, we use the following two complexity measures. The first is the amount of space that each processor uses. It is important to note that even for a simple task such as token management, it was shown in [IJ90] that no system can solve the problem with a finite number of states if there is no bound on the number of processors. From an algorithmic point of view an infinite state protocol might be interesting, but from a practical point of view, a self-stabilizing protocol with infinite amount of space does not make a lot of sense. Note that if there is no known bound on the number of processors in the system, then a bounded protocol does not exist for almost any non trivial problem (in fact this is true for all the problems considered here). Therefore we assume some bound N on the number of processors (n) for the benefit of finite state protocols. This approach has been taken in almost all self-stabilizing algorithms which perform a non trivial task. The number of states is, of course, a function of N. The second interesting complexity measure is the (expected) stabilization time. Since our system is totally asynchronous, we should first explain and define the notion of asynchronous *distributed* time. In the spirit of [AFL83], the distributed time is best explained through the notion of computational *rounds*. Informally, a round is any computation (sequence of configurations) within which every processor is activated at least once. This should capture the fact that the system has done some minimal and "balanced" amount of work. The subtlety here is that an activation of a processor in the case of a read/write demon, can fail to capture this notion because the minimal contribution of a single processor to the system occurs usually just after completing one *cycle* of reading the content of all neighbours' registers, changing state, and writing the content of all registers. A computation is said to be a *round* if every processor is activated enough times within it to complete a cycle (for a central or a distributed demon this is achieved by a single activation for each processor) and no prefix of that computation has this property. Given a computation, we can uniquely partition it into segments such that every segment (except possibly the last) is a round. The *run time* of a finite computation is defined to be the number of segments in this partition to rounds.

3   The Unique Naming Protocol (UNP)

We start by giving some intuition for our algorithm. Let us ignore for the moment the complications evolving from the unstable environment and assume first that the system is not disturbed by any faults and second that it starts from some proper initial configuration. The first step is to randomly assign an ID to each node. Since we know a bound N on the size of the communication graph, we can choose these IDs from a large space and thus, the probability of assigning the same ID to two distinct nodes can be low. However, our goal is to assign distinct names to distinct nodes with probability that tends to 1 as time goes on. This means that some non-trivial coordination is needed. This coordination is done via the use of random walks. Each node sends a packet that contains its candidate ID to a tour on the graph. We call this packet a *token*. The tokens travel in a random fashion along the graph and, therefore, are very likely to visit every node after a short random walk ([AKL+79], [BK89], [CRR+89], [BKRU89]). These walks achieve simultaneously two goals. First they spread the information that certain IDs are already taken, and second, if two nodes have chosen the same ID as their candidates, their tokens will sooner or later clash with each other at some node along their (intersecting) trajectories. Aleliunas *et al.* [AKL+79] show that such collisions happen sooner rather than later. Once clashed at a certain node, these tokens compete for the right to use this ID as a candidate. While the loser runs back to its initiator, to inform it of the bad news, the winner resumes its (random) walk, continuing to inform other nodes that the ID in its packet is still taken, and at the same time, hunting for other opponents. The loser node chooses a new candidate for its ID, but this time it avoids any "already taken" IDs that it knows about. Since we want every loser in such a competition to return immediately to its creator, we provide every token with a "thread" that is "tied" to its creator. This thread is simply a map describing the way back. This map is accumulated while traveling forward simply by attaching port numbers. Note that the way back using this map, is most probably not a short way (not to mention shortest) but this is probably the best we can hope for since the system is anonymous. The use of such maps also means that if we want to prevent the packets of winning tokens from growing boundlessly, we must send them back after some bounded length walk. We call such walks that travel randomly "forward" for some bounded length and then return to their initiator, *Mitos-walks*[3]. Obviously, the bound on the walk must be larger than the number of nodes of the graph, if we want to give it a positive chance of covering all the nodes of the graph. Results on random walks (e.g. [AKL+79], [CRR+89]) show that a polynomial length (in N) walk covers the graph with very high probability. Although several subtle points have to be finessed, the description above provides a simple unique naming algorithm for a nonfaulty distributed system (although slower than the fast naming algorithms of Schieber and Snir [SS89]).

The next step is to "upgrade" this algorithm to be self-stabilizing. In essence, the resulting self-stabilizing algorithm uses the same idea of Mitos-walks but in addition it must incorporate some mechanism to detect and wash out any faulty information from the system. To do this we provide every node with some data structures which hold sufficient information to detect local inconsistencies. Upon detecting local inconsistencies in some environment consisting of neighbouring nodes, some of these nodes will be able to erase the inconsistent data from their data structures and thus making their contribution to the global stabilization process.

3.1 A Communication Primitive

Since the system is asynchronous, the processors need some coordination in order to send and receive data from their neighbours. Following the idea in [IJ90], we introduce a self-stabilizing mini-protocol to ensure a reliable and deadlock-free information passing. The mini-protocol we offer is best described if it is further decomposed into two subprotocols each of which controls a unidirectional token passing. Let us describe the subprotocol for passing tokens from processor p_1 to processor p_2. Assume that p_2 is to the "right" of p_1. The subprotocol P_{right} is used to pass a token from p_1 to p_2 and has four ordered configurations: $(Idle, Idle), (Send, Idle), (Send, Receive)$, and $(Idle, Receive)$. When p_1 wants to send a set of tokens T to p_2 it changes its bit from $Idle$ to $Send$ (which results in the configuration $(Send, Idle)$) and at the same time it writes its state[4] in $R_{1,2}$. When p_2 reads p_1's link register it discovers the new configuration together with the set of tokens that p_1 wants to pass to it. Now p_2 receives the tokens and changes its bit to $Receive$

---

[3] *Mitos* = the name of the thread in the Greek maze myth.

[4] As we will later, the state of p_1 indicates (among other things) which tokens p_1 wants to pass to which of its neighbours.

(which results in the configuration $(Send, Receive)$). When p_1 is activated again it changes its bit to $Idle$ which indicates that this transaction has completed from p_1's point of view. p_2 then changes to $Idle$. The subprotocol P_{left} is defined similarly.

Clearly, the protocols P_{left} and P_{right} can be composed to a single protocol P which controls token passing in both directions. This protocol has 16 configurations and can be implemented by using two bits for each processor. When processor p_1 wants to pass some set of tokens to p_2 but the protocol P_{right} is not in the configuration $(Idle, Idle)$, i.e. it is "busy", then p_1 keeps doing all its other activities and when P_{right} is not "busy" anymore, p_1 initiates the delivery. Proving that this protocol is self-stabilizing, reliable and deadlock-free is straightforward.

3.2 High Level Description of the UNP

If u and v are neighbours then $P_u(v)$ denotes the port number of u through which u is connected to v. For a network path $p_0, p_1, \ldots p_k$ of $k+1$ processors such that the port number of p_{i-1} within p_i's hardware is n_i ($n_i = P_{p_i}(p_{i-1})$), the corresponding port path is $n_1 \circ n_2 \circ \ldots \circ n_k$. For a processor u, d_u denotes the number of neighbours of u. Note that if u and v are neighbours, then u knows $P_v(u)$ and v knows $P_u(v)$. This is implemented by instructing each processor u to always write the corresponding port number in each of its link-registers.

Each processor creates exactly one token. If by some initial fault a processor discovers that it has generated more than one token it eliminates arbitrarily all but one of them. If it discovers that it has not created any tokens then it creates one. Hence, with each processor p we associate one and only one token that was created by p. p is called the $initiator$ of the token. Every processor sends its token for a random walk of some bounded length W_N. The token then returns to its initiator, carrying information about the network and according to this information the initiator takes some action before it sends the token for another random walk. Informally a token can be seen as the front end of a random walk, while the rear end of this walk is always at the initiator. Formally a $token$ is a 5-tuple of the form

$$\langle id, \ rb, \ pp, \ dir, \ comp \rangle$$

where $id \in \{1, 2, \ldots, N'\}$ and $N' = cN^{d+1}$ for some $c > 0$ and some $d \geq 1$ such that $N' \geq N$. This id represents the name that the initiator of this token currently has. rb is a random bit. The initiator of this token uniformly chooses this random bit at the beginning of each random walk. rb is used to $resolve$ collisions between different tokens that have the same id. pp is a port path that gives the (unique) path to go back to the initiator following the port numbers of the nodes that this token has visited. dir is the direction bit and $dir \in \{F, B\}$. When $dir = F$ the token is going (randomly) forward. When $dir = B$ the token is travelling back to its initiator. $comp \in \{winner, loser\}$ is a bit indicating whether this token has lost a competition to another token with the same id. If $comp = winner$ (this is the initial value when it starts the random walk) the token is called a $winner$. If $comp = loser$, then the token is called a $loser$. For convenience we sometimes use 1 for winner and 0 for loser. When a loser token goes back to its initiator u it informs u that it should change its (candidate) ID. During a walk the attributes id and rb remain constant, pp and dir change and $comp$ may change. Note also that due to initial inconsistencies it is possible that more than n tokens exist in the system. Exactly n of them though have initiators, i.e., correspond to processors. When a processor u sends a token t for a random walk, t has the form $\langle id_u, rb, \emptyset, F, winner \rangle$, where id_u is the current candidate name for processor u and rb is a random bit that u chose for this walk.

Each processor u has a table Tab_u with T_N entries (the exact value of T_N will be estimated later, but it is always greater than NW_N). Each entry of the table is called a *table-tuple*. Each table-tuple in Tab_u keeps information about a single token that has passed through node u but has not come back. It is possible that a specific token t visits node u more than once. Tab_u has a separate entry for each one of these visits. When the token starts to go back to its initiator, each processor it passes through deletes from its table the corresponding entry.

A *table-tuple* is defined to be a triple $\langle t, pn, f \rangle$ where t is a token, pn is some port number, and f is a boolean flag. If the table-tuple $\langle t, pn, 1 \rangle$ is in the table Tab_u of some processor u, it means that processor u currently has the token t and is going to give it to the neighbour which is connected through port number pn. If $f = 0$ it means that processor u has already given the token t to its neighbour at port pn. For convenience we sometimes refer to a table-tuple $\langle t, pn, f \rangle$ as a 7-tuple where the first five attributes are the token attributes and the sixth and seventh attributes are pn and f respectively. An attribute x of a tuple tt will be denoted by $tt.x$. For example if we have the table-tuple $tt = \langle 15, 0, 3 \circ 7 \circ 1, F, winner, 6, 0 \rangle$ then $tt.pp = 3 \circ 7 \circ 1$, $tt.pn = 6$, $tt.dir = F$, etc. If an attribute of some tuple has the value X then it means that it is a *DON'T CARE* value. Besides the table Tab_u, every processor u maintains a list containing the IDs that u is not allowed to use and a variable id_u with its current name.

Each processor u performs the following tasks.

Initiates a Random Walk: When u initiates a walk, it randomly and uniformly chooses a port number pn from the set $\{1, \ldots, d_u\}$ and sends the token $t = \langle id_u, b, \emptyset, F, winner \rangle$ to its neighbour that is connected through port pn. id_u is the candidate name of u, and b is a bit chosen randomly and uniformly for this walk. When u initiates this walk it writes the table-tuple $\langle id_u, b, \emptyset, F, winner, pn, 1 \rangle$ in its table Tab_u. When u delivers this token to its neighbour at port pn it assigns the flag f of this table-tuple to 0.

Receives and Sends Tokens: When u receives a token $t = \langle id, b, p, d, c \rangle$ on its way forward ($d = F$) from a neighbour v, it first checks to see whether the token has completed its walk, i.e. if $|p| = W_N - 1$. If this is the case, u changes the *direction* bit to B and sends the token $\langle id, b, p, B, c \rangle$ back to v, without keeping an entry for this token in its table. Later we show that u may also change the $comp$ attribute of this token.

If t has not completed its walk, u attaches $P_u(v)$ (the number of the port from which it received t) to $t.pp$, selects randomly $pn \in \{1, \ldots, d_u\}$ and sends t to its neighbour on port pn. u also writes the table-tuple $\langle id, b, p \circ P_u(v), F, c, pn, 1 \rangle$ in its table Tab_u. After u delivers the token to its neighbour on port pn the flag f is assigned to 0.

If a processor u receives a token $t = \langle id, b, p, d, x_1 \rangle$ on its way backward $(d = B)$ from a neighbour v, it checks to see if the table-tuple $\langle id, b, p, F, x_2, P_u(v), 0 \rangle$ exists in Tab_u. If such an entry does not exist, then u ignores t. Otherwise, let $p = n_1 \circ n_2 \circ \ldots \circ n_k \neq \emptyset$. Process u sends to its neighbour on port n_k the token $t' = \langle id, b, p', B, x_1 \cdot x_2 \rangle$ where $x_1 \cdot x_2$ is the bitwise multiplication of x_1 with x_2, $p' = n_1 \circ n_2 \circ \ldots \circ n_{k-1}$, and erases from Tab_u the corresponding tuple. If $p = \emptyset$, then u is the initiator of t and now t has just came back. In this case u immediately initiates another random walk with the token $t' = \langle id', b', \emptyset, F, winner \rangle$ where b' is a new random bit and $id' = id$ iff u was not forced to change name (this happens as we shall see when $x_1 \cdot x_2 = 0$). Otherwise, id' is a new (random) name from the list of allowable names that u maintains.

We provide now the exact details of token passing. Assume that u wants to pass the token $t = \langle id, b, p, X, X \rangle$ to its neighbour v at port pn. This means that Tab_u has the table-tuple $tt = \langle id, b, p, X, X, pn, 1 \rangle$. When u initiates the mini-protocol (see section 3.1) for the delivery of t, u writes tt and the other table-tuples of Tab_u that are "related" to v into $R_{u,v}$, enters the $(Send, Idle)$ stage of the mini-protocol, and assigns the flag f of tt to 0. This means that u no longer holds this token t. If the token is going backwards then u also deletes the table-tuple from Tab_u. The table-tuples of Tab_u that are *related* to v are the ones that refer to tokens that u has received from v, the ones that u has delivered to v or the ones that u is going to deliver to v. This, as we shall see later, will provide the means for local consistency checking between neighbours. Note that when v reads $R_{u,v}$ it can extract the information about the tokens that u wants to pass to it by examining the pn and f fields of the table-tuples. Also notice that it is possible for u to deliver more than one token to v with one invocation of the mini-protocol.

For two neighbours u and v we say that the table-tuple $tt_i \in Tab_u$ is *consistent* with a table-tuple $tt_j \in Tab_v$ iff $tt_i = \langle id, b, p, F, X, P_u(v), 0 \rangle$ and $tt_j = \langle id, b, p \circ P_v(u), X, X, X, X \rangle$. We denote this by $tt_i \vdash tt_j$. The meaning is that a token, on its way forward, visited u (table-tuple tt_i corresponds to this visit) and then visited v (table-tuple tt_j corresponds to this visit). Note that this relation is not reflexive.

For any configuration C, a *segment* is a maximal sequence $tt_0, tt_1, \ldots, tt_k$ of table-tuples such that $\forall i, 1 \leq i \leq k, tt_{i-1} \vdash tt_i$ in C. Notice that every segment of k table-tuples induces a path with $k - 1$ edges (k nodes) on the communication graph. Let $tt_k \in Tab_u$. If $tt_k.f = 1$ then tt_k is called the *physical token* of this segment, and the processor u is said to *hold* the physical token. If $tt_k.f = 0$ then the physical token is either in the register between u and the neighbour of u that u is passing this token to, and in this case the token is said to be *in transit*, or it does not exist at all. The latter case can occur only in faulty configurations. A *walk* is defined to be a segment $tt_0, \ldots, tt_k$ if $tt_0.pp = \emptyset$. The processor that holds the table-tuple tt_0 is the *initiator* of the walk. If a segment has a physical token tt_k and $tt_k.dir = F$ then the segment is said to go forward, otherwise (this includes the case where the segment has no physical token) it is going *backward*.

If a node u holds one physical token t_1 and any other token t_2 such that $t_1.id = t_2.id$ then we say that a *collision* between these two physical tokens occurs. A collision is *successful* if

$t_1.rb \neq t_2.rb$ and $t_1.comp = t_2.comp = winner$. By defining it so we extend the definition of a collision to cover the case where a physical token t_1 visits a node u that has been visited by another token t_2 with the same ID id, but u does not hold now the physical token t_2. In the latter case we say that we have a collision of a token (t_1) with a path (trace of another token, t_2). The fact that a collision occurs even in the absence of one of the physical tokens speeds up the stabilization time of the protocol considerably and also simplifies the analysis.

Detects and Resolves Collisions: Assume that u detects a successful collision between tokens t_1 and t_2 such that $t_1.rb = 1$ and $t_2.rb = 0$. In this case u makes token t_1 to be winner by leaving in the bit $tt_1.comp$ the value $winner$, and makes token t_2 loser by assigning $t_2.comp = loser$. u marks the information that t_2 is a loser in the table-tuple corresponding to t_2 and the physical token corresponding to t_2 will acquire the loser bit when it comes back to u. If the collision was not successful then no special action is taken.

Updates the queue with the forbidden IDs, FI_u: Every processor u has a queue FI_u of size $N - 1$ that holds the names that u is not allowed to use. When a token $t = \langle id, * \rangle$ visits a node u such that $id \notin FI_u$ and id is different than the candidate name that u currently has, u writes the value id in (the front of) FI_u. As mentioned earlier, the other case when a processor u updates its forbidden set FI_u, is when a token comes back as a loser. The queue FI_u is of size $N - 1$. If it becomes full, then the next ID that enters shifts out the last one. This guarantees that the queue will always hold the most "fresh" forbidden ids.

Resolves Inconsistencies: It is possible that some of the processors are initially faulty. For this reason each processor u uses the contents of its table Tab_u and the contents of the tables of its neighbours to detect and resolve inconsistencies. An *inconsistency* occurs when a processor u has in its table Tab_u some tuple $tt_i = \langle id_u, b, p, F, X, pn, 0 \rangle$ such that there is no other tuple tt_j with $tt_i \vdash tt_j$ and there is no corresponding token in transit between u and the neighbour of u at port pn. Hence u is waiting for a token that will never arrive. In this case u eliminates tt_i from its table. In this way every faulty segment will eventually be eliminated from the system. If u was the initiator of that token, then u generates another token and sends it again for a walk.

Flushes the table Tab_u: If processor u receives a set of tokens and has no sufficient space in its table to accommodate all of them, then when Tab_u becomes full it "*flushes*" Tab_u by deleting all the entries, and accommodates the remaining tokens that it has received. The intuition behind this action is that by flushing the table, a processor eliminates more inconsistent entries than the number of inconsistent entries that will be created as a result of the flush.

4   Correctness Proof and Analysis

We now prove that the Unique Naming protocol is correct. We do that in two stages. In the first stage we prove that starting in any arbitrary configuration, the protocol stabilizes to a closed set of configurations that is called *semi-legal*. When the system is in one of these semi-legal configurations, it is guaranteed to be globally consistent. Informally, in any of these configurations, the memories contain no garbage or meaningless information. In the next stage, it is shown that starting from any semi-legal configuration, the system will stabilize into a legal configuration and therefore, since this set of configurations is proved to be closed too, the system will remain there forever (as long as new faults will not be introduced).

We define a *semi-legal* configuration C to be a configuration where exactly n walks and no other segments exist, and where no ID is abandoned in C. An ID id is *abandoned* in a configuration C if there exists a token t in C such that $t.id = id$, $t.comp = looser$, and there is no other token t' in C having $t'.id = id$ and $t'.comp = winner$. Let $\mathcal{SL}$ be the set of semi-legal configurations. A configuration which is not *semi-legal* is called *illegal*. A configuration C is called *legal* if C is in $\mathcal{SL}$ and all the processors have unique IDs. The set of legal configurations is denoted by $\mathcal{L}$. A configuration that is not legal could be either semi-legal or illegal. It is obvious that after the system reaches a configuration where exactly n walks exist, the requirement for no abandoned IDs will be satisfied as soon as the first n walks (one for each processor) come back. As discussed in section 2 the scheduler can be central, distributed or Read/Write and can be fair or proper. From the six combinations the Read/Write proper demon is the most powerful and therefore our proofs will assume this scheduler.

In the following two sections we will need the following important lemma.

Lemma 1 *The protocol satisfies the no-livelock requirement.*

Proof: Sketch. Let us assume the contrary. Let F be a set of "frozen" nodes (nodes that are not activated) and let A be the set of activated nodes. From the properties of the UNP and the fact that the mini-protocol of the previous section has no wait configurations we get that eventually, all the physical tokens that "circulate" in the subgraph induced by A will have to pass through nodes in F. Therefore, since all the processors in A will be disabled and some processors in F will be enabled, the (proper) scheduler will be forced to activate processors in F. Contradiction. $\square$

We now prove that the system will stabilize in $\mathcal{SL}$, and in the next section we prove that the system will stabilize in $\mathcal{L}$.

4.1 Stabilizing to Semi-Legal Configurations

We examine the system after the processors start to work correctly. In other words we look at the system where the processors are non-faulty but the initial configuration is arbitrary and even illegal. We need the following two lemmata. Both are easily proved by inspection of the protocol.

Lemma 2 *After the first round there will be exactly n walks in the system.*

Lemma 3 *The set $\mathcal{SL}$ is closed under the protocol operations.*

Even though the number of walks is n, it is possible that that there exist some additional (faulty) segments, that are not walks. The remaining of this section is devoted to proving that the system will reach a configuration in $\mathcal{SL}$. For this purpose, we first prove that the number of flushes in the system is bounded. After showing that, it will follow that the "self-cleaning" process of inconsistencies from the tables will be achieved. We use a potential function argument. We will define a function $\Phi : C \to \mathbb{N}$ for which the following hold: (1) If C_j is reachable from C_i by activating once one processor u, and u flushed its table in this transition from C_i to C_j, then $\Phi(C_i) > \Phi(C_j)$. (2) For any C_i, C_j such that C_j is reachable from C_i by activating one processor once, $\Phi(C_i) \geq \Phi(C_j)$.

To facilitate the definition of Φ we classify all possible segments into the following four types: (i) Segment of type wf: A walk that is going forward. (ii) Segment of type wb: A walk that is going backward. (iii) Segment of type sf: A segment that is not a walk and is going forward. (iv) segment of type sb: A segment that is not a walk and is going backward. It is straightforward to verify that in any configuration C, the previous *classification* provides a *partition* of the set of segments in C.

If x is a segment, we denote by $|x|$ the length of x (i.e. the number of table-tuples in its sequence). If $x = tt_0, tt_1, \ldots, tt_k$ is a segment of type sf and $tt_0 = \langle X, X, p, F, X, X, X \rangle$ we define the potential length of x ($PL(x)$) to be $W_N - |p|$ where W_N is the length of every Mitos-walk. This indicates the length of segment x if it grows as much as it can. Note that it may not grow so much, but it has the potential for such a growth.

We now extend the definition of an sf type segment to include some objects that are not "segments" as defined above. A motivation for this extension comes from the following scenario. Assume that a node u has the physical token tt_k of some segment s, $s = tt_1, tt_2, \ldots tt_k$, that is going forward (i.e. it is of type sf or wf). u wants to pass this token to its neighbour v and therefore writes the table-tuple tt_k in its register $R_{u,v}$. Before v reads this register (and receives this token) u flushes its table. At this point segment s is broken into one segment s', $s' = tt_1, tt_2, \ldots tt_{k-1}$ and the physical token tt_k which does not reside in any table but it is in the register $R_{u,v}$ and is going to start growing after v reads this register. We define a segment of type sf' to be a segment that has no table-tuples in any of the tables but, as described above, has one table-tuple on some register. Note that this situation can not happen for a segment that is going backwards because u would have deleted the corresponding table-tuple from Tab_u at the same time when u wrote it on the register.

If x is a segment of type sf' such that the register table-tuple of x has the form $tt = \langle X, X, p, F, X, X, 1 \rangle$, we define the potential length of such a segment to be $PL(x) = W_N - (|p| + 1)$.

We call the segments of type wf and wb, *good* segments (these are the n walks), and the segments of type sf, sf' and sb, *bad* segments (these are the faulty ones). It is easy to see that a configuration is semi-legal iff the system has *only* good segments. Let WF, WB, SF, SF' and SB be the sets of segments of type wf, wb, sf, sf' and sb respectively in a configuration C. We define the *potential* Φ of the configuration C to be :

$$\Phi(C) = (|WF| + |WB|)W_N + \sum_{s \in SF} PL(s) + \sum_{s \in SF'} PL(s) + \sum_{s \in SB} |s|$$

Intuitively it means that we take into account: (i) the n walks with a multiplicative factor of W_N (their maximum potential length, because even if they are going backwards they will start growing again), (ii) the lengths of all the segments of type sb (the shrinking bad segments that will not grow any more), and (iii) for each segment of type sf or sf' its potential length. This notion captures the "amount" of bad segments in the system because a configuration C is good iff $\Phi(C) = nW_N$. We will prove that eventually the system will get rid of all the bad segments.

Lemma 4 *The function Φ satisfies the conditions (1) and (2).*

Proof: Sketch. Proving that Φ satisfies condition (1) is by calculating for each processor u how each segment x with x_u entries in Tab_u before a flush, changes the potential Φ after the flush. It follows that if we choose the size of tables to be $T_N = NW_N + c$, where $c > 0$, then each flush decreases the potential by at least c. By checking all possible transitions that do not involve flushes, the reader may verify that Φ satisfies condition (2). $\square$

Notice that the potential is maximized when each walk consists solely of its initial table-tuple, each *table* is full, and each faulty segment is of type sf or sf' and has potential $W_N - 1$. Therefore we have:

Lemma 5 *For every configuration C,*

$$nW_N \leq \Phi(C) \leq nT_NW_N + 2nT_NW_N.$$

The first term of the previous upper bound represents the potential due to the processor tables and the second term represents the potential due to the register tables. Recall that each register $R_{u,v}$ has table tuples only for the tokens that passed from u to v and vice versa.

From lemmata 4 and 5 we conclude that only a bounded number of flushes can occur. In particular the following holds.

Corollary 1 *If C_0 is the initial configuration and if we choose T_N to be $2NW_N$, then at most f flushes can occur where*

$$f \leq \frac{\Phi(C_0) - nW_N}{c} = \frac{\Phi(C_0) - nW_N}{NW_N} \leq 3T_N \frac{n}{N}$$

Theorem 1 *The system will reach a semi-legal configuration after at most $O(fW_N)$ rounds.*

Proof: Sketch. The protocol satisfies the no-livelock property (Lemma 1). By inspection of the protocol it is easy to see that every segment of type sb is reduced by at least one table-tuple in each round and every segment of type sf or sf' is increased by at least one table-tuple in each round until it reaches a length of W_N and then it becomes a segment of type sb. So every bad segment (type sb, sf or sf') will be eliminated within at most $2W_N$ rounds unless a flush occurs and creates new bad segments. But only f flushes can occur. This proves that after at most $(f + 1)2W_N$ rounds the system will reach a semi-legal configuration. $\square$

Hence we have:

Corollary 2 *The system will reach a semi-legal configuration within $O(\frac{n}{N}T_NW_N)$ rounds.*

It is important to notice that the proofs of this section, except for lemma 1, do not use any probabilistic arguments and that they are indeed the (deterministic) worst case bounds (as opposed to expected number of rounds of the next section). We will need this observation in section 4.3 where we will prove the same bound for a much stronger adversary.

4.2 Stabilizing to Legal Configurations

We will show now that the system will enter a legal configuration and that it will remain in $\mathcal{L}$ forever. First we state the following lemma.

Lemma 6 *The set of configurations $\mathcal{L}$ is closed under the protocol operations.*

Proof: Clearly if all the processors have different IDs, then no processor will change name forever and no table will ever be full so no flushes will ever occur. $\square$

In light of Theorem 1 we assume that the system is now in a semi-legal configuration, meaning that there exist exactly n walks and no other segments of any other type. Informally, this means that all the garbage has been washed out from all the *tables* (memories), and that no garbage will be introduced to the system. Therefore, the nature of the proofs in this section has the nature of a "common" distributed computing analysis without any need to regard the self-stabilizing subtleties.

The argument is as follows. We first show that no processor can change ID more than n times. We then argue that no matter how the system is scheduled, in any sufficiently long computation a change of ID by at least one processor must occur, or the system must have reached a legal configuration. Since no more than n^2 ID changes can occur overall, it follows that after some finite computation the system must enter a legal configuration.

From now on we assume that the system has started to work from some semi-legal configuration which, from Corollary 2, occurs anyway after some bounded number of rounds. We say that a processor p *uses* an ID x, if p initiates a walk of a token t such that $t.id = x$. We say that a processor p has *relinquished* an ID x, if that processor has lost in a competition on x to some other processor.

Lemma 7 *In every semi-legal configuration, the subset $S \subseteq \{1, \ldots, N'\}$ of IDs that are used or were used by any of the processors is of size less than or equal to n.*

Proof: Sketch. If some processor u has used more than n IDs then this implies that some ID was relinquished by every processor in the system. This is a contradiction since for every successful competition there is always one winner. $\square$

Corollary 3 *Any processor can not change ID more than n times.*

We need the following claim:

Claim 1 *If the system is in some semi-legal but not legal configuration at some time r then a collision will occur between time r and time $r + 4W_N$ with probability $\geq \frac{1}{8}$ assuming that a random walk of length W_N covers the graph with probability $\geq \frac{1}{2}$.*

Proof: Sketch. Let u and v be two processors with the same IDs. Since each walk has length W_N it will come back after at most $2W_N$ rounds. The proof is completed by using the fact that the random walk covers the graph with probability $\geq \frac{1}{2}$ and the fact that one of the processors will choose a random bit different than the random bit of the other with probability $\frac{1}{2}$. □

The main result of this section is the following theorem:

Theorem 2 *Starting in any semi-legal but not legal configuration the expected number of rounds before the system reaches a legal configuration is bounded by $O(W_N \log N)$.*

Proof: Sketch. First note that if a processor chooses a new name then this name will be unique with very high probability ($\frac{N^d - N}{N^d}$, for some $d > 2$). If at some round r a set of l processors share the same ID, then after $O(W_N)$ rounds the expected number of processors that will still have this ID is $\frac{l}{2}$. The proof follows by linearity of expectation □

4.3 Correctness Under A More Powerful Scheduler

We will prove in this section that our protocol stabilizes even in the presence of a scheduler that has the following extra privileges:

1. It can choose *any* constant R (independent of the size of the network), at the beginning of the computation such that at any configuration it knows R future random tosses (of every random number generator used in the system).

2. It can choose (before hand) any bounded set of ϵ_is such that $0 < \epsilon_i < \frac{1}{2}$ for all i, and it can bias *any* random choice *individually* such that the probability of 1 is now $\frac{1}{2} \pm \epsilon_i$.

In order to beat this powerful adversary we slightly change the algorithm. We provide each processor u, having d_u neighbours, with $d_u + 3$ random number generators. One generator for choosing the random bits rb, one for choosing the candidate IDs and $d_u + 1$ to choose ports. When u gets a token from the ith neighbour it invokes the ith random number generator that is associated with this port. The $d_u + 1^{st}$ generator is used to produce the port number for the token when it is initiated.

We first prove that the adversary can not cause livelock; i.e., it can not block (freeze) for ever any processor. This proof is similar to the one of lemma 1 and is omitted.

Lemma 8 *The protocol satisfies the no-livelock requirement even under the strongest adversary.*

Note that all the proofs of section 4.1 make no use of any probabilistic arguments or any properties of the scheduler but they only use the no-livelock property. Therefore, combining lemma 8 with exactly the same proofs as in section 4.1, we have the following.

Corollary 4 *Starting from any configuration, the system will enter a semi-legal configuration after at most $O(T_N W_N)$ rounds even in the presence of the strongest adversary.*

The major difficulty is to prove that the system will stabilize to some legal configuration starting from any semi-legal configuration. The reason is that the adversary can now prevent collisions by using its extra privileges. The major result of this section is the following theorem.

Theorem 3 *Starting from some semi-legal but not legal configuration the system will stabilize to some legal configuration with probability that tends to 1 as time goes on.*

Proof: Sketch. Let u and v have the same ID. The idea is to show that for every configuration there exists a bounded size sequence S of favorable (for the protocol) random choices that will drive the system to a successful collision no matter what the adversary does. It will follow that in every configuration the probability ρ of successful collision after some bounded number of rounds is greater than zero and in fact it is bounded below by some positive constant. Let $T = \{u_0, u_1, \ldots, u_m, u_0\}$ be a DFS tour of some arbitrary spanning tree of the network. Informally, the sequence S has for every random port number generator of each processor a "sufficiently" large number of port choices that will create a a large continuous flow of tokens along T, and also a large number of 1s for the random bits of u and a large number of 0s for the random bits of v. This sequence S is bounded and when consumed by the system will necessarily create a successful collision no matter what the adversary does. $\square$

Corollary 5 *The system stabilizes to some legal configuration starting from any configuration even in the presence of the strongest adversary.*

5 The Topology Problem

We now present a self stabilizing solution for the topology problem. In this problem, each node in an anonymous network computes the same (labeled) description of the network's topology. It turns out that a simple variant of the UNP achieves this goal while preserving the properties of the UNP. The only major change is to add an additional, bounded size field to every token. This field carries the names of the neighbours of its initiator. That is, every initiator sends within its token a list with the IDs of its neighbours.

Each node knows the names of its neighbors by reading their registers. If G is the system's communication graph, it is clear that once the naming algorithm has stabilized, every processor p_i can compute a graph G_i such that $G \subseteq G_i$. The possible additional parts in any G_i might have evolved from early (prior to stabilization of the UNP) inconsistent information. It remains to detect and erase these faulty parts from every G_i. This task becomes easy if we remember a very trivial "property" of any graph: every edge connects *two* nodes. This means that a processor p_i needs only to remove from G_i all inconsistent edges. An *inconsistent edge* is an edge whose existence was not "declared" by exactly two nodes. After these removals, the resulting G_i might contain several components such that all but one are faulty. However, since G is connected it is guaranteed that one of these components in G_i is exactly G. Moreover, p_i's ID eventually labels exactly one node in G_i (see the Naming Protocol) and therefore, the right component is easy to detect; it is the component which contains p_i. This is a complete description of the Topology Protocol.

Needless to mention (but we mention!) that almost exactly the same time/space results for the Naming Protocol hold for the Topology Protocol.

6 Other Results

It turns out that the Naming and the Topology protocols provide simple solutions to many other self-stabilizing problems. Surprisingly, we are able to solve either by simple reductions, or by using the Mitos-walk technique, *all* the self stabilizing problems that we know of! Let us mention a few. For every problem mentioned, we provide an informal description of its solution.

Ranking

"After" knowing all the names each processor can deterministically determine its rank in the name set.

Leader Election

Once we assign a unique ID for every node and the subset (of the initial ID space) of IDs used is known, we can simply take the processor with the maximal ID as the leader. This goal is trivialy achieved by the Topology Protocol.

Spanning Tree

For every edge (u, v) in the graph, we assign any unique weight (e.g. $(min\{ID(u), ID(v)\}, max\{ID(u), ID(v)\})$). After the Topology Protocol has stabilized, every processor computes locally (and deterministically) an MST for the graph. Obviously, the spanning tree is the same for each process.

Mutual Exclusion

We use the Leader Election Protocol (Topology Protocol). The Mitos-walk of the leader provides a token management and a graph traversal scheme that solves the Mutual Exclusion ([IJ90]). Interestingly, we can solve the Mutual Exclusion problem using another approach. After we compute the Spanning Tree (Topology Protocol), we can use the result in [AET91] to reduce the Mutual Exclusion problem on that tree to the Mutual Exclusion on rings [Jal90].

7 UNP-The Improved Version

In this section we derive a faster protocol than the previous one. This protocol uses Mitos-walks and a *reset* subroutine to obtain an $O(C_G)$ expected stabilization time for the Unique Naming problem and consequently for the Topology and the other problems. A reset subroutine is a procedure that after invoked by any processor in the system makes the whole system start the underlying protocol from scratch. For this improved version of the UNP we can use either the new $O(D)$-time reset subroutine (where D is a bound on the network's diameter) in [AEH91], or the recent reset protocol of Awerbuch *et al.* [APV91].

The algorithm works similarly to the previous one in the sense that each processor chooses an ID and sends a token for a random walk in the graph. The difference, however, is the treatment of inconsistencies and the detection of multiple IDs. When a processor u discovers some inconsistency between itself and some neighbour, including the case where its table becomes full, or when u discovers that two processors have the same ID, then u executes the reset subroutine. The response (by every processor) for a reset command is to flush its table. Therefore, any invocation of the reset subroutine will clear all the tables of all the processors and will force each processor to choose a new random ID from the set of IDs. It turns out that each processor will choose unique ID with probability greater than $(e^{-1})^{\frac{1}{N^d}}$ where N^{d+1} is the size of the name space (note that this probability converges rapidly to one). If two processors have the same IDs then after $O(C_G)$ expected number of rounds this will be detected and some processor will invoke the reset procedure. It can be easily seen that this algorithm stabilizes within $O(C_G)$ rounds. The space complexity is the same as before.

8 Conclusions and Discussion

We have presented uniform and bounded space self-stabilizing solutions for many problems in the area. These problems include the Unique Naming, the Topology, the Spanning Tree, the Ranking, the Token Management and other.

We are now completing a work that uses the same idea and yields self-stabilizing solutions for most of the problems mentioned above on the more restrictive message passing model even with unbounded channel capacities. Nevertheless our solution still has bounded space and time complexities. As we mentioned in the introduction our space and time complexities do not measure up with other faster protocols such as [AKY91] and [DIM91]. However, our contribution is of fairly simple protocols which are robust against some very powerful adversaries. More importantly, we have introduced here a simple paradigm of self-checking via the use of tables. The usefulness and effectiveness of this paradigm has already proved itself in another new result by Anagnostou et al. [AEH91] which uses it to derive an almost time-optimal ($O(D)$) Naming and Topology protocols, where D is a known bound for the diameter of the network.

Acknowledgements

We are grateful to Vassos Hadzilacos and Hisao Tamaki for helpful discussions and many comments that improved this work both on style and on content. We thank Baruch Schieber and Moti Yung for pointing out an error that appeared in a previous version of this paper. We thank Shay Kutten and the anonymous referees for mentioning us several important works that we were not aware of. Finally, the first author wishes to thank Pat Donnelly for a lot of help during this work. The second author wishes to thank Michael Loftus for letting him using his computing facilities.

References

[AEH91] E. Anagnostou, R. El-Yaniv, and V. Hadzilacos. Fast and Simple Self-Stabilizing Algorithms, 1991. In preparation.

[AET91] E. Anagnostou, R. El-Yaniv, and H. Tamaki. A Self Stabilizing Reduction and Applications, 1991. Unpublished Manuscript.

[AFL83] E. Arjomandi, M. Fisher, and N. Lynch. Efficiency of Synchronous Versus Asynchronous Distributed Systems. *Journal of the ACM*, 30 (3):449–456, 1983.

[AKL+79] R. Aleliunas, R. Karp, R. Lipton, L. Lovász, and C. Rackoff. Random walks, Universal traversal sequences, and the complexity of maze problems. In *Proc. of the 20th Annual Symposium on Foundations of Computer Science*, pages 218–223, 1979.

[AKY90] Y. Afek, S. Kutten, and M. Yung. Memory-Efficient Self Stabilizing Protocols for General Networks. In *4th IWDAG*, pages 15–28, Bari, Italy, September, 1990.

[AKY91] Y. Afek, S. Kutten, and M. Yung, August, 1991. Personal Communication.

[APV91] B. Awerbuch, B. Patt-Shamir, and G. Varghese. Self-Stabilization by Local Checking and Correction. In *32nd FOCS*, October, 1991.

[AV91] B. Awerbuch and G. Varghese. Distributed Program Checking: a Paradigm for Building Self-Stabilizing Distributed Protocols. In *32nd FOCS*, October, 1991.

[BGW89] G. Brown, M. Gouda, and C. Wu. Token Systems that Self-Stabilize. *IEEE Transactions on Computers*, 38, 6:845–852, 1989.

[BK89] A. Broder and A. Karlin. Bounds on Cover Times. *Journal of Theoretical Probability*, 2:101–120, 1989.

[BKRU89] A. Broder, A. Karlin, P. Raghavan, and E. Upfal. Trading Space for Time in Undirected s-t Connectivity. In *Proc. of the 21st ACM Symposium on Theory of Computing*, pages 543–549, Seattle, WA, 1989.

[BP89] L. E. Burns and J. Pachl. Uniform Self-Stabilizing Rings. *ACM Transactions on Programming Languages and Systems*, 11, 2:330–344, 1989.

[CRR+89] A. Chandra, P. Raghavan, W. Ruzzo, R. Smolensky, and P. Tiwari. The Electrical Resistance of a Graph Captures its Commute and Cover Times. In *Proc. of the 21st ACM Symposium on Theory of Computing*, pages 574–586, Seattle, WA, 1989.

[Dij74] E. W. Dijkstra. Self-stabilizing systems in spite of distributed control. *Comm. of the ACM*, 17(11):643–644, 1974.

[DIM90] S. Dolev, A Israeli, and S. Moran. Self Stabilization of Dynamic Systems Assuming Only Read/Write Atomicity. In *Proc. of the 9th ACM Symposium on Principles of Distributed Computing*, pages 103–117, Quebec City, Canada, 1990.

[DIM91] S. Dolev, A. Israeli, and S. Moran. Uniform Dynamic Self-Stabilizing Leader Election. In *5th IWDAG*, Delphi, Greece, October, 1991.

[IJ90]   A. Israeli and M. Jalfon. Token Management Schemes and Random Walks Yield Self Stabilizing Mutual Exclusion. In *Proc. of the 9th ACM Symposium on Principles of Distributed Computing*, pages 119–131, 1990.

[Jal90] Marc Jalfon. Randomized Self Stabilizing Uniform Protocols for Distributed Systems. Master's thesis, Department of Computer Science, Technion, 1990.

[KP90]   S. Katz and K. J. Perry. Self-stabiling Extensions for Message-passing Systems. In *Proc. of the 9th ACM Symp. on Principles of Distr. Computing*, pages 91–101, Quebec City, Canada, 1990.

[Kru79] H. S. Kruijer. Self-stabilization (in spite of distributed control) in tree-structured systems. *Inf. Proc. Letters*, 8, 2:91–95, 1979.

[SS89]   B. Schieber and M. Snir. Calling Names on Nameless Networks. In *Proc. of the 8th ACM Symp. on Prenciples of Distr. Computing*, pages 319–328, Edmonton, Canada, 1989.

Pseudo Read-Modify-Write Operations: Bounded Wait-Free Implementations
(Extended Abstract)

James H. Anderson*   Bojan Grošelj†

Department of Computer Science
The University of Maryland at College Park
College Park, Maryland 20742

Abstract

We define a class of operations called *pseudo read-modify-write* (PRMW) operations, and show that nontrivial shared data objects with such operations can be implemented in a bounded, wait-free manner from atomic registers. A PRMW operation is similar to a "true" read-modify-write (RMW) operation in that it modifies the value of a shared variable based upon the original value of that variable. However, unlike an RMW operation, a PRMW operation does not return the value of the variable that it modifies. We consider a class of shared data objects that can either be read, written, or modified by a commutative PRMW operation, and show that any object in this class can be implemented without waiting from atomic registers. The implementations that we present are polynomial in both space and time and thus are an improvement over previously published ones, all of which have unbounded space complexity.

1   Introduction

The implementation of shared data objects is a subject that has received much attention in the concurrent programming literature. A *shared data object* is a data structure that is shared by a collection of processes and is accessed by means of a fixed set of operations. Traditionally, shared data objects have been implemented by using mutual exclusion,

---

*Work supported in part by an award from the General Research Board, University of Maryland, and by an NSF Research Initiation Award. E-mail: jha@cs.umd.edu.

†On leave from: The Center for Advanced Computer Studies, University of Southwestern Louisiana, Lafayette, Louisiana 70504. E-mail: bojan@cs.umd.edu.

with each operation corresponding to a "critical section." Although conceptually simple, such implementations suffer from two serious shortcomings. First, they are not very resilient: if a process experiences a halting failure while accessing such a data object, then the data object may be left in a state that prevents subsequent accesses by other processes. Second, such implementations restrict parallelism. This may be undesirable, especially if operations are time-consuming to execute (e.g., file transfers).

As a result of these two shortcomings, there has been much interest recently in wait-free implementations of shared data objects. An implementation of a shared data object is *wait-free* iff the operations of the data object are implemented without any unbounded busy-waiting loops or idle-waiting primitives. Wait-free shared data objects are inherently resilient to halting failures: a process that halts while accessing such a data object cannot block the progress of any other process that also accesses that same data object. Wait-free shared data objects also permit maximum parallelism: such a data object can be accessed concurrently by any number of the processes that share it since one access does not have to wait for another to complete.

One of the major objectives of researchers in this area has been to characterize those shared data objects that can be implemented without waiting in terms of single-reader, single-writer, single-bit atomic registers. An *atomic register* is a shared data object consisting of a single shared variable that can either be read or written in a single operation [20]. An N-reader, M-writer, L-bit atomic register consists of an L-bit variable that can be read by N processes and written by M processes. It has been shown in a series of papers that multi-reader, multi-writer, multi-bit atomic registers can be implemented without waiting in terms of single-reader, single-writer, single-bit atomic registers [6, 9, 10, 16, 17, 20, 21, 22, 24, 25, 26, 27, 28]. This work shows that, using only atomic registers of the simplest kind, it is possible to solve the classical readers-writers problem without requiring either readers or writers to wait [13].

Another shared data object of interest is the composite register, a data object that generalizes the notion of an atomic register. A *composite register* is an array-like shared data object that is partitioned into a number of components. As illustrated in Figure 1, an operation of such a register either writes a value to a single component, or reads the values of all components. Afek et al. [2] and Anderson [3, 4] have shown that composite registers can be implemented from atomic registers without waiting. This work shows that, using only atomic registers of the simplest kind, it is possible to implement a shared memory that can be read in its entirety in a single "snapshot" operation, without resorting to mutual exclusion.

In this paper, we consider the important question of whether there exist other nontrivial shared data objects that can be implemented from atomic registers without waiting. We define a class of operations called *pseudo read-modify-write* (PRMW) operations and consider a corresponding class of shared data objects called *PRMW objects*. This class of

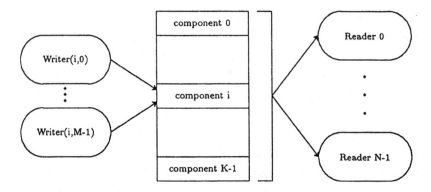

Figure 1: Composite register structure (only writers for component i are depicted).

objects includes such fundamental objects as counters, shift registers, and multiplication registers. A PRMW object consists of a single shared variable that can either be read, written, or modified by a commutative PRMW operation. The PRMW operation takes its name from the classical read-modify-write (RMW) operation as defined in [18]. The RMW operation has the form "$temp$, $X := X$, $f(X)$," where X is a shared variable, $temp$ is a private variable, and f is a function. Executing this operation has the effect of modifying the value of X according to f, and returning the original value of X in $temp$. The PRMW operation has the form "$X := f(X)$," and differs from the RMW operation in that the value of X is not returned.

We prove that any PRMW object can be implemented from atomic registers in a wait-free manner. We establish this result by first considering the problem of implementing a counter without waiting. A *counter* is a PRMW object whose value can be read, written, or incremented by an integer value.[1] We first show that counters can be implemented from composite registers without waiting, and then show that our implementation can be generalized to apply to any PRMW object. Given the results of [2, 3, 4], this shows that any PRMW object can be implemented without waiting using only atomic registers. Our results stand in sharp contrast to those of [5, 14], where it is shown that RMW operations cannot, in general, be implemented from atomic registers without waiting.

The problem of implementing PRMW objects without waiting has been studied independently by Aspnes and Herlihy in [7]. Aspnes and Herlihy give a general, wait-free implementation that can be used to implement any PRMW object. A counter implementation, which is obtained by optimizing the general implementation, is also given. Both of these implementations have unbounded space complexity: the first uses a graph of unbounded size to represent the history of the implemented data object, and the sec-

---

[1]Note that decrementing can be defined in terms of incrementing; thus, a counter actually supports four different operations: read, write, increment, and decrement.

ond uses unbounded timestamps. Our counter implementation and its generalization are polynomial in space and time.

The rest of the paper is organized as follows. In the next section, we formally define the problem of implementing a counter from composite registers. The counter implementation mentioned above is described in Section 3. The correctness proof, while only sketched here, is given in its entirety in the full paper using the UNITY programming theory of Chandy and Misra [11]. In Section 4, we discuss several issues pertaining to our implementation, and show that the implementation can be generalized to implement any PRMW object. Concluding remarks appear in Section 5.

2   Problem Definition

In this section, we consider the problem of implementing a counter from composite registers, and give the conditions that such an implementation must satisfy to be correct. An implementation consists of a set of procedures along with a set of variables. Each procedure has the following form:

> **procedure** *name(inputs)*
> > *body*;
> > **return(***outputs***)**
> **end**

where *name* is the name of the procedure, *inputs* is an optional list of input parameters, *outputs* is an optional list of output parameters, and *body* is a program fragment comprised of atomic statements. We assume that the counter is shared by N processes, and that each process has three resident procedures, *Read*, *Write*, and *Increment*. A Read procedure is invoked by a process to read the value of the counter; the value read is returned as an output parameter. A Write procedure is invoked to write a new value to the counter; the value to be written is specified as an input parameter. An Increment procedure is invoked to increment the value of the counter; the value to add is given as an input parameter. We assume that each process invokes its resident procedures in a serial manner. An example of an implementation is given in Figures 2 and 3.

Each variable of an implementation is either private or shared. A *private variable* is defined only within the scope of a single process, whereas a *shared variable* is defined globally and may be accessed by more than one process. The procedures and variables of an implementation are required to satisfy the following two restrictions.

- *Atomicity Restriction:* Each shared variable is required to correspond to a component of a composite register. Therefore, each statement may either write a single shared variable or read one or more shared variables, but not both.

- *Wait-Freedom Restriction*: As mentioned in the introduction, each procedure is required to be "wait-free," i.e., idle-waiting primitives and unbounded busy-waiting loops are not allowed. (A more formal definition of wait-freedom is given in [5].)

We now define several concepts that are needed to state the correctness condition for an implementation. These definitions apply to a given implementation. A *state* is an assignment of values to the variables of the implementation. (Each process's "program counter" is considered to be a private variable of that process.) One or more states are designated as *initial states*. An *event* is an execution of a statement of a procedure. We use $s \xrightarrow{e} t$ to denote the fact that state t is reached from state s via the occurrence of event e. A *history* of the implementation is a sequence (either finite or infinite) $s_0 \xrightarrow{e_0} s_1 \xrightarrow{e_1} \cdots$ where s_0 is an initial state. Note that the definition of a history abstracts away from the activities of each process that do not directly affect the implemented counter.

The set of events in a history corresponding to a single procedure invocation is called an *operation*. An operation of a Read (respectively, Write or Increment) procedure is called a *Read operation* (respectively, *Write operation* or *Increment operation*).[2] Each history defines an irreflexive partial order on operations: an operation p precedes another operation q in this ordering iff each event of p occurs before all events of q in the history.

As mentioned above, each Read procedure has an output parameter that returns the value read from the counter; the value returned by a Read operation is called the *output value* of that operation. As also mentioned above, each Write (Increment) procedure has an input parameter that specifies the value to be written (added) to the counter; the value written (added) to the counter by a Write (Increment) operation is called the *input value* of that operation.

An operation of a procedure P in a history is *complete* iff the last event of the operation occurs as the result of executing the **return** statement of P. A history is *well-formed* iff each operation in the history is complete.

Given this terminology, we are now in a position to define what it means for a history of an implementation to be "linearizable." Linearizability provides the illusion that each operation is executed instantaneously, despite the fact that it is actually executed as a sequence of events. It can be shown that the following definition is equivalent to the more general definition of linearizability given by Herlihy and Wing in [15], when restricted to the special case of implementing a counter.

Linearizable Histories: A well-formed history h of an implementation is *linearizable* iff the partial order on operations defined by h can be extended to a total order $\prec$ such that for each Read operation r in h, the following condition is satisfied.

---

[2] In order to avoid confusion, we capitalize the terms "Read," "Write," and "Increment" when referring to the operations of the implemented counter, and leave them uncapitalized when referring to the variables used in the implementation.

- If there exists a Write operation w such that $w \prec r \wedge \neg \langle \exists v : v$ is a Write operation :: $w \prec v \prec r \rangle$,[3] then the output value of r equals that obtained by adding the input value of w to the sum of the input values of all Increment operations ordered between w and r by $\prec$.

- If no such w exists, then the output value of r equals that obtained by adding the initial value of the counter to the sum of the input values of all Increment operations ordered before r by $\prec$. □

An implementation of a counter is *correct* iff it satisfies the Atomicity and Wait-Freedom restrictions and each of its well-formed histories is linearizable.

3   Counter Implementation

In this section, we present our counter implementation. For now, we assume that the counter stores values ranging over the integers. Later, in Section 4, we consider the case in which the counter stores values over some bounded range. (In the latter case, overflow is a problem.)

The shared variable declarations for the implementation are given in Figure 2 and the procedures for process i, where $0 \le i < N$, are given in Figure 3. Central to the proof of the implementation is the "history variable" H, which is defined in Figure 2. H is used to totally order the operations of the implemented counter, and is one of several auxiliary variables that are used to facilitate the proof of correctness. H is a sequence of tuples of the form $(op, pnum, id, val)$, where op ranges over {READ, WRITE, INC}, $pnum$ ranges over $0..N - 1$, and id and val are integers. Each tuple records the "effect" of a specific operation. The type of the particular operation is identified by the op field, the process invoking the operation is identified by the $pnum$ field ($pnum$ stands for "process number"), and the id field is used to differentiate between operations of the same type by the same process. The val field is used to record the output value of a Read operation or the input value of a Write or Increment operation. The following notational conventions regarding history variables will be used in the remainder of the paper.

Notational Conventions for History Variables: If X and Y are sequences of tuples, then $X \cdot Y$ denotes the sequence obtained by appending Y onto the end of X. If X is a prefix of Y, then we write $X \sqsubseteq Y$. If z is a tuple, then $X - z$ denotes the sequence obtained by removing all occurrences of z from X. (Note that there may be no occurrences of z in X, in which case $X = X - z$.) The symbol $\emptyset$ is used to denote the empty sequence. □

---

[3]The notation of [11] is used for quantification. Thus, for example, $\langle \sum j : B(j) :: E(j) \rangle$ denotes the sum $\sum_{j \text{ s.t. } B(j)} E(j)$.

type

$Tagtype$ = **record** $seq : 0..N + 1$; $pnum : 0..N - 1$ **end**;

$Qtype$ = **record** val: **integer**; $tag : Tagtype$ **end**;

$Htype$ = **record** op: (READ, WRITE, INC); $pnum : 0..N - 1$;

$\quad\quad\quad\quad\quad$ id : **integer**; val : **integer** **end**

shared var

Q : **array**$[0..N]$ **of** $Qtype$; $\quad\quad\quad\quad$ { $(N + 1)$-component composite register }

H, S : **sequence of** $Htype$; $\quad\quad\quad$ { Auxiliary history variables }

$Ovlap$: **array**$[0..N - 1]$ **of boolean** $\quad$ { Auxiliary flag: indicates if an Increment is

$\quad\quad\quad\quad\quad\quad\quad\quad\quad\quad\quad\quad\quad\quad\quad\quad\quad\quad\quad\quad$ overlapped by a Write }

initially

$\langle \forall j : 0 \leq j < N :: Q[j].val = 0 \ \wedge \ Q[j].tag = (0,j) \rangle \ \wedge \ Q[N].val = init \ \wedge$
$Q[N].tag = (0,0) \ \wedge \ H = S = \emptyset$

Figure 2: Shared variable declarations.

According to the semantics of a counter, Write and Increment operations change the value of the implemented counter, whereas Read operations do not. This is reflected in the definition of the function Val, given next. This function gives the "value" of the implemented counter as recorded by a sequence of READ, WRITE, and INC tuples.

Definition of Val: Let i range over $0..N - 1$, let n and v be integer values, let $init$ be the initial value of the implemented counter, and let α be a sequence of READ, WRITE, and INC tuples. Then, the function Val is defined as follows:

$$Val(\alpha \cdot (\text{READ}, \ i, \ n, \ v)) \ \equiv \ Val(\alpha)$$
$$Val(\alpha \cdot (\text{WRITE}, \ i, \ n, \ v)) \ \equiv \ v$$
$$Val(\alpha \cdot (\text{INC}, \ i, \ n, \ v)) \ \equiv \ Val(\alpha) + v$$
$$Val(\emptyset) \ \equiv \ init \quad\quad\quad\quad\quad\quad \square$$

The proof of correctness is based upon the following lemma.

Lemma: If the following two conditions hold, then each well-formed history of the implementation is linearizable.

- *Ordering*: During the execution of each operation (i.e., between its first and last events), an event occurs that appends a unique tuple for that operation to H, and this tuple is not subsequently removed from H.

private var   { Private variables for process i }

  $x : Qtype$;   { Local copy of Q }

  $seq : 0..N + 1$;   { Sequence number }

  $tag : Tagtype$;   { Tag value for Write or Increment operation }

  $outval$, sum, id : integer;   { Output value; sum of increment input values;
               auxiliary operation id }

  $tuple : Htype$ { Auxiliary variable for recording INC tuple }

procedure $Read()$ **returns integer**

 0: **read** $x := Q$;   { Take snapshot and compute output value }

  $outval := x[N].val + \langle \sum j : 0 \leq j < N \ \wedge \ x[j].tag = x[N].tag :: x[j].val\rangle$;

  H, $id := H \cdot (\text{READ}, i, id, outval)$, $id + 1$;

 1: **return**($outval$)

end { $Read$ }

procedure $Write(inval : \text{integer})$ { Input value passed as a parameter }

 2: **read** $x := Q$;   { Take snapshot; compute sequence no., tag }

  select seq **such that** $\langle \forall j : 0 \leq j \leq N :: seq \neq x[j].tag.seq\rangle$; { Such a seq exists }

  $tag := (seq, i)$;

 3: **write** $Q[N] := (inval, tag)$;   { Write base component }

  H, S, id, $Ovlap[0]$, $\ldots$, $Ovlap[N-1] :=$

        $H \cdot S \cdot (\text{WRITE}, i, id, inval)$, $\emptyset$, $id + 1$, $true$, $\ldots$, $true$;

 4: **return**

end { $Write$ }

Figure 3: Procedures for process i.

- *Consistency*: The tuples in H are consistent with the semantics of a counter. More precisely, the following assertion is an invariant: $\langle \forall \alpha :: \alpha \cdot (\text{READ}, i, n, v) \sqsubseteq H \Rightarrow v = Val(\alpha)\rangle$.   □

To see why this lemma holds, consider a well-formed history h. Let α denote the "final" value of H in h: i.e., if h is finite, then $H = \alpha$ in the final state of h, and if h is infinite, then α is infinite and every finite prefix of α is a prefix of H for some infinite sequence of states in h. Define a total order $\prec$ on the operations in h as follows: $p \prec q$ iff p's tuple occurs before q's tuple in α. By the Ordering condition, $\prec$ extends the partial precedence ordering on operations defined by h. By the Consistency condition and the definition of Val, $\prec$ is consistent with the semantics of a counter. That is, the output value of each Read operation in h equals that obtained by adding the input value of

procedure *Increment*(*inval* : **integer**) { Input value passed as a parameter }

{ First phase: read Q and write $Q[i]$ }
5: **read** $x := Q$;   { Take snapshot and copy tag }
   $tag := x[N].tag$;
   $S, tuple, id := S \cdot (\text{INC}, i, id, inval), (\text{INC}, i, id, inval), id + 1$;

{ Compare new tag with tag of last Increment operation }
6: **if** $x[i].tag = tag$ **then** $sum := x[i].val$ **else** $sum := 0$ **fi**;
7: **write** $Q[i] := (sum, tag)$;   { Write increment component }

{ Second phase: read Q and write $Q[i]$ }
8: **read** $x := Q$;   { Take snapshot and copy tag }
   $tag := x[N].tag$;
   $Ovlap[i] := false$;

{ Compare tag from first phase with tag from second phase }
9: **if** $x[i].tag = tag$ **then** $sum := x[i].val + inval$ **else** $sum := 0$ **fi**;
10: **write** $Q[i] := (sum, tag)$;   { Write increment component }
    if $tuple \notin H \lor (\neg Ovlap[i] \land x[i].tag = tag)$ **then**
       $H, S := (H - tuple) \cdot tuple, S - tuple$
    fi;
11: **return**
end { *Increment* }

Figure 3: Procedures for process i (continued).

the most recent Write operation according to $\prec$ (or the initial value of the implemented counter if there is no preceding Write operation) to the sum of the values of all intervening Increment operations according to $\prec$. This implies that h is linearizable.

We justify the correctness of the implementation below by informally arguing that the Ordering and Consistency conditions are satisfied; a formal proof is presented in the full paper. Before proceeding, several comments concerning notation are in order.

Notational Conventions for Implementations: Each initial state of the implementation is required to satisfy the **initially** assertion given in Figure 2. (If a given variable is not included in the **initially** assertion, then its initial value is arbitrary. Note that each private variable has an arbitrary initial value.) As in the definition of *Val*, we use *init* to denote the initial value of the implemented counter. To make the implementation easier to understand, the keywords **read** and **write** are used to distinguish reads and writes of (nonauxiliary) shared variables from reads and writes of private variables. To simplify the implementation, each labeled sequence of statements is assumed to be a single atomic statement. (Each such sequence can easily be implemented by a single multiple-assignment.) □

Each of the labeled atomic statements in Figure 3 satisfies the Atomicity restriction of Section 2. In particular, no statement writes more than one (nonauxiliary) shared variable, and no statement both reads and writes (nonauxiliary) shared variables. The Wait-Freedom restriction is also satisfied, since each procedure contains no unbounded loops or idle-waiting primitives. With regard to the Atomicity restriction, it should be emphasized that auxiliary variables are irrelevant. These variables are not to be implemented, but are used only to facilitate the proof of correctness: observe that no auxiliary variable's value is ever assigned to a nonauxiliary variable.

We continue our description of the implementation by considering the shared variables as defined in Figure 2. There is only one nonauxiliary shared variable, namely the $(N+1)$-component composite register Q. Each process i, where $0 \leq i < N$, may read all of the components of Q, and may write components $Q[i]$ and $Q[N]$. Each component of Q consists of two fields, *val* and *tag*. The *val* field is an integer "value," and is used to record the input value of an Increment or Write operation. The *tag* field consists of two fields, *seq* and *pnum*. The *seq* field is a "sequence number" ranging over $0..N+1$, and the *pnum* field is a "process number" ranging over $0..N - 1$.

As mentioned previously, a number of shared auxiliary variables are also included in the implementation. The most important of the auxiliary variables is the history variable H, described above. Another shared history variable S is used to hold INC tuples for "pending" Increment operations. The role of S is explained in detail below. The shared auxiliary boolean variable $Ovlap[i]$ indicates whether an Increment operation of process i is "overlapped" by a Write operation.

The components of Q are used in the following manner. The last component $Q[N]$ is the "base component," and is updated whenever a Write operation is performed by any process. Each other component $Q[i]$, where $0 \leq i < N$, is an "increment component," and is updated when process i performs an Increment operation. Informally, the value of the counter is defined to be that obtained by adding the input value of the most recent Write operation — which is stored in the base component $Q[N]$ — to the sum of the input values of all subsequent Increment operations — which are stored in the increment components $Q[0]$ through $Q[N - 1]$. In this context, "recent" and "subsequent" are interpreted with respect to the total order $\prec$, which is defined using the history variable H as explained above. The value of the counter is formally defined by the following expression.

$$Q[N].val + \langle \sum j : 0 \leq j < N \ \wedge \ Q[j].tag = Q[N].tag :: Q[j].val \rangle \qquad (1)$$

In this expression, $Q[N].val$ represents the input value of the most recent Write operation, and $\langle \sum j : 0 \leq j < N \ \wedge \ Q[j].tag = Q[N].tag :: Q[j].val \rangle$ gives the sum of the input values of all subsequent Increment operations.

With expression (1) in mind, we now consider how Read, Write, and Increment operations are executed in the implementation. A Read operation simply computes the sum defined by (1). Note that, because Q is a composite register, this sum can be computed by reading Q only once.

A Write operation first computes a new tag value, and then writes its input value and tag value to $Q[N]$. The tag value consists of a sequence number and process number. The sequence number is obtained by first reading Q, and then selecting a value differing from any sequence number appearing in the components of Q. Note that, because there are $N+1$ sequence numbers appearing in the $N+1$ components of Q, and because each sequence number ranges over $0..N+1$, such a value exists.

Each Increment operation of process i is executed in two phases. In both phases, Q is read and then $Q[i]$ is written. In each phase, the tag value written to $Q[i]$ is obtained by copying the value read from $Q[N].tag$. If several successive Increment operations of process i obtain the same tag value, then their input values are accumulated in $Q[i].val$ (see statement 9). It can be shown that the value assigned to $Q[i].val$ by an Increment operation equals the sum of the input values of all Increment operations of process i that are ordered by $\prec$ to occur after the most "recent" Write operation.

To formally establish the correctness of the implementation, it suffices to prove that the Ordering and Consistency conditions hold. It is straightforward to show that the Ordering condition holds. Each Read and Write operation appends a unique tuple for itself to H (statements 0 and 3) and such tuples are never removed from H. For Increment operations, the situation is slightly more complicated. When an Increment operation p of process i reads Q in its first phase (statement 5), a unique tuple for that operation is appended to the history variable S. S contains tuples for "pending" Increment operations. The tuple for p is subsequently appended to H either by p itself or by an "overlapping" Write operation. In particular, if a write to $Q[N]$ (statement 3) occurs between the execution of statements 5 and 10 by p, then the first such write removes p's tuple from S and appends it to H. On the other hand, if no write to $Q[N]$ occurs in this interval, then p's tuple is removed from S and appended to H when p executes statement 10. To complete the proof, note that an INC tuple can be removed from H only by the operation that appended that tuple to H (statement 10), and in this case, the tuple is reappended. This implies that Ordering is satisfied.

In the remainder of this section, we outline the proof of Consistency. To see that this condition holds, first observe that the tuples in H may be reordered only when an INC tuple is removed and reappended (see statement 10). However, prior to being removed, such an INC tuple must be part of a sequence of INC tuples followed by a WRITE tuple (see statement 3). By the definition of *Val*, removing and reappending such an INC does not invalidate the value of any READ tuple.

To complete the proof, we must show that each READ tuple has a valid value when first appended to H. That is, we must prove that $outval = Val(H)$ whenever such a tuple is appended to H by the execution of statement 0 by any process. This can be established by proving the invariance of the following assertion.

$$Val(H) = Q[N].val + \langle \sum i : 0 \leq i < N \ \wedge \ Q[i].tag = Q[N].tag :: Q[i].val \rangle \quad (2)$$

Establishing the invariance of (2) is the crux of the proof.

By the definition of the initial state, assertion (2) is initially true. Thus, to prove that it is an invariant, we must show that it is not falsified by the execution of any statement by any operation. Showing that (2) is not falsified by any Increment operation is the most difficult part of the proof. The main thrust of this part of the proof is to show that the following two conditions hold for each Increment operation p: first, if the execution of statement 10 by p increments the right-side of (2) by the input value of p, then it also appends an INC tuple for p to H; second, if the execution of statement 10 by p leaves the right-side of (2) unchanged, then either p's input value is 0 (in which case it really does not matter how p is linearized) or H is not modified (in which case p's INC tuple has already been appended to H by an "overlapping" Write operation).

Showing that (2) is not falsified by the statements of Read and Write operations is somewhat simpler. The statements that must be considered are those that may modify H or Q. For Read and Write operations, there are two statements to check, namely 0 and 3.

Statement 0 does not modify Q, but appends a READ tuple to H. However, appending a READ tuple to H does not change the value of $Val(H)$. Therefore, statement 0 does not falsify (2).

Statement 3 assigns "$Q[N].val := inval$" and "$H := H \cdot S \cdot (\text{WRITE}, i, id, inval)$," and thus, by the definition of *Val*, establishes the assertion $Val(H) = Q[N].val$. Thus, to prove that (2) is not falsified, it suffices to show that the following assertion is established as well.

$$\langle \forall j : 0 \leq j < N :: Q[j].val \neq 0 \ \Rightarrow \ Q[j].tag \neq Q[N].tag \rangle \quad (3)$$

This is the second key invariant of the proof. The importance of (3) should not be overlooked. By the semantics of a counter, each Write operation should have the effect of completely overwriting the previous contents of the implemented counter. Assertions (1) and (3) imply that this is indeed the case.

Given the results of [2, 3, 4] and the results of this section, we have the following theorem.

Theorem: Counters can be implemented in a bounded, wait-free manner from atomic registers. □

4 Discussion

In the following subsections, we discuss several issues pertaining to our counter implementation.

4.1 Handling Overflows

In Section 3, we assumed that the value of the implemented counter ranges over the integers. To implement a counter that stores values over some bounded range, our implementation must be modified to prevent overflows. An overflow may result if an Increment operation is performed when the value of the counter is "very close" to the maximum allowed. Overflows can be dealt with in two ways: we can either modify the Increment procedure so that potential overflows are detected and avoided; or we can allow the value of the counter to "wrap around" when an overflow occurs. Incorporating the latter approach into our implementation is straightforward. For example, to implement a counter whose value ranges over $0..L - 1$, we need only modify statements 0 and 9 in Figure 3 so that when *outval* and *sum* are computed, addition is performed modulo L.

Overflows can be detected and avoided by testing the value assigned to *sum* by the Increment procedure. Suppose, for example, that each *val* field in Q ranges over $-L..L$. In this case, if $|sum| \leq L$ holds following the execution of statement 9, then $Q[i]$ is updated as before. However, if $|sum| > L$, then an error code is returned to process i and $Q[i]$ is not modified. This approach has the disadvantage that the counter does not have a single "maximum value": for a given value of the counter, an Increment operation by one process may cause an overflow error, while an Increment operation with the same input value by another process does not. This inconsistency results from the fact that overflow for process i depends only on the value of $Q[i].val$, and not on the value of any other component.

4.2 Complexity

The time complexity of an implementation is defined to be the number of reads and writes of composite registers required to execute an operation of the implemented counter. It is easy to see that the time complexity of each Read, Write, and Increment operation in our implementation is $O(1)$. The space complexity of an implementation is defined to be the number of single-reader, single-writer, single-bit atomic registers required to realize the implementation. If the implemented counter stores values ranging over $\{-L, \ldots, L\}$, then by the results of [2, 3, 4], the space complexity is polynomial in L and N.

4.3   Generalizing the Implementation

By the correctness of the implementation, the partial order on operations in any well-formed history h can be extended to a total order $\prec$ that is consistent with the semantics of a counter. In particular, the output value of each Read operation r in h equals $y + z$, where y and z are defined as follows.

- If there exists a Write operation w such that $w \prec r \wedge \neg \langle \exists v : v$ is a Write operation :: $w \prec v \prec r \rangle$, then y equals the input value of w and z equals the sum of the input values of all Increment operations ordered between w and r by $\prec$.

- If no such w exists, then y equals the initial value of the implemented counter, and z equals the sum of the input values of all Increment operations ordered before r by $\prec$.

In more general terms, the protocol followed in the implementation allows each Read operation to determine two values y and z, where y is the most recently written value according to $\prec$, and z is a function over the input values of all intervening Increment operations according to $\prec$. Note that this protocol does not enable a Read operation to determine the relative ordering of the intervening Increment operations. However, this ordering is irrelevant in determining the value of the counter because Increment operations are defined in terms of addition, which is commutative. This protocol can be generalized to yield the following theorem.

Theorem: Any shared register X that can either be read, written, or modified by a PRMW operation of the form "$X := X \circ v$," where $\circ$ is a commutative operator and v is an integer value, can be implemented in a bounded, wait-free manner from atomic registers. □

   As an example, consider the problem of implementing a "multiplication register," i.e., one that can either be read, written, or multiplied by an integer value. We can implement such a register by defining the value of the implemented register to be

$$Q[N].val \times \langle \textstyle\prod j : 0 \leq j < N \wedge Q[j].tag = Q[N].tag :: Q[j].val \rangle$$

and by modifying statement 0 of the Read procedure accordingly. The procedure used to multiply the value of the register would be similar to the Increment procedure in Figure 3, except that in statements 6 and 9, "$sum := 0$" would be replaced by "$sum := 1$," and in statement 9, "$sum := x[i].val + inval$" would be replaced by "$sum := x[i].val \times inval$" (actually, $prod$ would be a more suitable variable name than sum in this case). The initialization of the register would be the similar to that given in Figure 2, except for the requirement $\langle \forall j : 0 \leq j < N :: Q[j].val = 1 \rangle$.

4.4 Implementing More Powerful Shared Data Objects

One may wonder whether atomic registers can be used to implement even more powerful shared data objects in a wait-free manner, i.e., ones that may be modified by means of numerous PRMW operations. In order to partially address this question, we consider the problem of implementing a register that combines the operations of both counters and multiplication registers; we call such a register an *accumulator register*. In the remainder of this subsection, we show that the following theorem holds.

Theorem: Accumulator registers cannot be implemented from atomic registers without waiting. □

The proof is based upon the problem of two-process consensus. In the consensus problem, two processes are required to agree on a common boolean "decision value"; trivial solutions in which both processes agree on a predetermined value are not allowed. It has been show both by Herlihy [14] and by Anderson and Gouda [5] that two-process consensus cannot be solved in a wait-free manner using only atomic registers. Therefore, to prove that accumulator registers cannot be implemented from atomic registers without waiting, it suffices to prove that accumulator registers can be used to solve two-process consensus in a wait-free manner.

Figure 4 depicts a program that solves two-process consensus without waiting by using a single shared accumulator register X. To see that this program solves the consensus problem, consider Figure 5. This figure depicts the possible values of X; each arrow is labeled by the statement that causes the change in value. Based on this figure, we conclude that if statement 0 is executed before statement 3, then the final value of y equals 4 or 8 and the final value of z differs from 2 and 5, in which case both processes decide on "true." On the other hand, if statement 3 is executed before statement 0, then the final value of z equals 2 or 5 and the final value of y differs from 4 and 8, in which case both processes decide on "false." Thus, this program solves the consensus problem.

5 Concluding Remarks

We have shown that there exist nontrivial shared data objects with PRMW operations that can be implemented from atomic registers in a bounded, wait-free manner. In particular, we have presented an implementation that can be generalized to implement any shared data object that can either be read, written, or modified by a commutative PRMW operation. Our implementation is polynomial in both space and time, and thus is an improvement over the unbounded implementations of Aspnes and Herlihy [7].

The results of this paper provide yet another example of an unbounded wait-free implementation that can be made bounded. An interesting research question is whether

shared var $X : 1..8$
initially $X = 1$

process 0
private var
   decide : boolean;
   $y : 1..8$
begin
  0: **increment** $X := X + 3$;
  1: **read** $y := X$;
  2: *decide* := $(y = 4 \lor y = 8)$
end

process 1
private var
   decide : boolean;
   $z : 1..8$
begin
  3: **multiply** $X := X \cdot 2$;
  4: **read** $z := X$;
  5: *decide* := $(z \neq 2 \land z \neq 5)$
end

Figure 4: A solution to the consensus problem.

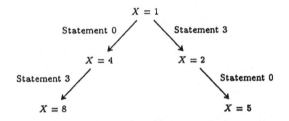

Figure 5: Possible values of X.

it is possible to develop a general mechanism for converting any unbounded wait-free implementation into a bounded one, provided the data object under consideration is syntactically bounded.[4] This question was noted previously by Afek et al. [2].

The correctness proof for our implementation (which has only been outlined here, but which is given in its entirety in the full paper) is noteworthy because of the fact that it is assertional, rather than operational. Most proofs of wait-free implementations that have been presented in the literature are based upon operational concepts such as histories and events, and rely on extensive case analysis to cover the many different interleavings of events that may occur in a history. Such proofs require one to mentally execute the program at hand (e.g., "... if process i does this, then process j does that ..."), and thus are quite often error prone and difficult to understand. In our proof, auxiliary history variables are used to record the effect of each operation; these history variables serve as a basis for stating the required invariants.

---

[4]For example, a counter whose value ranges over the integers is syntactically unbounded.

The use of auxiliary history variables in correctness arguments is, of course, not new. Early references include the work of Clint [12] and also Owicki and Gries [23]. More recent references include Abadi and Lamport's work on refinement mappings [1] and Lam and Shankar's work on module specifications [19]. Our use of history variables was, in fact, motivated by the latter paper, where history variables are used to formally specify database serializability. We believe that history variables better facilitate the development of assertional proofs of wait-free implementations than do shrinking functions and their variants [8].

Acknowledgements: We would like to thank A. Udaya Shankar for several helpful discussions on the subject of this paper, and for bringing reference [19] to our attention. We would also like to thank John Tromp for his comments on a draft of this paper.

References

[1] M. Abadi and L. Lamport, "The Existence of Refinement Mappings," *Theoretical Computer Science*, Vol. 82, No. 2, May 1990, pp. 253-284.

[2] Y. Afek, H. Attiya, D. Dolev, E. Gafni, M. Merritt, and N. Shavit, "Atomic Snapshots of Shared Memory," *Proceedings of the Ninth Annual Symposium on Principles of Distributed Computing*, 1990, pp. 1-14.

[3] J. Anderson, "Multiple-Writer Composite Registers," Technical Report TR.89.26, Department of Computer Sciences, University of Texas at Austin, September 1989.

[4] J. Anderson, "Composite Registers," *Proceedings of the Ninth Annual Symposium on Principles of Distributed Computing*, 1990, pp. 15-30.

[5] J. Anderson and M. Gouda, "The Virtue of Patience: Concurrent Programming With and Without Waiting," Technical Report TR.90.23, Department of Computer Sciences, University of Texas at Austin, July 1990.

[6] J. Anderson and M. Gouda, "A Criterion for Atomicity," *Formal Aspects of Computing: The International Journal of Formal Methods*, scheduled to appear in 1992.

[7] J. Aspnes and M. Herlihy, "Wait-Free Data Structures in the Asynchronous PRAM Model," *Proceedings of the Second Annual ACM Symposium on Parallel Architectures and Algorithms*, July, 1990.

[8] B. Awerbuch, L. Kirousis, E. Kranakis, P. Vitanyi, "On Proving Register Atomicity," Report CS-R8707, Centre for Mathematics and Computer Science, Amsterdam, 1987.

[9] B. Bloom, "Constructing Two-Writer Atomic Registers," *IEEE Transactions on Computers*, Vol. 37, No. 12, December 1988, pp. 1506-1514. Also appeared in *Proceedings of the Sixth Annual Symposium on Principles of Distributed Computing*, 1987, pp. 249-259.

[10] J. Burns and G. Peterson, "Constructing Multi-Reader Atomic Values from Non-Atomic Values," *Proceedings of the Sixth Annual Symposium on Principles of Distributed Computing*, 1987, pp. 222-231.

[11] K. Chandy and J. Misra, *Parallel Program Design: A Foundation*, Addison Wesley, 1988.

[12] M. Clint, "Program Proving: Coroutines," *Acta Informatica*, Vol. 2, p. 50-63, 1973.

[13] P. Courtois, F. Heymans, and D. Parnas, "Concurrent Control with Readers and Writers," *Communications of the ACM*, Vol. 14, No. 10, October 1971, pp. 667-668.

[14] M. Herlihy, "Impossibility and Universality Results for Wait-Free Synchronization," *Proceedings of the Seventh Annual Symposium on Principles of Distributed Computing*, 1988, pp. 276-290.

[15] M. Herlihy and J. Wing, "Linearizability: A Correctness Condition for Concurrent Objects," *ACM Transactions on Programming Languages and Systems*, Vol. 12, No. 3, 1990, pp. 463-492.

[16] A. Israeli and M. Li, "Bounded time-stamps," *Proceedings of the 28th IEEE Symposium on Foundations of Computer Science*, 1987, pp. 371-382.

[17] L. Kirousis, E. Kranakis, and P. Vitanyi, "Atomic Multireader Register," *Proceedings of the Second International Workshop on Distributed Computing*, Springer Verlag Lecture Notes in Computer Science 312, 1987, pp. 278-296.

[18] C. Kruskal, L. Rudolph, M. Snir, "Efficient Synchronization on Multiprocessors with Shared Memory," *ACM Transactions on Programming Languages and Systems*, Vol. 10, No. 4, October 1988, pp. 579-601.

[19] S. Lam and A. Shankar, "Specifying Modules to Satisfy Interfaces: A State Transition System Approach," Technical Report CS-TR-2082.3, University of Maryland at College Park, 1988.

[20] L. Lamport, "On Interprocess Communication, Parts I and II," *Distributed Computing*, Vol. 1, pp. 77-101, 1986.

[21] M. Li, J. Tromp, and P. Vitanyi, "How to Construct Wait-Free Variables," unpublished manuscript, Centrum voor Wiskunde en Informatica, Amsterdam, The Netherlands, 1989.

[22] R. Newman-Wolfe, "A Protocol for Wait-Free, Atomic, Multi-Reader Shared Variables, *Proceedings of the Sixth Annual Symposium on Principles of Distributed Computing*, 1987, pp. 232-248.

[23] S. Owicki and D. Gries, "An Axiomatic Proof Technique for Parallel Programs I," *Acta Informatica*, Vol. 6, pp. 319-340, 1976.

[24] G. Peterson, "Concurrent Reading While Writing," *ACM Transactions on Programming Languages and Systems*, Vol. 5, 1983, pp. 46-55.

[25] G. Peterson and J. Burns, "Concurrent Reading While Writing II: The Multi-Writer Case," *Proceedings of the 28th Annual Symposium on Foundations of Computer Science*, 1987.

[26] A. Singh, J. Anderson, and M. Gouda, "The Elusive Atomic Register, Revisited," *Proceedings of the Sixth Annual Symposium on Principles of Distributed Computing*, 1987, pp. 206-221.

[27] J. Tromp, "How to Construct an Atomic Variable," unpublished manuscript, Centrum voor Wiskunde en Informatica, Amsterdam, The Netherlands, 1989.

[28] P. Vitanyi and B. Awerbuch, "Atomic Shared Register Access by Asynchronous Hardware," *Proceedings of the 27th IEEE Symposium on the Foundations of Computer Science*, 1986, pp. 233-243.

Maintaining Digital Clocks In Step

Anish ARORA[1]   *Shlomi DOLEV*[2]   *Mohamed GOUDA*[1]

1. Department of Computer Sciences, The University of Texas at Austin, USA
2. Department of Computer Science, Technion, ISRAEL

Abstract

We design a stabilizing system of simultaneously triggered clocks: if the clock values ever differ, then the system is guaranteed to converge to a state where all clock values are identical, and are subsequently maintained to be identical. Our design of an N-clock system uses N registers of $2 \log N$ bits each and is guaranteed to converge to identical values within N^2 "triggers".

Keywords: stabilization, reliability, distributed algorithms, digital clocks, convergence.

1 Introduction

Digital systems are often designed to be synchronous; that is, a system-wide clock pulse is used to ensure that system parts operate in "lock-step". One building block that is commonly used in the design of such systems is a *digital clock*. Operationally, the task of a digital clock is to maintain a count of the number of clock pulses as they occur.

The use of digital clocks in circuit design is frequent; for example,

- *Synchronization* : A circuit may include some parts that need to synchronize with other parts periodically [1]. The periods, in this case, can be measured using a digital clock.

- *Mode Change* : A circuit may be designed to operate in several modes. For instance, a mode can be allocated to gate asynchronous inputs into registers, and then the register values can be operated upon in a second mode [1]. In such cases, the signal to change the current mode of execution can be generated using a digital clock.

- *Periodic Reset* : A circuit may be designed such that a global signal is periodically generated so as to reset the state of the circuit to some predefined state. In this case, the reset signal can be generated using a digital clock.

In designing circuits where many parts rely on digital clocks, it is preferable to use a number of digital clocks rather than to use only one. This is because using one digital clock may result in long wires which (possibly) traverse multiple chips. It may also result in high fan-out. Consequently, the propagation delay is significant, extra power is needed to drive the signals, and chip space is wasted [2].

The problem of using multiple digital clocks, however, is the added obligation on the circuit designer to ensure that all digital clocks maintain identical values. This obligation is complicated by situations that may set the digital clocks to different values; for instance, powering up the circuit, transient failures, or circuit reconfiguration can all cause different clocks to have different values.

One approach to ensure that all digital clocks maintain identical values is to *compare-and-reset* ; that is, the values of all clocks can be repeatedly compared and, when they differ, all clocks can be reset to some fixed value. One problem of this approach is that it can itself result in long wires and high fan-in and fan-out. Recall that these were the problems that motivated the introduction of multiple clocks in the first place. Another problem is that all clocks are reset even if only one clock differs from the rest. Finally, since the approach is centralized, the overall circuit is no more robust than the reset circuitry itself.

An alternate approach is to design the circuit to be *stabilizing* ; that is, upon starting from an arbitrary state, the circuit is guaranteed within a finite number of pulses to reach a state where all clocks have the same value and are incremented at every subsequent pulse. Based on this approach, we develop in this paper two designs for maintaining clocks in step. In both designs, we avoid the problems of the previous approach: global comparison and global reset of clock values are not used. Moreover, we show that both designs can tolerate the fail-stop failures of individual clocks and interconnecting wires (provided the overall circuit remains connected).

The problem of maintaining clocks in step has been studied earlier by Even and Rajsbaum [3]. They show how clocks can operate in step in a synchronous system where all clocks have identical values initially, but different clocks start counting at different instants. In their design, 'neighboring' clock values may differ by one during early stages of execution, but eventually a state is reached where in step operation resumes. There are, however, two drawbacks in their design. First, if neighboring clock values ever differ by two or more, e.g., due to a transient failure, in step operation cannot be regained. Second, their design employs clocks whose values range over the natural numbers. More recently, Gouda and

Herman have presented a design [4] that overcomes the first drawback, but still employs clocks whose values range over an infinite domain. By way of contrast, both of our designs employ finite-state clocks and so can be used in digital circuits.

The reader should note that the problem of maintaining clocks in step differs from the problem of clock synchronization [5,6]—in the latter, different clocks are updated with respect to different triggering signals whose rate of drift is within known bounds, and it is required to maintain the clocks so that their values are never too far apart. In some respect, the clock maintenance problem is strictly weaker than the clock synchronization problem: all clocks are updated simultaneously in the former whereas different clocks may be updated at different times in the latter. However, clock synchronization solutions presented thus far perform their task under the assumption of some condition holding initially; hence, these solutions are not stabilizing.

2   In-Step Digital Clocks

Consider an undirected and connected graph. Associated with each node u in the graph is

- a *register-set* whose value ranges over a predefined, but finite domain $Q.u$, and
- a *function* whose domain is the cartesian product of $Q.u$ and all $Q.v$ such that the pair (u, v) is an edge in the graph, and whose range is $Q.u$.

A *state* of this system is defined by a value for each register-set.

An *execution* of this system is an infinite sequence of transitions; in each transition, all functions associated with nodes in the graph are simultaneously executed starting from the current state, thereby yielding the next state.

The problem of maintaining digital clocks in step can be stated as follows:

It is required to design a register-set and a function for each node u such that the register-set associated with u includes a register $x.u$ whose range is the interval $0 .. m-1$, where m is a natural number greater than 1, and the resulting system satisfies the following two properties.

1. *Unison* : Starting from any *consensus state*, i.e., one where all x registers have equal values, the value of each x register is incremented by one modulo m in every following transition of the system. (This implies that all states following a consensus state are also consensus states.)

2. *Stabilization* : Starting from an arbitrary state, the system is guaranteed to reach a consensus state.

In Sections 3 and 4, we present two designs that solve the problem posed above. Both designs enjoy low atomicity: in each transition of the resulting system, every x register is updated based on its value and the x value associated with some neighboring node. Hence, the combinational logic needed to implement the function is easily designed in a modular fashion. Moreover, both designs are symmetric: associated with all nodes are identical registers and identical functions (modulo the exact set of neighbors accessed). Hence, both designs can tolerate fail-stop failures of nodes and edges in the system graph provided the remaining graph is connected. Of the two designs, however, the one discussed in Section 4 has better space and time complexity, i.e., it uses fewer bits of memory and attains unison faster, respectively.

3 A Catch Up Design

In our first design, we associate with each node u a new register $r.u$ whose value ranges over the nodes adjacent to u in the system graph. In each transition, $x.u$ is updated to $max(x.u, x.(r.u)) +_m 1$, where $+_m$ denotes addition modulo m. Also, $r.u$ is updated to $nxt.(r.u)$, where nxt is a function over the nodes adjacent to u such that upon successive application to $r.u$, nxt returns nodes in some fixed cyclic order. Thus, associated with each node u we have :

> **register-set of node u ::**
> $x.u : 0.. \ m-1$
> $r.u$: whose value ranges over the set
> $\quad \{ v \mid (u, v)$ is an edge in the graph$\}$
>
> **function of node u ::**
> $x.u \ , \ r.u \ := \ max(x.u, x.(r.u)) +_m 1 \ , \ nxt.(r.u)$

Operationally speaking, the design ensures that each node u "scans" the clocks adjacent to it in a fixed cyclic order and, in each transition, $x.u$ "catches up" with the clock $x.(r.u)$ being scanned if $x.u < x.(r.u)$; else, it counts normally.

We prove below that our construction solves the problem of maintaining digital clocks in step provided the following inequality holds:

$$m > n \times deg \times dia$$

where n is the number of nodes in the graph, *dia* is the diameter of the graph, and *deg* is the maximal degree of a node in the graph. We also show that the convergence span, i.e, the maximum number of transitions needed to reach in-step operation from an arbitrary state, is m.

Proof of Correctness

Unison. At a consensus state all x's are equal and, hence, $max(x.u, x.(r.u)) +_m 1 = x.u +_m 1$ for all nodes u and $r.u$. Executing the function of each node in parallel increments each x by 1 modulo m, thereby yielding the next consensus state.

Stabilization. We prove the stabilization property in two parts:
  (i) Starting from any state, the system is guaranteed to reach a state that satisfies ($\forall$ nodes $u : x.u < m - (deg \times dia)$).
  (ii) Starting from any state that satisfies ($\forall$ nodes $u : x.u < m - (deg \times dia)$), the system is guaranteed to reach a consensus state.

Proof of (i). Consider an arbitrary state q. By the pigeon-hole principle, there exists a number l, $0 \leq l < m$, such that no x has a value strictly within the circular interval l .. $l +_m \lceil m/n \rceil$ at q.

Executing a transition starting from q yields a state where no x has a value strictly within the circular interval $(l +_m 1)$ .. $(l +_m \lceil m/n \rceil +_m 1)$ since all x's are assigned the value of some x incremented by one modulo m. By repeating this argument $m - \lceil m/n \rceil - l$ times, we observe that the system is guaranteed to reach a state where no x has a value strictly within the circular interval $m - \lceil m/n \rceil$ .. 0 . That is, a state where ($\forall$ nodes $u : x.u \leq m - (\lceil m/n \rceil)$). Since $m > n \times deg \times dia$, this state satisfies ($\forall$ nodes $u : x.u < m - (deg \times dia)$) (end proof of (i)).

Proof of (ii). Let k be an arbitrary node such that $x.k$ is largest among all x's at a state q where ($\forall$ nodes $u : x.u < m - (deg \times dia)$). We observe that if $x.k < m$ at q then, upon executing a transition, $x.k$ remains largest among all x's and is incremented by one. Thus, if $x.k < m - (deg \times dia)$ at q, then $x.k$ is largest among all x's at the states resulting from the next $(deg \times dia)$ transitions.

Consider any node u that is adjacent to k in the system graph. If q satisfies $x.k < m - (deg \times dia)$ and the system starts executing from q, then within the first deg transitions register $r.u$ is set to k at least once (due to our choice of the function nxt of node u). Therefore, at some state in the first deg states following q, $x.u$ equals $x.k$ and is hence also largest among all x's. Since $x.k < m$ holds at each of these states, once $x.u$ is largest it remains largest.

By repeating the above argument, in $deg \times dia$ transitions $(\forall u : x.u = x.k)$ holds; i.e, the system is in a consensus state   (end proof of (ii)).

Convergence Span. We note from the proof of part (i) that, upon starting from an arbitrary state, the maximum number of transitions needed to reach a state where $(\forall$ nodes $u : x.u < m - (deg \times dia))$ holds is $m - \lceil m/n \rceil$. We note from the proof of part (ii) that, upon starting from a state where $(\forall$ nodes $u : x.u < m - (deg \times dia))$ holds, a consensus state is reached in at most $deg \times dia$ transitions. Hence, the number of transitions needed to reach in-step operation from an arbitrary state is bounded by $m - \lceil m/n \rceil + (deg \times dia)$, and, therefore, by m.

4 A Fall Back Design

The solution presented in the previous section requires m to be greater than $n \times deg \times dia$. In this section, we present a strikingly similar solution that solves the problem of maintaining digital clocks in step for even smaller values of m. The only difference is that instead of assigning $max(x.u, x.(r.u)) +_m 1$ to $x.u$, we now assign $min(x.u, x.(r.u)) +_m 1$ to $x.u$. That is, associated with each node u we have :

---

register-set of node u ::
  $x.u$: $0.. \ m-1$
  $r.u$: whose value ranges over the set
          $\{\, v \,|\, (u, v) \text{ is an edge in the graph} \,\}$

function of node u ::
  $x.u$, $r.u$:= $min(x.u, x.(r.u)) +_m 1$, $nxt.(r.u)$

Operationally speaking, the design ensures that each node u scans the clocks adjacent to it in a fixed cyclic order and, in each transition, $x.u$ "falls back" with the clock $x.(r.u)$ being scanned if $x.u > x.(r.u)$; else, it counts normally.

We prove below that our construction solves the problem of maintaining digital clocks in step provided the following inequality holds:

$$m > 2 \times deg \times dia$$

We also show that the convergence span is $3 \times deg \times dia$, and invite the reader to observe that $2 \times deg \times dia < n^2$ for $n > 1$.

Proof of Correctness

Unison. The proof of the unison property is identical to the corresponding proof in the previous section with *max* replaced by *min*.

Stabilization. We prove the stabilization property by considering two cases for the first $deg \times dia$ states in the system execution following an arbitrary state q:

(i) At each of these states, no x has the value 0.

(ii) At some of these states, some x has the value 0.

Proof of (i). Let k be an arbitrary node such that $x.k$ is smallest among all x's at q. We observe that at each of the first deg states following q, $x.k$ is smallest among all x's.

Consider any node u that is adjacent to k in the system graph. Within the first deg transitions starting from q, register $r.u$ is set to k at least once (due to our choice of the function nxt of node u). Therefore, at some state in the first deg states following q, $x.u$ equals $x.k$ and is hence also smallest among all x's. Since no x has the value 0 at these states, once $x.u$ is smallest it remains smallest.

By repeating the above argument, after $deg \times dia$ transitions $(\forall u : x.u = x.k)$ holds; i.e, the system is in a consensus state $\hspace{4cm}$ (end proof of (i)).

Proof of (ii). Let k be an arbitrary node such that $x.k$ has value 0 at some state s in the first $deg \times dia$ states following q. After deg transitions following s, the value of $x.k$ is at most deg.

Consider any node u that is adjacent to k in the system graph. Within deg transitions following s, register $r.u$ is set to k at least once (due to our choice of the function nxt of node u). Therefore, after deg transitions following s, the value of $x.u$ is also at most deg.

By repeating this argument, after $deg \times dia$ transitions following s, the system is at a state t that satisfies $(\forall u : x.u \leq deg \times dia)$. Hence, at each of the $deg \times dia$ transitions following t, no x has the value 0. Thus, by the first case, the system will be at a consensus state after $deg \times dia$ transitions following t $\hspace{3cm}$ (end proof of (ii)).

Convergence Span. We note from the proof of part (i) that, within $deg \times dia$ transitions from an arbitrary state, the system either reaches a consensus state or a state where some x register is set to 0. In the latter case, we note from the proof of part (ii) that, after $deg \times dia$ subsequent transitions, the system reaches a state where $(\forall$ nodes $u : x.u \leq deg \times dia)$ holds; a consensus state is guaranteed within the next $deg \times dia$ transitions. In all, therefore, at most $3 \times deg \times dia$ transitions are needed to attain unison.

5 Linear Arrays of Digital Clocks

The catch-up and fall-back designs presented in Sections 3 and 4 respectively can be easily implemented in hardware. In this section, we describe a hardware implementation of the two designs for the special case of a linear array of clocks.

Shown below is a schematic diagram of the circuit associated with an intermediate node u. The circuit has a $(\log m)$-bit register that stores $x.u$ (we assume that m is a power of two) and a toggle flip flop $t.u$ that is used as a selector. *Mplus* is a combinational circuit that computes the maximum (minimum, respectively, for the fall back design) of two $(\log m)$-bit input operands and increments the result by one modulo m. We assume for correct operation that the delay of the combinational logic exceeds the clock width and is less than the clock period [2].

Each intermediate circuit u operates as follows. At the leading edge of each clock pulse, if $t.u = 1$ holds then $x.u$ is updated to the output produced by the *Mplus* circuit of $x.u$ and $x.(u-1)$; else, $t.u = 0$ holds and $x.u$ is updated to the *Mplus* output of $x.u$ and $x.(u+1)$. Simultaneously, $t.u$ is toggled. In the circuit operation of end nodes, their x register is updated to the *Mplus* of their x value and their (only) neighbor's x value.

pulse

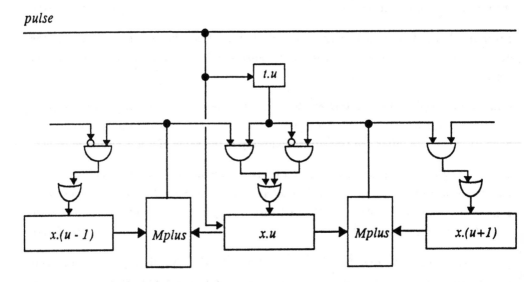

As an example, we note that with 8-bit registers clock arrays of upto 11 clocks can be built with the first design (since $deg = 2$, $dia = 10$, and $m = 2^8 > n \times deg \times dia$), whereas clock arrays of upto 64 clocks can be built with the second design (since $deg = 2$, $dia = 63$, and $m = 2^8 > 2 \times deg \times dia$).

6 Conclusions

We have presented, in this paper, two designs for maintaining digital clocks in step. These designs are based on the notion of *self-stabilization*. Consequently, the resulting systems are self-initializing, do not require global state detection and are robust to transient failures and reconfiguration.

More generally, the concepts discussed in this paper can be extended to reason about the correctness of several classes of reliable systems in a formal manner. In [7], Arora and Gouda present such a formalism based on two notions: closure (which is a safety property analogous to unison) and convergence (which is a progress property analogous to stabilization).

Issues for further research include: What is the least m such that there is a solution for the problem posed in Section 2 in which at most one adjacent clock is "accessed" per clock pulse? Likewise, what is the lower bound on the convergence span for achieving unison? We conjecture that the second solution is optimal in these respects.

Acknowledgements. We thank Shimon Even, Jayadev Misra, and Sergio Rajsbaum for helpful discussions concerning this work.

References

[1]   J.B. Peatman, *Digital Hardware Design,* Addison-Wesley (1980).

[2]   C. Mead and L. Conway, *Introduction to VLSI Systems,* Addison-Wesley (1980).

[3]   S. Even and S. Rajsbaum, "Lack of a global clock does not slow down the computation in distributed networks", TR522, Computer Science Department, Technion (1988) ; the first part of this paper, "Unison in distributed networks", appeared in *Sequences, Combinatorica, Compression, Security, and Transmission,* E.M. Capocelli (ed.), Springer-Verlag.

[4]   M.G. Gouda and T. Herman, "Stabilizing unison", *Information Processing Letters,* Vol. 35, No. 4, pp. 171-175 (1990).

[5]   F. Cristian, "Probabilistic clock synchronization", *Distributed Computing,* Vol. 3, pp. 146-158 (1989).

[6]   F.B. Schneider, "Understanding protocols for byzantine clock synchronization", Technical Report TR-87-859, Cornell University, Department of Computer Science, Ithaca, NY 14853 (1987).

[7]   A. Arora and M.G. Gouda, "Closure and convergence: A formal basis for fault-tolerance", Manuscript (1991).

Implementing FIFO Queues and Stacks

(EXTENDED ABSTRACT)

Hagit Attiya[*]
Department of Computer Science
The Technion
Haifa 32000, Israel

Abstract

The cost of implementing FIFO queues and stacks is studied under two consistency conditions for shared memory multiprocessors, *sequential consistency* and *linearizability*. The cost measure is the worst-case response time in distributed implementations of virtual shared memory supporting one of the two conditions. The worst-case response time is very sensitive to the assumptions that are made about the timing information available to the system. All the results in this paper assume that processes have clocks that run at the same rate as real time and that all message delays are in the range $[d - u, d]$ for some known constants u and d, $0 \le u \le d$.

If processes have perfectly synchronized clocks or if every message has delay exactly d, then the response time of a dequeue operation is at least d, for any sequentially consistent implementation of FIFO queues. This matches exactly an upper bound in which an enqueue operation is performed instantaneously and the response time of a dequeue operation is d; this upper bound implements linearizability. If clocks are not perfectly synchronized and if message delays are variable, i.e., $u > 0$, then, for any linearizable implementation of a queue, the response time of an enqueue operations is at least $\Omega(u)$. In contrast, we present sequentially consistent implementation for this weaker timing model in which an enqueue operation is performed instantaneously, and the worst-case response time of a dequeue operation is $2d$. (This algorithm is completely asynchronous and does not rely on any timing information.)

Similar results are proved for implementing stacks, with the pop operation playing the role of the dequeue operation, and the push operation playing the role of the enqueue operation.

1 Introduction

A fundamental problem in concurrent computing is how to provide programmers with a useful model of logically shared data that can be accessed atomically, without sacrificing performance. The model must specify how the data can be accessed and what guarantees are provided about the results. To enhance performance (e.g., response time, availability, or fault-tolerance), many implementations employ multiple copies of the same logical piece of shared data (*caching* or *replication*). Also, several user programs must be able to access data "concurrently." More complications arise because at some level, each access to shared data has duration in time; it is not instantaneous. Therefore,

---

[*]Electronic mail address: hagit@TECHSEL.BITNET or hagit@cs.technion.ac.il. A complete version of this paper appeared as Technical Report #672, Department of Computer Science, The Technion, Haifa, Israel, May 1991.

the illusion of atomic operations on single copies of objects must be supported by a *consistency mechanism*. The consistency mechanism guarantees that although operations may be executed concurrently on various copies and have some duration, they will *appear* to have executed atomically, in some sequential order that is consistent with the order seen at individual processes.[1] When this order must preserve the global (external) ordering of non-overlapping operations, this consistency guarantee is called *linearizability* ([11]); otherwise, the guarantee is called *sequential consistency* ([12]). Obviously, linearizability implies sequential consistency.

Attiya and Welch ([3]) compare the cost of implementing sequential consistency and linearizability, in non-bused distributed system. The comparison was based on worst-case time complexity — the inherent response time of the best possible distributed implementation supporting each consistency condition. Several upper and lower bounds were given for the case where the shared memory consists of a collection of *read/write objects*. It was shown that, under certain timing assumptions, implementing linearizable read/write objects requires more time than implementing sequentially consistent ones.

How do these results change when other objects are implemented? Since read/write objects do not provide an expressive and convenient abstraction for concurrent programming (cf. [10]), many multiprocessors now support more powerful concurrent objects, e.g., FIFO queues, stacks, test-and-set, fetch-and-add ([8]). In this paper, we study the time complexity of implementing sequential consistency and linearizability for FIFO queues and stacks; that is, we present upper and lower bounds on the worst-case response time for performing an operation on an object.

1.1 Results

We consider a collection of application programs running concurrently and communicating via virtual shared memory. The application programs are running in a distributed system consisting of a collection of nodes and a complete communication network.[2] The shared memory abstraction is implemented by a *memory consistency system* (mcs), which uses local memory at the various nodes and some protocol executed by the mcs processes (one at each node). (Nodes that are dedicated storage can be modeled by nullifying the application process.) Figure 1 illustrates a node, on which is running an application process and an mcs process. The application process sends calls to access shared data to the mcs process; the mcs process returns the responses to the application process, possibly based on messages exchanged with mcs processes on other nodes.

The correctness conditions are defined at the interface between the application processes (written by the user) and the mcs processes (supplied by the system). Thus, the mcs must provide the proper semantics when the values of the responses to calls are considered, throughout the network. We assume that on each node there is a real-time clock readable by the mcs process at that node, that runs at the same rate as real-time. We assume that every message incurs a delay in the interval $[d - u, d]$, for some known constants u and d, $0 \leq u \leq d$ (u stands for *uncertainty*). If $u = 0$, then the message delays are constant.

Most of the paper deals with implementations of FIFO queues, which can be accessed by enqueue and dequeue operations.

---

[1]This condition is similar in flavor to the notion of *serializability* from database theory ([6, 17]); however, serializability applies to *transactions* which aggregate many operations.

[2]The assumption of a complete communication network can be omitted and is made here only for clarity of presentation.

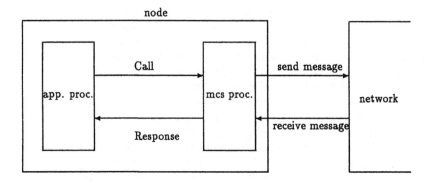

Figure 1: System Architecture

If processes have perfectly synchronized clocks and message delays are constant, we show that the worst-case response time for a dequeue operation is at least d. The result is proved for sequential consistency, and thus, holds also for linearizability. We show that this bound is tight by presenting an algorithm in which an enqueue operation returns instantaneously, while a dequeue operation returns within time d. The algorithm achieves linearizability, and hence, sequential consistency.

For the more realistic case of approximately synchronized clocks and uncertain message delays, we present a sequentially consistent implementation of FIFO queues in which an enqueue operation returns instantaneously, and a dequeue operation returns within time $2d$. (This algorithm is completely asynchronous and does not rely on any timing information.) In contrast, we show that for any mcs that supports linearizability, an enqueue operation cannot return instantaneously, regardless of the response time for the dequeue operation. Specifically, we show that the worst-case response time of an enqueue operation is at least $u/2$; note that u can be as big as d.

Similar results are proved for implementing stacks, with the pop operation playing the role of the dequeue operation, and the push operation playing the role of the enqueue operation.

1.2   Related Work

Some bounds are known on the worst case time to implement operations on read/write objects. For the case of perfectly synchronized clocks and constant message delay, Lipton and Sandberg ([14]) prove a lower bound of d on the sum of the response times of read and write operations, for any implementation that supports sequential consistency. This tradeoff is tight for this model: Attiya and Welch ([3]) present linearizable implementations in which one operation (either read or write) is performed instantaneously and the response time of the other operation is d. In [3], the model where clocks are not perfectly synchronized and message delays are variable, i.e., $u > 0$, is studied. It is shown that such a tradeoff cannot be achieved by linearizable implementations: the response time for both read and write operations is at least $\Omega(u)$. In contrast, [3] present sequentially consistent implementations for this weaker timing model in which one operation (either read or write) is performed instantaneously, and the worst-case response time of the other operation is $O(d)$.

As in [3], we show that, in the presence of mild timing uncertainty, implementing sequential consistency is more efficient than implementing linearizability. While for read/write objects, there is a tradeoff between the response time of the two operations (read and write), in the case of FIFO queues and stacks there is one operation (dequeue or pop, resp.) whose response time must be big and cannot be traded with the response time for the other operation (enqueue or push, resp.).

Several papers exploited the relative weakness of sequential consistency to achieve more efficient implementation of read/write objects ([1, 2, 3]). To the best of our knowledge, this is the first time such improvements are achieved for other objects.

The implementations in [3] were based on the *atomic broadcast* protocol of Birman and Joseph ([7]). Similarly, our implementations of sequentially consistent FIFO queues and stacks in the case of imperfect clocks also embed a protocol for atomic broadcast. However, they rely on a different, simpler algorithm for atomic broadcast, which yields better constants.

2 Correctness Conditions

The formal model is very similar to the one in [3], with minor modifications to accommodate FIFO queues (these definitions will be adapted for stacks in Section 5); the model is only outlined here.

A memory consistency system (mcs) consists of a collection of processes, one on each node of a distributed system. Process p interacts with the application program using *call* events, $\text{Enq}_p(Q)$ and $\text{Deq}_p(Q, v)$ for all objects Q and values v, and *response* events, $\text{Ack}_p(Q)$ (for Enq) and $\text{Ret}_p(Q, v)$ (for Deq). A special value, $\perp$, represents the empty queue. A process communicates with other processes using *message-send* and *message-receive* events. It sets timers for itself (to go off at some future clock time) and responds to them using *timer-set* and *timer* events, respectively. The process is modeled as an automaton with states and a transition function that takes as input the current state, clock time, and a call or message-receive or timer event, and produces a new state, a set of response events, a set of message-send events, and a set of timer-set events. A *history* of a process describes what steps (i.e., applications of the transition function) the process takes at what real times; it must satisfy the obvious conditions.

An *execution* of a set of processes is a set of histories, one for each process, satisfying the following two conditions: (1) A timer is received by p at clock time T if and only if p has previously set a timer for T. (2) There is a one-to-one correspondence between the messages sent by p to q and the messages received by q from p, for any processes p and q. We use the message correspondence to define the *delay* of any message in an execution to be the real time of receipt minus the real time of sending. (The network is not explicitly modeled, although the constraints on executions imply that the network reliably delivers all messages sent.) An execution is *admissible* if the delay of every message is in the range $[d - u, d]$, for fixed nonnegative integers d and u, $u \leq d$, and for every p, at any time at most one call at p is *pending* (i.e., lacking a matching subsequent response).

Every object is assumed to have a *serial specification* (cf. [11]) defining a set of *operations*, which are ordered pairs of call and response events, and a set of *operation sequences*, which are the allowable sequences of operations on that object. In the case of a FIFO queue, the ordered pair of events $[\text{Deq}_p(Q), \text{Ret}_p(Q, v)]$ forms an *operation* for any p, Q, and v, as does $[\text{Enq}_p(Q, v), \text{Ack}_p(Q)]$. The set of operation sequences consists of all sequences that obey the usual FIFO queue semantics. That is, with a sequence of operations we associate a sequence of *queue states*, starting with an initial empty state and continuing with a state for each operation (representing the state of the queue *after* the operation). We require that each enqueue operation adds an item to the end of the queue, and each dequeue operation removes an item from the head of the queue, or return $\perp$ if the queue is empty.[3]

A sequence τ of operations for a collection of processes and objects is *legal* if, for each object Q, the restriction of τ to operations of Q is in the serial specification of Q. Given an execution σ, let $ops(\sigma)$ be the sequence of call and response events appearing in σ in real-time order, breaking ties by ordering all events at the same process by their order at that process and then using process ids.

Definition 2.1 (Sequential consistency, Lamport) *An execution σ is* sequentially consistent *if there exists a legal sequence τ of operations such that, for each process p, the restriction of $ops(\sigma)$ to operations of p is equal to the restriction of τ to operations of p.*

Definition 2.2 (Linearizability, Herlihy and Wing) *An execution σ is* linearizable *if there exists a legal sequence τ of operations such that, for each process p, the restriction of $ops(\sigma)$ to operations of p is equal to the restriction of τ to operations of p, and furthermore, whenever the response for operation op_1 precedes the call for operation op_2 in $ops(\sigma)$, then op_1 precedes op_2 in τ.*

An mcs is a *sequentially consistent* implementation of a set of objects if any admissible execution of the mcs is sequentially consistent; similarly, an mcs is a *linearizable* implementation of a set of objects if any admissible execution of the mcs is linearizable.

We measure the efficiency of an implementation by the worst-case response time for any operation on the object. Given a particular mcs and a FIFO queue Q implemented by it, we denote by $|E(Q)|$ the maximum time taken by an enqueue operation on Q and by $|D(Q)|$ the maximum time taken by a dequeue operation on Q, in any admissible execution. Denote by $|E|$ the maximum of $|E(Q)|$, and by $|D|$ the maximum of $|D(Q)|$, over all objects Q implemented by the mcs.

3 Perfect Clocks

We start by considering the case in which processes have perfectly synchronized (*perfect*) clocks and message delay is constant and known.[4] We first show that the worst-case response time of a dequeue operation is at least d, even in this strong model, and even under sequential consistency. We then show that the lower bound is tight for this model by describing an algorithm that matches the lower bound exactly; in the algorithm, enqueues return instantaneously, while the worst-case response time for a dequeue operation is d. The algorithm implements linearizability, which is a stronger condition than sequential consistency.

---

[3]This specification is operational. It is possible to give an algebraic (axiomatic) specification (cf. [11]); the operational specification is used here for simplicity.

[4]We remark that the assumptions that processes have perfect clocks and that message delays are constant (and known) are equivalent.

3.1 Lower Bounds for Sequential Consistency

In this section we prove our main lower bound result:

Theorem 3.1 *For any memory-consistency system that is a sequentially consistent implementation of a FIFO queue Q, $|D(Q)| \geq d$.*

Proof: Let p and q be two processes that access Q. Assume by way of contradiction that there exists a sequentially consistent implementation of Q for which $|D(Q)| < d$. Denote $T = |D(Q)|$. By definition, the queue Q is initially empty.[5] By the specification of Q, there is some admissible execution α'_1 such that $ops(\alpha'_1)$ is

$$\text{Enq}_q(Q, 1) \ \text{Ack}_p(Q) \ \text{Deq}_p(Q) \ \text{Ret}_p(Q, v_1) \ldots \text{Deq}_p(Q) \ \text{Ret}_p(Q, v_i) \ldots$$

$\text{Enq}_q(Q, 1)$ occurs at real time 0 and $\text{Ack}_p(Q)$ occurs at time t; the first $\text{Deq}_p(Q)$ occurs at time t, while the jth $\text{Deq}_p(Q)$ occurs at time $t + (j - 1)T$ (see Figure 2(a)[6]). Consider now the infinite sequence $v_1, \ldots, v_i, \ldots$. It is possible that many of them are $\perp$, however, since only a finite number of Deq operations can be serialized before the Enq operation, we have:

Lemma 3.2 *There exists some i such that $v_i \neq \perp$.*

Fix this particular i, and note that $v_i = 1$ and, for all j, $1 \leq j < i$, $v_j = \perp$. Let α_1 be α'_1 truncated after the ith Deq operation by p. More precisely, $ops(\alpha_1)$ is

$$\text{Enq}_q(Q, 1) \ \text{Ack}_p(Q) \ \text{Deq}_p(Q) \ \text{Ret}_p(Q, \perp) \ldots \text{Deq}_p(Q) \ \text{Ret}_p(Q, \perp) \ \text{Deq}_p(Q) \ \text{Ret}_p(Q, 1)$$

$\text{Enq}_q(Q, 1)$ occurs at real time 0 and $\text{Ack}_p(Q)$ occurs at time t; the first $\text{Deq}_p(Q)$ occurs at time t, while the ith $\text{Deq}_p(Q)$ occurs at time $t + (i - 1)T$ (see Figure 2(b)). It is clear that the v_j's are exactly as in α'_1. By assumption, the real time at the end of α is less than $t + (i - 1)T + d$. Thus, no message sent after $t + (i - 1)T$ is received during α_1.

We now consider the execution where the ith (and last) dequeue by p is replaced with a dequeue by q. More precisely, by the specification of Q, there is some admissible execution α_2 such that $ops(\alpha_2)$ is

$$\text{Enq}_q(Q, 1) \ \text{Ack}_p(Q) \ \text{Deq}_p(Q) \ \text{Ret}_p(Q, \perp) \ldots \text{Deq}_p(Q) \ \text{Ret}_p(Q, \perp) \ \text{Deq}_q(Q) \ \text{Ret}_q(Q, u)$$

$\text{Enq}_q(Q, 1)$ occurs at real time 0 and $\text{Ack}_p(Q)$ occurs at time t; the first $\text{Deq}_p(Q)$ occurs at time t, while the $i - 1$st $\text{Deq}_p(Q)$ occurs at time $t + (i - 2)T$, and Deq_q occurs at time $t + (i - 1)T$ (see Figure 2(c)). Since α_2 is sequentially consistent, it follows that $u = 1$. By assumption, the real time at the end of α_2 is less than $t + (i - 1)T + d$. Thus, no message sent after $t + (i - 1)T$ is received during α_2.

Since no message sent after time $t + (i - 1)T$ is ever received in α_1 and α_2, and since α_1 and α_2 are identical until time $t + (i - 1)T$, the execution α obtained from α_1 by replacing q's history with q's history in α_2 is admissible. Then $ops(\alpha)$ is

---

[5]If we allow queues to be initially non-empty, the proof of the lower bound becomes much simpler; we leave the details to the interested reader.

[6]In the figures, time runs from left to right, and each line represents events at one process. Important time points are marked at the bottom.

$$\text{Enq}_q(Q,1)\,\text{Ack}_p(Q)\,\text{Deq}_p(Q)\,\text{Ret}_p(Q,\bot)\ldots\text{Deq}_p(Q)\,\text{Ret}_p(Q,1)\,\text{Deq}_q(Q)\,\text{Ret}_q(Q,1)$$

(see Figure 2(d)). By assumption, α is sequentially consistent. Thus, there is a legal sequential execution τ, which is a permutation of the above operations. However, in τ the element "1" is enqueued once, and dequeued twice. A contradiction. ■

We remark that it is possible to prove the following variant on a theorem of Lipton and Sandberg ([14]) by mimicking the proof of this theorem presented in [3]. (Details are left to the reader.)

Theorem 3.3 *For any memory-consistency system that is a sequentially consistent implementation of two FIFO queues Q_1 and Q_2, $|E| + |D| \geq d$.*

The proof of this tradeoff is omitted, as it is superseded by Theorem 3.1 and the algorithms presented below.

3.2 Upper Bounds for Linearizability

In this section we show that the lower bound given in Theorem 3.1 is tight for the model with perfect clocks. Specifically, we present an algorithm in which an enqueue operation returns instantaneously, while a dequeue operation returns within time d. The algorithm ensures the stronger condition of linearizability. It assumes that clocks are perfect and message delays are constant.

Informally, the algorithm works as follows. Each process keeps a copy of all queues in its local memory. When an $\text{Enq}_p(Q, v)$ occurs, p sends $enqueue(Q, v)$ messages to all other processes (including a message to itself which is delayed d time units) and does an Ack immediately. When a $\text{Deq}_p(Q)$ occurs, p sends $dequeue(Q)$ messages to all other processes (including a message to itself which is delayed d time units). After waiting d time units, p handles its own message and does a $\text{Ret}_p(Q, v)$. Whenever a process receives an $enqueue(Q, v)$ or $dequeue(Q)$ message, it makes the appropriate update to the copy of Q in its local memory. (If it receives several messages at the same time, it "breaks ties" using sender ids, that is, it handles them by increasing order of process ids.)

Theorem 3.4 *There exists a linearizable implementation of FIFO queues with $|ENQ| = 0$ and $|DEQ| = d$.*

In the proof, we serialize each operation to occur d time after it is called. Since all processes update their local copies at these serialization times, the claim follows. (Proof omitted.)

4 Imperfect Clocks

Obviously, the assumptions of the previous section are unrealistically strong. In this section we relax them, and assume a system in which clocks run at the same rate as real time but are not initially synchronized, and in which message delays are in the range $[d - u, d]$ for some $u > 0$.

process p — Deq(X, v_1) Deq(X, v_2) ...... Deq(X, v_{i-1}) Deq(X, v_i) ......

process q — Enq($X, 1$)

Time 0 t $t + T$ $t + (i-2)T$ $t + (i-1)T$

(a) The execution α_1'.

process p — Deq($X, \perp$) Deq($X, \perp$) ...... Deq($X, \perp$) Deq($X, 1$)

process q — Enq($X, 1$)

Time 0 t $t + T$ $t + (i-2)T$ $t + (i-1)T$

(b) The execution α_1.

process p — Deq($X, \perp$) Deq($X, \perp$) ...... Deq($X, \perp$)

process q — Enq($X, 1$) Deq(X, u)

Time 0 t $t + T$ $t + (i-2)T$ $t + (i-1)T$

(c) The execution α_2.

process p — Deq($X, \perp$) Deq($X, \perp$) ...... Deq($X, \perp$) Deq($X, 1$)

process q — Enq($X, 1$) Deq($X, 1$)

Time 0 t $t + T$ $t + (i-2)T$ $t + (i-1)T$

(d) The execution α.

Figure 2: Executions used in the proof of Deq lower bound.

Under these assumptions, the lower bound of Theorem 3.1 still holds, but the algorithm of Theorem 3.4 is no longer correct. (It is possible to construct an execution that is neither sequentially consistent nor linearizable.) We first show that if only sequential consistency is required, then this can be remedied. Specifically, we present a sequentially consistent implementation of FIFO queues in which enqueue operations return instantaneously while the worst-case response time for a dequeue operation is $2d$. This algorithm matches (within constant factors) the lower bound of Theorem 3.1. We then show that there does not exist a similar algorithm for implementing linearizability — in any linearizable implementation of a FIFO queue the worst-case response time of an enqueue operation must depend on u, the message delay uncertainty (which can be as big as d).

4.1 Upper Bounds for Sequential Consistency

Inspecting the algorithm for fast enqueues (Theorem 3.4) reveals that the key point of its correctness is the fact that updates (either enqueues or dequeues) are handled by all processes in the same order and at the same time. In order to guarantee sequential consistency, it suffices for processes to update their local copies in the same order (not necessarily at the same time). A simple way to achieve this property is for a centralized controller to collect update messages and broadcast them. This idea can be developed into an algorithm in which each enqueue operation is performed instantaneously and the response time for dequeue is $O(d)$. We now present an algorithm that is completely distributed and does not rely on a centralized controller. The algorithm uses a variant of atomic broadcast ([7]) to guarantee that all messages are handled at the same order at all processes. The algorithm does not use any timing information and is completely asynchronous.

The atomic broadcast algorithm, embedded in our implementation, is based on assigning timestamps to messages, and hence to operations. Each process maintains a local timestamp and a vector with (conservative) estimates of the timestamps of all other processes. A process keeps its timestamp bigger than or equal to timestamps of all other processors (according to its estimates). Each operation generates an update message; this message is tagged with the process's current timestamp, and its id. Processes handle update messages by performing the appropriate change to their local copies. A process handles an update message with timestamp x only when it is certain that all other update messages with timestamp x have arrived at it. This is done by waiting to learn that all processes have increased their timestamp to be at least $x + 1$.[7] Once it learns that all processes have increased their timestamp beyond x, it handles all pending updates with timestamps less than or equal to x, breaking ties using process ids.

A more detailed description of the algorithm follows. Each process keeps a local copy of every object, a vector of timestamps (counters) – one for each process, and a set of updates that it is waiting to make to its local copies.

When a request to enqueue v to Q comes in to p, p sends a message $update(t_p, enqueue, Q, v)$ to all processes,[8] where t_p is p's current timestamp. It then increases its own timestamp by one. The operation returns immediately. When a request to dequeue from Q comes in to p, p sends a message $update(t_p, dequeue, Q, \perp)$ to all processes. It then increases its own timestamp by one and waits until it can handle this update locally.

---

[7]For simplicity, we assume FIFO channels, but this assumption can be removed if sequence numbers are employed.

[8]We assume that p sends each message also to itself.

When a process q receives an update message (either *dequeue* or *enqueue*) with timestamp t_p, it saves it in a list of pending updates, sorted by timestamp and process id. It then increases its timestamp to be at least as large as $t_p + 1$ and sends a timestamp increase message $timestamp(t_q, q)$.

When a process receives a timestamp increase message, it updates the timestamp entry for the sender, and checks to see if there are any pending updates whose timestamp is strictly less than all processes' timestamps (saved in its local vector). It handles these updates in increasing timestamp order, breaking ties using process ids. Handling an update to an object is done by performing the appropriate change (enqueue or dequeue) on the local copy of the object. If the update is a dequeue by the same process, the dequeue operation that is currently waiting returns the value that was dequeued from the local copy. (Note that by well-formedness, there is only one pending dequeue operation.)

The complete code of the algorithm appears in Figure 3.

We start with an outline of the correctness proof. To show sequential consistency, we must demonstrate, for any admissible execution σ, a sequential order for all operations in σ which implies a legal sequential execution. The ordering of operations is done by timestamps (breaking ties with process ids). The resulting sequence respects the order at each process by construction and because of the way timestamps are assigned. The sequence satisfies the specification of FIFO queues (is legal) since updates are done at each process in timestamp, process id order.

We now present the details of the formal proof. Fix some admissible execution of the algorithm, σ. With each operation on an object Q we associate, in a natural manner, an update by each process to its local copy of Q. The next lemma guarantees that each such update is performed within time $2d$ from the initiation of the operation. (Proof omitted.)

Lemma 4.1 *Suppose an operation op on object Q is invoked by process p at time T, then each process performs the update associated with op to its local copy of Q no later than time $T + 2d$.* ∎

The next claim follows immediately from the code of the algorithm:

Lemma 4.2 *Let p be some process. Then every operation of p in σ is given a unique timestamp in increasing order.*

Which immediately implies:

Lemma 4.3 *The timestamps assigned to operations in σ, together with process ids, form a total order.*

This total order is called *timestamp order*. The next lemma is the key to showing sequential consistency.

Lemma 4.4 *For any process p, p performs all changes to local copies in timestamp order.*

The algorithm uses the following data types:

> timestamp = integer
> update = record with fields
>> obj : name of an object (object to be updated)
>> op : name of an operation (enqueue or dequeue)
>> val : value of obj (value to be enqueued)
>> ts : timestamp (assigned by initiator)
>> id : process id (of the initiator)

The state of each process consists of the following components:

> ts : array[1..n] of integer, all initially 0 (estimate (from below) timestamps of all processes)
> *updates* : set of updates, initially empty (set of updates waiting to be made to local copies)
> copy of every object Q, initially equal to its initial value

The transition function of process p is as follows:

> $\text{Enq}_p(Q, v)$
>> send $update(ts[p], enqueue, Q, v)$ to all processes
>> $ts[p] := ts[p] + 1$
>> generate $\text{Ack}_p(Q)$

> $\text{Deq}_p(Q)$
>> send $update(ts[p], dequeue, Q, \perp)$ to all processes
>> $ts[p] := ts[p] + 1$

> receive $update(t, op, Q, v)$ from q
>> add (Q, op, v, t, q) to *updates*
>> if $t + 1 > ts[p]$ then
>>> $ts[p] := t + 1$
>>> send $timestamp(ts[p], p)$ to all processes
>> endif

> receive $timestamp(t)$ from q
>> $ts[q] := t$
>> repeat
>>> let E be element with smallest (ts,id) pair in *updates*
>>> if for some q, $ts[q] \leq E.ts$ then exit
>>> make appropriate modification to local copy of E.obj
>>> if E.id=p and E.op = deq then generate $\text{Ret}_p(Q, v)$
>>> remove E from *updates*
>> endrepeat

Figure 3: An asynchronous implementation of FIFO queues.

Proof: Let (t_1, q_1) be the timestamp of the update corresponding to op_1, and let (t_2, q_2) be the timestamp of the update corresponding to op_2. Suppose, by way of contradiction, that $(t_1, q_1) <$ (t_2, q_2) but p performs op_2's update before op_1's update.

When p performs op_2's update, it cannot yet have the update message for op_1, because otherwise it would perform op_1's update before op_2's update. By the code, in order to perform op_2's update, it must be that $ts_p[q_1] > t_2$. But then p must have received a *timestamp* message from q_1 with a timestamp $t \geq t_2 + 1$. Since $(t_1, q_1) < (t_2, q_2)$ it must be that $t_1 \leq t_2$, and hence $t > t_1$. By the code, the update message corresponding to op_1 was sent before the *timestamp* message. But then the FIFO property of the communication system implies that p has already received the update message corresponding to op_1. A contradiction. ∎

We can now prove:

Theorem 4.5 *There exists a sequentially consistent implementation of FIFO queues with $|E| = 0$ and $|D| = 2d$.*

Proof: Clearly, the time for an enqueue is 0. By Lemma 4.1, a process p performs the update corresponding to a dequeue operation it initiated at most $2d$ time after it was initiated. Thus, by the code, p generates Ret_p for that operation at most $2d$ time after it receives Deq_p. Hence, the time for any dequeue is at most $2d$.

The remainder of the proof is devoted to showing sequential consistency. For every admissible execution σ, we must specify a legal sequence of operations τ, such that for every process p, $ops(\sigma)|p = \tau|p$. Fix some admissible execution σ.

Define the sequence of operations $\tau = \text{serial}(\sigma)$ as follows: Order the operations in σ by timestamp order. From Lemma 4.2 it follows that operations by p are ordered in τ as they were ordered in σ, and thus $ops(\sigma)|p = \tau|p$, for all processes p.

It remains to show that τ is legal, i.e., that for every FIFO queue Q, the restriction of τ to Q is in the serial specification of Q. Pick any Q and consider $\tau|Q = op_1 op_2 \ldots$. Suppose op_i is $[\text{Deq}_p(Q), \text{Ret}_p(Q, v)]$. Since the local updates at p occur at the timestamp order (Lemma 4.4) it follows that exactly the updates associated with operations that precede op_i occur in p before the update associated with op_i occurs in p. The claim follows. ∎

Theorem 4.6 (presented below) implies that this algorithm does not guarantee linearizability. It is also possible to construct an explicit scenario which violates linearizability.

4.2 Lower Bound for Linearizability

We show that in any linearizable implementation of a FIFO queue the worst-case time for an enqueue is $u/2$ (assuming that at least two processes can enqueue to the same FIFO queue). The proof uses the technique of *shifting*. Shifting is used to change the timing and the ordering of events in the system while preserving the local views of the processes. It was originally introduced in [15] to prove lower bounds on the precision achieved by clock synchronization algorithms. Here we describe the technique and its properties informally; the formal details can be found in [3].

Given an execution with a certain set of clocks, if process p's history is changed so that the real times at which the events occur are shifted by some amount s and if p's clock is shifted by the same amount, then the result is another execution in which every process still "sees" the same events happening at the same clock time. The intuition is that the changes in the real times at which events happen at p cannot be detected by p because its clock has changed by a corresponding amount. The resulting changes to message delays in the new execution are as follows: the delay of any message to p is s less, the delay of any message from p is s more, and the delay of any message not involving p has the same delay as in the original execution.

Theorem 4.6 *If Q is a FIFO queue with at least two enqueuers then any linearizable implementation of Q must have $|E(Q)| \geq \frac{u}{2}$.*

Proof: Let p and q be two processes that can enqueue to Q and r be a process that dequeues from Q. Assume in contradiction that there is an implementation with $|E(Q)| < \frac{u}{2}$. Initially, Q is empty. By the specification of Q, there is an admissible execution α such that

- $ops(\alpha)$ is $Enq_p(Q, 1)$ $Ack_p(Q)$ $Enq_q(Q, 2)$ $Ack_q(Q)$ $Deq_r(Q)$ $Ret_r(Q, 1)$;

- $Enq_p(Q, 1)$ occurs at time 0, $Enq_q(Q, 2)$ occurs at time $\frac{u}{2}$, and $Deq_r(Q)$ occurs at time u; and

- the message delays in α are d from p to q, $d - u$ from q to p, and $d - \frac{u}{2}$ for all other ordered pairs.

Let $\beta = shift(shift(\alpha, p, -\frac{u}{2}), q, \frac{u}{2})$. I.e., we shift p later by $\frac{u}{2}$ and q earlier by $\frac{u}{2}$. The result is still an admissible execution, since the delay of a message from p or to q becomes $d - u$, the delay of a message from q or to p becomes d, and all other delays are unchanged. But $ops(\beta)$ is

$$Enq_q(Q, 2) \ Ack_q(Q) \ Enq_p(Q, 1) \ Ack_p(Q) \ Deq_r(Q) \ Ret_r(Q, 1)$$

which violates linearizability, because r's dequeue should return 2, not 1 (by the FIFO property). ∎

The assumption about the number of enqueuers made in Theorem 4.6 is crucial to the results, since it can be shown that the algorithm of Theorem 3.4 is correct if there is only one enqueuer.

5 Stacks

The results presented for FIFO queues can be extended in a straightforward manner to apply for stacks.

The specification of a stack is similar to the specification of a queue: The operations are $[Pop_p(S), Ret_p(S, v)]$ and $[Push_p(S, v), Ack_p(S)]$, for any p, S, and v. The set of operation sequences consists of all sequences that obey the usual stack semantics. Given a particular mcs and a stack S implemented by it, we denote by $|PUSH(S)|$ the maximum time taken by push operation on S and by $|POP(S)|$ the maximum time taken by a pop operation on S, in any admissible execution. Denote by $|PUSH|$ the maximum of $|PUSH(S)|$, and by $|POP|$ the maximum of $|POP(S)|$, over all objects S implemented by the mcs.

As the algorithms and the proofs are very similar, the following theorems are stated without proof.

Theorem 5.1 *In the model with perfect clocks, for any sequentially consistent implementation of a stack S, $|POP(S)| \geq d$.*

Theorem 5.2 *In the model with perfect clocks, there exists a linearizable implementation of stacks with $|PUSH| = 0$ and $|POP| = d$.*

Theorem 5.3 *In the model with imperfect clocks, there exists a sequentially consistent implementation of stacks with $|PUSH| = 0$ and $|POP| = 2d$.*

Theorem 5.4 *In the model with imperfect clocks, for any linearizable implementation of a stack S, $|PUSH(Q)| \geq \frac{u}{2}$, if more than two processes can push to the same stack.*

6   Conclusions and Further Research

This paper addresses a simplification of the problem of memory coherence in loosely-coupled multi-processors ([4, 5, 8, 9, 13, 16, 18, 19]). Our formal model ignores several important practical issues, e.g., limitations on the size of local memory storage, network topology, clock drift and "hot-spots". Since our lower bounds are proved in a very strong model, they clearly hold for more practical systems. We believe our algorithms can be adapted to work in more realistic systems.

Our work leaves open many interesting questions. Obviously, it is desirable to narrow the gaps between our upper and lower bounds. It will be interesting to understand how practical issues such as local memory size influence the bounds; in particular, it would be interesting to incorporate directory-based consistency schemes. The cost measure we have chosen to analyze is response time, but there are other interesting measures, including throughput and network congestion. It would be interesting to develop a better understanding of the relationship between the semantics of an object and the efficiency of implementing it.

Like the results of [3], our results suggest that linearizability may be too strong a condition, and that relaxing it yields some performance benefits. On the other hand, it is clear that efficiency, in general, and response time, in particular, are not the only criterion for choosing consistency guarantees. The ease of designing, verifying, programming, and debugging algorithms using these consistency conditions is also of great importance, and should be studied.

Acknowledgements: I would like to thank Roy Friedman and Jennifer Welch for many stimulating discussions and comments.

References

[1] S. Adve and M. Hill, "Implementing Sequential Consistency in Cache-Based Systems," *Proc. Int. Conf. on Parallel Processing*, 1990.

[2] Y. Afek, G. Brown, and M. Merritt. "A Lazy Cache Algorithm," *Proc. 1st ACM Symp. on Parallel Algorithms and Architectures*, 1989, pp. 209–222.

[3] H. Attiya and J. L. Welch, "Sequential Consistency versus Linearizability," *Proc. 3rd ACM Symp. on Parallel Algorithms and Architectures*, 1991, pp. 305–315.

[4] J. Bennett, J. Carter, and W. Zwaenepoel, "Munin: Distributed Shared Memory Based on Type-Specific Memory Coherence," *Proc. 2nd ACM Symp. on Principles and Practice of Parallel Programming*, 1990, pp. 168–176.

[5] R. Bisiani, A. Nowatzyk, and M. Ravishankar, "Coherent Shared Memory on a Distributed Memory Machine," *Proc. Int. Conf. on Parallel Processing*, 1989, pp. I-133–141.

[6] P. Bernstein, V. Hadzilacos, and H. Goodman, *Concurrency Control and Recovery in Database Systems*, Addison-Wesley, Reading, MA, 1987.

[7] K. Birman and T. Joseph, "Reliable Communication in the Presence of Failures," *ACM Trans. on Computer Systems*, Vol. 5, No. 1 (February 1987), pp. 47–76.

[8] W. Brantley, K. McAuliffe, and J. Weiss, "RP3 Processor-Memory Element," *Proc. Int. Conf. on Parallel Processing*, 1985, pp. 782–789.

[9] M. Dubois and C. Scheurich, "Memory Access Dependencies in Shared-Memory Multiprocessors", *IEEE Trans. on Software Engineering*, vol. 16, no. 6 (June 1990), pp. 660–673.

[10] M. Herlihy, "Wait-Free Implementations of Concurrent Objects," *Proc. 7th ACM Symp. on Principles of Distributed Computing*, 1988, pp. 276–290.

[11] M. Herlihy and J. Wing, "Linearizability: A Correctness Condition for Concurrent Objects," *ACM Trans. on Programming Languages and Systems*, Vol. 12, No. 3 (July 1990), pp. 463–492.

[12] L. Lamport, "How to Make a Multiprocessor Computer that Correctly Executes Multiprocess Programs," *IEEE Trans. on Computers*, Vol. C-28, No. 9 (September 1979), pp. 690–691.

[13] K. Li and P. Hudak, "Memory Coherence in Shared Virtual Memory Systems," *ACM Trans. on Computer Systems*, vol. 7, no. 4, pp. 321–359.

[14] R. Lipton and J. Sandberg, *PRAM: A Scalable Shared Memory*, Technical Report CS-TR-180-88, Princeton University, September 1988.

[15] J. Lundelius and N. Lynch, "An Upper and Lower Bound for Clock Synchronization," *Information and Control*, Vol. 62, Nos. 2/3 (August/September 1984), pp. 190–204.

[16] S. Min and J. Baer, "A Timestamp-Based Cache Coherence Scheme," *Proc. Int. Conf. on Parallel Processing*, 1989, pp. I-23–32.

[17] C. Papadimitriou, *The Theory of Concurrency Control*, Computer Science Press, Rockville, MD, 1986.

[18] U. Ramachandran, M. Ahamad, and M. Y. Khalidi, "Coherence of Distributed Shared Memory: Unifying Synchronization and Data Transfer," *Proc. Int. Conf. on Parallel Processing*, 1989, pp. II-160–169.

[19] C. Scheurich and M. Dubois, "Correct Memory Operation of Cache-Based Multiprocessors," *Proc. 14th Int. Symp. on Computer Architecture*, 1987, pp. 234–243.

Optimal Amortized Distributed Consensus *

(Extended Abstract)

Amotz Bar-Noy [†]   Xiaotie Deng [‡]   Juan A. Garay [†]   Tiko Kameda [§]

Abstract

Are randomized consensus algorithms more powerful than deterministic ones? Seemingly so, since randomized algorithms exist that reach consensus in expected constant number of rounds, whereas the deterministic counterparts are constrained by the $r \geq t+1$ lower bound in the number of communication rounds, where t is the maximum number of faults to be tolerated.

In this paper, however, we study the behavior of deterministic algorithms when consensus is repeatedly needed, say k times. We show that it is possible to achieve consensus with the optimal number of processors ($n > 3t$), and optimal amortized cost in all other measures: the number of communication rounds r^*, the maximal message size m^*, and the total bit complexity b^*.

More specifically, we achieve the following amortized bounds for k consensus instances: $r^* = O(1 + \frac{t}{k})$, $b^* = O(t^2 + \frac{t^4}{k})$, and $m^* = O(1 + \frac{t^2}{k})$. When $k \geq t^2$, then r^* and m^* are $O(1)$ and $b^* = O(t^2)$ which is optimal.

---

*Work of second and fourth authors partially supported by the NSERC (Natural Sciences and Engineering Research Council of Canada) International Postdoctoral Fellowship and research grants from the NSERC and the Advanced Systems Institute of British Columbia.

[†]IBM T. J. Watson Research Center, P. O. Box 704, Yorktown Heights, NY 10598.

[‡]Department of Computer Science, York University, North York, Ontario, Canada M3J 1P3. This work was carried out while this author was at the School of Computing Science, Simon Fraser University.

[§]School of Computing Science, Simon Fraser University, Burnaby, B.C., Canada V5A 1S6.

1 Introduction

A major task in fault tolerant distributed systems is to agree on common values. The *Consensus* problem (and its variations, e.g., the *Byzantine Generals* problem) [PSL] is related to this task and has received much attention in the literature. The problem is defined as follows. Given is a set P of n processors, a subset T ($|T| = t$) of which may be *faulty*. Each processor starts off with an initial binary value. The goal is to find an algorithm where each non-faulty (correct) processor *decides* on a value under the following two conditions:

- **Agreement:** All the non-faulty processors decide on the same binary value.

- **Validity:** If the initial value is identical for all the non-faulty processors, then this will be the decision value.

We assume the standard distributed model. Processors are synchronized by rounds of communication in which each sends messages, receives messages, and performs some local computation. By the last round of the algorithm, after the local computation is completed, the processors must decide. The communication is carried out by a complete and reliable network, i.e., each processor can send a message to every other processor and messages arrive unaltered in the same round they are sent. The processors are deterministic and no randomized operations are allowed. The faulty processors might behave in a malicious (*Byzantine*) way, and collude in order to prevent reaching a valid agreement. For ease of the exposition, processors' names are assumed to be the numbers $1, \ldots, n$.

Three complexity measures determine the efficiency of a solution: the ratio between n and t; the number of rounds required in the worst case, denoted by r; and the communication complexity, alternatively given by the upper bound on the size of any message, denoted by m, or the total number of transmitted bits, denoted by b. The known lower bounds for these complexity measures are: $n \geq 3t + 1$ [PSL], $r \geq t + 1$ [FL], $m \geq 1$ (obvious), and $b \geq n(t+1)/4$ [DR]. These lower bounds are independent in the sense that they are valid no matter what the values of the other parameters are. The number of rounds lower bound, in particular, asserts the difficulty of overcoming potential faults.

One approach to reduce this (high) price for fault tolerance is that of Dolev, Reischuk and Strong [DRS]. By permitting processors to terminate at different times (rounds), instead of simultaneously, they show that *early-stopping* consensus protocols are possible. This type of agreement is called *eventual*, and they prove a lower bound of two rounds of termination uncertainty for this case. Such protocols take a number of rounds proportional to the actual number of failures that actually occur, rather than the worst-case t. (However, if simultaneous termination is required, then the algorithm must run for $t + 1$ even if no faults occur.)

Randomization is another way to circumvent the lower bound for the number of rounds: there exist probabilistic algorithms that reach consensus in expected *constant* number of rounds. Feldman and Micali [FM], in particular, have a probabilistic algorithm that achieves this together with the optimal number of processors (at the expense, however, of non-trivial communication).

In this paper, we study whether this constant number of rounds can also be achieved on the average in situations where we need to perform consensus repeatedly, say k times. In addition, we would like to achieve this constant average number of rounds while obtaining an optimal average performance in the other parameters. Here we concentrate on the case where the consensus instances have to be performed far apart, i.e. the next consensus does not start until the previous one has already finished. We call this problem the *multi-consensus* problem.

Denote by $R(k)$, $M(k)$, and $B(k)$ the overall round complexity, message size complexity, and bit complexity over k instances of the consensus problem, respectively ($M(k)$ is the sum of the maximal size of the messages required at each instance). We introduce the *amortized* counterpart of the efficiency parameters: $r^* = R(k)/k$, $m^* = M(k)/k$, and $b^* = B(k)/k$, the average value of the corresponding parameters over the k consensus instances.

The faulty processors can force one of the instances, not known to the correct processors, to last $t + 1$ rounds using at least t^2 bits. Therefore, for the rest of the instances we need at least one round and t^2 bits per instance. The second part of the last claim is true since any processor must send its input bit to at least $t + 1$ other processors, as otherwise it can be obstructed by the faulty processors and n is no longer greater than $3t$. Consequently, $R(k) = \Omega(t + k)$, implying $r^* = \Omega(1 + \frac{t}{k})$, and $B(k) = \Omega(t^2 k)$, implying $b^* = \Omega(t^2)$. The third lower bound is obviously $m^* \geq 1$.

A related problem is when the inputs of all instances are given in advance. Then by concatenating all the input bits together it is possible to achieve $r^* = \frac{r}{k}$, $b^* = b$, and $m^* = km$, for any algorithm with complexity: r rounds, b bits, and m maximum size of a message. However, this is a very unrealistic solution because of the enormous size of the messages for large k. Bar-Noy and Dolev [BD] considered solving many instances of the consensus problem in a pipeline fashion by using only single-bit messages ($m^* = 1$). Their generic algorithm achieves $r^* = 1 + \frac{t}{k}$ and $b^* = O(t^3)$; however, the number of required processors is quadratic in t. Ben-Or and El-Yaniv [BE] presented an expected constant rounds algorithm for solving many instances together by generalizing the result of [FM] in a non-trivial fashion. However, as in the simple concatenating solution, they use messages the size of which is proportional to the number of instances.

An apparent main requirement to achieve the goal of optimal amortized parameters is that a constant number of rounds is invested per occurring fault. We note that previous to this paper, "technology" already existed to achieve this property, namely, the simultaneous application of the early-stopping and *fault detection* techniques. To the best of our knowledge, the only published consensus algorithm in which these techniques are used together to achieve consensus with the optimal number of processors is that of [BDDS].[1] However, using this algorithm as the basis for solving many instances of the consensus problem yields the following parameters: $r^* = O(1 + \frac{t}{k})$, $b^* = O(t^3 + \frac{t^4}{k})$, and $m^* = O(t)$ (i.e. non-optimal amortized message complexity).

The contribution of this paper is to achieve, for $n > 3t$, a constant amortized round complexity and to improve the result implied by [BDDS] in both message size and bit complexity. We thus obtain an optimal amortized solution in all four parameters:

- $n = 3t + 1$,

- $r^* = O(1 + \frac{t}{k})$,

- $b^* = O(t^2 + \frac{t^4}{k})$, and

- $m^* = O(1 + \frac{t^2}{k})$.

When $k \geq t^2$ we get: $r^* = O(1)$, $b^* = O(t^2)$, and $m^* = O(1)$. Note that when $k = o(t^2)$ the amortized bit complexity and the amortized message size complexity are not optimal. We leave it as an open problem whether an optimal solution is possible for all values of k. When $n > 4t$, we can do slightly better: $b^* = O(t^2 + \frac{t^3 \log t}{k})$, and $m^* = O(1 + \frac{t \log t}{k})$.

The reminder of the paper is organized as follows. In Section 2 we describe the early-stopping version of the *Phase King* protocol of [BGP]. In Section 3, we discuss some modifications to the *Phase King* protocol (needed for our multi-consensus algorithm). In particular, we address a more stringent version of the consensus problem. The solution for the multi-consensus problem and its complexity are presented in Section 4.

2 A Simple Early-Stopping Consensus Algorithm

In this section we present a simple consensus algorithm (*Early Stopping Phase King*—ESPK for short) that uses messages of constant size and satisfies the early-stopping property, i.e. it is able to terminate in a number of rounds proportional to the actual number of occurring faults. For ease of presentation, we discuss a version of the protocol that requires $n > 4t$, $r = \min\{2(t+1), 2(f+2)\}$, $b = 48t^2 \min\{f+2, t+1\}$, and $m = 2$. The protocol is shown in Figure 1.

---

[1] In fact the algorithm of [BDDS] uses only the fault detection technique. However, it is easy to see how the early-stopping technique can be added.

```
Procedure ESPK (V: {0,1});
begin
        for i := 1 to t + 1 do
        begin

                broadcast(V)     (* universal exchange *)
                for j := 0 to 1 do  C[j] := the number of received j's;
                V := C[1] > 2t;
                π := C[V] ≥ n - t;

                broadcast(⟨V, π⟩);     (* king's broadcast *)
                S := the number of received ⟨v, 1⟩'s, for any v;
                if S ≤ t then V := v, where ⟨v, w⟩ is the message received from i
                elseif S > 2t then return(V);     (* and halt *)

        end for
end;
```

Figure 1: The Early-Stopping Phase King algorithm for processor p.

A simple modification of the algorithm yields consensus with $n > 3t$, $r = \min\{3(t+1), 3(f+2)\}$, $b = 36t^2 \min\{f + 2, t + 1\}$ bits, and $m = 2$.

Algorithm ESPK executes at most $t+1$ phases each consisting of two communication rounds: the universal exchange, in which all processors broadcast their values, and the broadcast of the "phase king;" each phase has a distinct king. We assume that each processor p has local variables V (the subject of the consensus problem), π (indicating whether or not there is a "strong preference" for V), S (which counts the number of processors with strong preferences), and integer array $C[0..1]$ (the counts of 0's and 1's received). In the universal exchange each processor changes V to what it perceives as the majority value and sets π to true if V is strongly favored. During the king's broadcast each processor sends the value determined to be the majority along with an indication of the strength of its preference for this value. After counting the number of processors with strong preferences, a processor replaces its own V with the king's value if there are too few processors with a strong preference; should it find many processors with a strong preference, it halts. Because some correct processors may halt before others, a convention is adopted to account for messages that are not received during a phase: if no message is received from p during the universal exchange, the receiving processor assumes that the missing value is equal to the value of its own V of the previous phase and that the silent processor indicates a strong preference during the king's broadcast.

Theorem 2.1 *Distributed Consensus is achievable using* $\min\{2(f+2), 2(t+1)\}$ *exchange rounds,* $48t^2 \min\{f+2, t+1\}$ *bits, and two-bit messages, where* f *is the number of processors that fail during its execution, provided* $n > 4t$.

Proof: As a preliminary step, we show the following property of the ESPK, as shown in Figure 1.

Persistency: If for all $p \in P - T$, $V_p = v$ is true at the start of a phase, every processor decides on v and halts at the end of the phase.

To see this, observe that the universal exchange results in every correct processor p having $C[v]_p \geq n - t > 2t$, $C[\bar{v}]_p \leq t$, and π being true. Therefore it computes $V_p = v$ with a strong preference. During the king's broadcast, it detects that at least $n - t$ processors have strong preferences so it ignores the king's broadcast and halts.

Now we can address the correctness of the protocol.

Agreement: Since the number of phases $(t + 1)$ exceeds the number of faulty processors (t), there is some first phase g with a correct king. One of the following cases holds at the end of phase g:

1. $S \leq t$ for each $p \in P - T$. Then every correct processor assigns V_g to V during the king's broadcast of phase g, where $\langle V_g, w \rangle$ is the message sent by g.

2. $S > t$ for some $p \in P - T$. Then π is true for at least one correct processor p' so at the end of the universal exchange, $V_{p'} = v$ and $C[v]_{p'} > n - t$. Among the $n - t$ processors from which p' received value v, at most t are faulty so at least $n - 2t$ are correct. Thus, *every* correct processor received v at least $n - 2t > 2t$ times, and $\bar{v}$ at most $2t$ times. As a consequence, at the end of the universal exchange $V_q = v$ for every $q \in P - T$, including king g; therefore, the king's broadcast must result in $V_q = v$ for all $q \in P - T$.

Persistency assures that no correct processor changes its value subsequent to phase g.

Validity: Follows trivially from persistency.

As for the running time, *Agreement* shows that, if the king of phase i is correct, then for all $p \in P - T$, $V_p = v$ for some v. By *Persistency*, all processors that have not already halted will halt by the end of phase $i + 1$. If g is the phase with the first correct king, the number of faulty processors f is at least $g - 1$ (because all prior kings were faulty). Thus $g + 1 \leq f + 2$.

The message size and bit complexity follow easily from inspection of the algorithm.

□

The version of ESPK that we have just discussed is asymptotically optimal with respect to rounds and number of processors.[2] By means of a simple modification, it is possible to achieve exact optimality in the number of processors, i.e. $n > 3t$. The modification consists of one additional universal exchange, after which we can show that there is at most one value v for which its count is greater than t, for any processor $p \in P - T$. Therefore, a processor can compute a new value for V using fewer processors ($t + 1$ versus $2t + 1$) than in the version of ESPK of Figure 1. The additional exchange can be either one round using two-bit messages, or two rounds using single-bit messages. Details are given in [BGP]. This allows us to state:

Theorem 2.2 *Distributed Consensus is achievable using* $\min\{3(f + 2), 3(t + 1)\}$ *exchange rounds,* $36t^2 \min\{f+2, t+1\}$ *bits, and messages of size at most two, or* $\min\{4(f+2), 4(t+1)\}$ *exchange rounds,* $36t^2 \min\{f, t + 1\}$ *bits, using single-bit messages, where f is the number of processors that fail during its execution, provided $n > 3t$.*

3 Modifications to the *ESPK* Algorithm

In this section, we perform some modifications to the algorithm presented in the previous section in order to use it as a basis of the multi-consensus construction. The first modification concerns a different version of the consensus problem. In this version, the input values for the correct processors are x and $x + 1$, for some unknown $0 \le x < t$. In addition a stronger Validity condition is required:

* **Strong Validity:** The decision value must be one of the correct processors' initial value.

In the standard binary case both Validity conditions are equivalent. We call this problem the *unknown binary set* consensus problem (UBC for short). We note that a straightforward reduction of UBC to the 0-1 case fails, since the faulty processors could introduce the values $x - 1$ and $x + 2$.

We first show how to solve UBC for $n > 4t$ with messages of size $O(\log t)$. We add an initial round of universal exchange before the first round of MOD-ESPK. At the end of this round, each correct processor adopts the majority value. In future rounds correct processors believe only to values that had a support of at least $t+1$ processors in this first initial universal exchange round.

Lemma 3.1 *ESPK with the above modification solves* UBC *using $n > 4t$ processors.*

Proof: Clearly only x and $x + 1$ can get a support of $t + 1$ processors and therefore a faulty king cannot introduce a third value. Also, since a correct processor adopts either x or $x + 1$ it follows that a correct king suggests either x or $x + 1$. Now assume that at least $2t + 1$ correct

---

[2]It is not hard to modify the algorithm to be optimal in message size, i.e., $m = 1$.

processors started with x or $2t + 1$ correct processors started with $x + 1$. Then by the first phase all the correct processors decide on the same value since all correct processors adopt the same value. Otherwise, since $n > 4t$, at least $t + 1$ correct processors started with x and at least $t + 1$ correct processors started with $x + 1$. In this case, all the correct processors would believe the first correct king if by that phase agreement has not been achieved. □

The above algorithm uses messages of size $\log x \leq \log t$. Note also that the above algorithm works even if the two values to be agreed upon are $0 \leq x < y \leq t$ (in the above proof $y = x+1$).

When $3t < n \leq 4t$ we solve UBC with larger messages. Each correct processor replaces its input by its *unary* representation, i.e., x ones followed by a single zero, or $x + 1$ ones followed by a single zero. The processors run at most $x + 3$ instances of ESPK in parallel[3] on the new representation of their input. If a correct processor started with x it participates only in the first $x + 2$ instances where its $x + 2$nd input is zero. If a correct processor started with $x + 1$ it participates only in the first $x + 3$ instances and its $x + 3$rd input is zero. The decision value is the location of the first zero minus one.

Lemma 3.2 ESPK *with the above modification solves* UCB *using* $n > 3t$ *processors.*

Proof: On each instance the processors run the regular ESPK and therefore decide on either zero or one. Since all correct processors have input one in the first x instances it follows that the first zero location is at least in the $x + 1$st position. On the other hand, since all correct processors have input zero in the $x + 2$nd instance it follows that the first zero location is at most in the $x + 2$nd position. Consequently, all correct processors either find the first zero at location $x + 1$ or at location $x + 2$ and therefore decide on either x or $x + 1$. □

The above algorithm uses messages of size at most $x + 3$. Note that here the two binary values must be x and $x + 1$, otherwise the above algorithm guarantees only that the decision is no less than the smaller value and no larger than the larger value.

The second modification is the addition of a *fault detection* mechanism. One additional property of ESPK is that only faulty kings can cause the algorithm to run for many rounds. Moreover, doing this forces them to reveal themselves as faulty to all the correct processors. Let F be the set of processors known as faulty by all the correct processors. We say a processor in F has been *globally* detected. Assume now that processor p terminates the execution of ESPK at phase i. We claim that in this case processors $\{1, \ldots, i - 2\} \subseteq F$. This is true because whenever a processor has to proceed to the next phase, it realizes that the current king is faulty. Since all correct processors could not terminate before phase $i - 1$ it follows that all detected at least $i - 2$ faulty kings.

---

[3] It is easy to see how to achieve consensus on more than one value in parallel. This is achievable in the same number of rounds as for one value, at the cost of a multiplicative increase in communication.

Our third and last modification concerns the sequence of kings used by ESPK. The ascending order of kings is not necessary, and any set of $t+1$ kings (known to all processors) would yield the correctness of the algorithm.

We call ESPK including the above three modifications MOD-ESPK. In the next section, we will invoke MOD-ESPK with a set of $t+1$ kings, and with two input values—the first being binary and the second being a number between 0 and t. In addition, the output will include the number of processors detected as faulty.

4 Optimal Amortized Consensus

In this section we describe an algorithm for k consensus instances, which achieves optimal amortized parameters. We assume that there is enough time to finish each agreement before the next one starts.

Our goal is to use for each consensus instance an early-stopping algorithm which has the following two properties.

1. Its round complexity is $r = \beta_r + \alpha_r f$ where α_r and β_r are constants and f is the number of faulty processors that actually occurred in the run and these faulty processors were globally detected. In later consensus instances these f faulty processors will be ignored. Therefore, the round complexity of the multi-consensus algorithm is at most $R(k) = \beta_r k + \alpha_r t$ and hence $r^* \leq \beta_r + \alpha_r \frac{t}{k}$. When $k \geq t$ we have $r^* \leq \beta_r + \alpha_r = O(1)$.

2. Its bit complexity is $b = \beta_b t^2 + \alpha_b t f$ where α_b and β_b are constants. Again, since these f processors will be ignored in later consensus instances, it follows that the bit complexity of the multi-consensus algorithm is at most $B(k) = \beta_b t^2 k + \alpha_b t^2$ and hence $b^* \leq \beta_b t^2 + \alpha_b \frac{t^2}{k}$. For all k it implies that $b^* = O(t^2)$.

(The algorithm of [BDDS] achieves the first property but not the second: its bit complexity for globally detecting f faulty processors is $O(ft^3)$.) Algorithm MOD-ESPK achieves these two properties for a non-constant $\alpha_b = O(t)$, however, there is a problem in using it straightforwardly. Implicit in this algorithm is the assumption that the processors know in advance the sequence of kings to be used during the run of the algorithm. When there is only one or two consensus instances this can be done as all processors agree on the names $1, \ldots, n$. The first instance can use the kings $1, \ldots, t+1$ and the second instance can use the kings $n, \ldots, n-t$. However, already in the third instance there is no common agreement among the correct processors about the next set of kings. If faulty kings were reused, then the desired amortized complexity would not be achieved. On the other hand, since this is an eventual agreement situation the processors cannot immediately agree on which processor acted as the last king.

```
Algorithm MULTI-CONSENSUS;
begin
        l := 1;    u := n;    f := 0;
        for j := 1 to k do
        begin
            input(V);
            if j is even then begin
                Call MOD-ESPK with input (V, f) and kings l, l + 1, ...;
                u := u - f;
            end
            if j is odd then begin
                Call MOD-ESPK with input (V, f) and kings u, u - 1, ...;
                l := l + f;
            end;
            f := # of faulty processors detected by MOD-ESPK in current instance;
            output(V)
        end for
end;
```

Figure 2: The Multi-Consensus algorithm for processor p

To overcome this problem we use the following mechanism. In even consensus instances (for short, even consensus) the processors use the kings $1, \ldots, t + 1$, while in odd consensus instances they use the kings $n, \ldots, n - t$. In addition, in the jth agreement the processors agree upon the first king to be used in the $j + 1$st agreement. This first king starts an increasing sequence of kings if $j + 1$ is even, and a decreasing sequence of kings if $j + 1$ is odd. Initially, in the first two instances, the starting kings are 1 or n. In later instances, as some processors are globally detected as faulty, the starting kings could be any number between 1 and $t + 1$ or n and $n - t$. To obtain the agreement on the starting king, and therefore on the sequence of kings to be used, the input for each instance is a pair of numbers: the first is the agreement input bit and the second is the name of a would-be king.

Such a solution implies an amortized bit complexity of $b^* = O(t^2 \log t)$, because the size of each message is multiplied by $\log t$. To get rid of the $\log t$ factor, instead of agreeing on the first king to be used in the next instance, the processors agree on the number of globally detected faulty processors. This number can be t, but t is also an upper bound for the sum of these numbers over all instances. Later we will show that the overhead bit complexity for this mechanism is $O(kt^2 + t^4)$.

We now describe the algorithm more formally (see Figure 2). In the multi-consensus algorithm, for each instance j, processor p gets the input bit in V and returns its decision before getting the next input bit. Processor p maintains two parameters l and u. In the even consensus instances the kings will be $l, l + 1, \ldots$ and in the odd consensus instances the kings will be $u, u - 1, \ldots$. Denote the values of these parameters at the beginning of the jth instance by l_j and u_j. Initially $l_1 = 1$ and $u_1 = n$. Another parameter maintained by p is f which is the number of new faulty processors detected by p after the termination of any call to MOD-ESPK. Denote the value of f at the beginning of the jth instance by f_{j-1}. Initially $f_0 = 0$. The input for the jth instance of MOD-ESPK is the pair (V, f_{j-1}). At the end of an even consensus instance j, p sets $l_{j+1} = l_j$ and $u_{j+1} = u_j - f_{j-1}$. At the end of an odd consensus instance j, p sets $l_{j+1} = l_j + f_{j-1}$ and $u_{j+1} = u_j$.

The following lemma shows that before each call to MOD-ESPK all the processors agree on the set of kings to be used, which in turn implies the correctness of our multi-consensus algorithm. The proof of this lemma follows by a straightforward induction on j.

Lemma 4.1 *At the beginning of an even (respectively, odd) instance j, all the correct processors have the same value for l_j (respectively, u_j).*

By the fault detection property of MOD-ESPK we get the following lemma.

Lemma 4.2 *For all j, processors $1, \ldots, l_j - 1 \in F$ and $u_j + 1, \ldots, u_n \in F$.*

Proof: Algorithm MOD-ESPK guarantees that for each j, f_j can get at most two values x and $x + 1$. Also, $x + 1$ faulty processors were detected by some correct processors, but maybe only x were globally detected. In any case, the processors agree on either x or $x + 1$ detected faulty processors. These x detected faulty processors are the first x kings used by MOD-ESPK. If this protocol used the kings $l, l + 1, \ldots$, (respectively, $u, u - 1, \ldots$) then processors $l, \ldots, l + x - 1$ (respectively, $u, \ldots, u - x + 1$) are faulty processors. The proof is completed by a straightforward induction on j. $\square$

Now we are ready to assert the amortized complexity of our multi-consensus algorithm.

Lemma 4.3 *The round complexity of the multi-consensus algorithm is $R(k) = O(k + t)$, and therefore $r^* = O(1 + \frac{t}{k})$.*

Proof: The early-stopping property of MOD-ESPK implies that the jth instance took at most $r_j = 4f_j + 8$ rounds. By Lemma 4.1, a faulty processor can play the role of a king only once and therefore $\sum_{j=1}^{k} f_j \leq t$. Consequently, $\sum_{j=1}^{k} r_j \leq 8k + 4t$. $\square$

Lemma 4.4 *The bit complexity of the multi-consensus algorithm is $B(k) = O(t^2 k + t^4)$, and therefore $b^* = O(t^2 + \frac{t^4}{k})$.*

Proof: Recall that the input to MOD-ESPK is a pair of numbers: the first is the input bit and the second is a number $f, 0 \leq f < t.$. First we analyze the bit complexity of the multi-consensus concerning the first part of each such pair.

In algorithm MOD-ESPK the bit complexity of each phase is $36t^2$. The early-stopping property of MOD-ESPK implies that the jth instance took at most $f_j + 2$ phases. Therefore, the bit complexity of the jth instance is $b'_j = 36t^2 f_j + 72t^2$. By Lemma 4.1, a faulty processor can play the role of a king only once and therefore $\sum_{j=1}^{k} f_j \leq t$. Consequently,

$$\sum_{j=1}^{k} b'_j \leq 72t^2 k + 36t^3.$$

We now analyze the bit complexity concerning the second part of each pair. This bit complexity is the same as that of the first part for all cases when $f_j = 0$. Let $i_1, \ldots, i_h$ be the indices for which $f_{i_g} \geq 1$. The agreement on f_{i_g} is done in instance $i_g + 1$ the bit complexity of which is at most $b''_{i_g+1} = 36t^2(3 + f_{i_g})(f_{i_g+1} + 2)$. Therefore the bit complexity for the cases where $f_j > 0$ is at most

$$\sum_{g=1}^{h} b''_{i_g+1} \leq \sum_{g=1}^{h} ct^2 f_{i_g+1} f_{i_g} \leq ct^3 \sum_{g=1}^{h} f_{i_g+1} \leq ct^4,$$

for some constant c. The above inequalities rely on Lemma 3.2. □

Note that if $n > 4t$ we can use Lemma 3.1 instead of Lemma 3.2 for the last inequality to obtain a bit complexity $b^* = O(t^2 + \frac{t^3 \log t}{k})$.

Lemma 4.5 *The average message size is* $m^* = O(1 + \frac{t^2}{k})$.

Proof: Using the same reasoning as in the Lemma 4.4 we get that in most instances messages are of constant size and in at most t instances messages are of size $O(t)$. Therefore, $m^* = O(\frac{k+t^2}{k})$. □

Again note that if $n > 4t$ we get $m^* = O(1 + \frac{t \log t}{k})$.

Lemmas 4.1, 4.3, 4.4, and 4.5 yield our main result:

Theorem 4.1 *The multi-consensus algorithm is correct for all k instances, provided $n > 3t$, and has the following amortized complexity:*

1. $r^* = O(1 + \frac{t}{k})$,

2. $b^* = O(t^2 + \frac{t^4}{k})$,

3. $m^* = O(1 + \frac{t^2}{k})$.

Acknowledgments

We would like to thank Vassos Hadzilacos for pointing out a flaw in an early version of the paper.

References

[BD] A. Bar-Noy and D. Dolev, "Consensus Algorithms with One-Bit Messages," *Distributed Computing*, Vol 4, pp. 105-110, 1991.

[BDDS] A. Bar-Noy, D. Dolev, C. Dwork, and H. R. Strong, "Shifting Gears: Changing Algorithms on the Fly to Expedite Byzantine Agreement," *Proc. 6th ACM Symp. of Principles of Dist. Computing*, pp. 42-51, 1987.

[BE] M. Ben-Or and R. El-Yaniv, "Interactive Consistency in Constant Time," unpublished manuscript.

[BG] P. Berman and J. A. Garay, "Asymptotically Optimal Distributed Consensus," *Proc. ICALP 89*, LNCS Vol. 372, pp. 80-94, 1989.

[BGP] P. Berman, J. A. Garay, and K. J. Perry, "Asymptotically Optimal Early-Stopping Consensus," *IBM Research Report*, in preparation.

[DR] D. Dolev and R. Reischuk, "Bounds of Information Exchange for Byzantine Agreement," *JACM*, Vol. 32, pp. 191-204, 1985.

[DRS] D. Dolev, R. Reischuk, and H.R. Strong, "Eventual is Earlier than Immediate," *Proc. 23rd Symp. on Foundations of Comp. Science*, pp.196-203, 1982. Revised version appears under the title, "Early-Stopping in Byzantine Agreement," *JACM*, Vol. 37, pp. 720-741, 1990.

[FL] M.J. Fischer and N.A. Lynch, "A Lower Bound for the Time to Assure Interactive Consistency," *Information Processing Letters*, Vol. 14, pp.183-186, 1982.

[FM] P. Feldman, and S. Micali, "Optimal Algorithms for Byzantine Agreement," *Proc. 20th Symposium on Theory of Computing*, pp. 148-161, 1988.

[PSL] M. Pease, R. Shostak, and L. Lamport, "Reaching Agreement in the Presence of Faults," *Journal of the ACM*, Vol. 27, pp. 228-234, 1980.

Optimally Simulating Crash Failures in a Byzantine Environment[*]

Rida Bazzi[†]
Gil Neiger

College of Computing, Georgia Institute of Technology
Atlanta, Georgia 30332–0280 U.S.A.

Abstract

The difficulty of designing of fault-tolerant distributed algorithms increases with the severity of failures that an algorithm must tolerate. Researchers have simplified this task by developing methods that *automatically* translate protocols tolerant of "benign" failures into ones tolerant of more "severe" failures. In addition to simplifying the design task, these translations can provide insight into the relative impact of different models of faulty behavior on the ability to provide fault-tolerant applications. Such insights can be gained by examining the properties of the translations. The *round-complexity* of a translation is such a property; it is the number of *rounds* of communication that the translation uses to simulate one round of the original algorithm. This paper considers *synchronous* systems and examines the problem of developing translations from simple stopping (crash) failures to completely arbitrary behavior with round-complexities 2, 3, and 4, respectively. In each case, we show a lower bound on the number of processors that must remain correct. We show matching upper bounds for all of these by developing three new translation techniques that are each optimal in the number of processors required. These results fully characterize the optimal translations between crash and arbitrary failures.

1   Introduction

Distributed computer systems give algorithm designers the ability to write fault-tolerant applications in which correctly functioning processors can complete a computation despite the failure of others. It has been well-established that the complexity of writing such applications depends upon the type of faulty behavior that processors may exhibit. While simple stopping failures are relatively easy to tolerate, completely arbitrary behavior is especially difficult to overcome. To assist the designers of such applications, researchers have developed *translations* that automatically increase the fault-tolerance of distributed algorithms. Such translations automatically convert algorithms tolerant of relatively benign type of failure into ones that tolerate more severe faulty behavior. These tools simplify the design task: the designer can design his or her algorithm with the assumption that faulty behavior is benign; the translated algorithm can then be run correctly in a system with more severe failures. In addition to simplifying the design task, these translations can provide insight into the relative impact of different models of faulty behavior on the ability to provide fault-tolerant applications. Such insights can be gained by examining the properties of the translations.

---

[*]Partial support for this work was provided by the National Science Foundation under grants CCR-8909663 and CCR-9106627.

[†]This author was supported in part by a scholarship from the Hariri Foundation.

The following failures, which are the ones most commonly considered, form a hierarchy from most benign to most severe:

1. *Crash Failures.* Processors subject to crash failures fail by stopping prematurely [8]. Before they stop, they are completely correct. After they stop, they take no further actions.

2. *Omission Failures.* Processors subject to omission failures may stop, or they may intermittently fail to send or receive messages [17]. The messages they send are always correct. Because these processors can fail and yet continue to send messages, their failure is more difficult to detect and overcome.

3. *Arbitrary Failures.* Processors subject to arbitrary failures can take any action whatsoever [13]. They can stop, omit to send or receive messages, or send spurious messages and claim to have received messages they did not. Such failures are also called *malicious* or *Byzantine*.

(Other failures models have been considered but are beyond the scope of this paper [8,18,19].)

Many researchers have developed translations for failures within this hierarchy. Coan [6] considered systems with *asynchronous* message-passing and developed a "compiler" that converts algorithms that tolerate crash failures into ones that tolerate arbitrary failures. Other researchers have considered systems with *synchronous* message-passing. Hadzilacos [8] developed a technique to translate algorithms tolerant of crash failures into ones that tolerate omission-to-send failures; it was later shown that his translation was not general in that it could not be applied to all algorithms [12].[1] Srikanth and Toueg [19] showed how algorithms written for systems in which message authentication was used to mitigate arbitrary failures could be run in systems without this authentication. Neiger and Toueg [15] considered systems with *synchronous* message-passing and developed a family of translations; some translate from crash to omission failures, while others translates from omission failures to arbitrary failures.

Most translations require simulating one round of communication of the original protocol by some number of rounds in the new protocol. Some translations require that no more than a certain fraction of processors fail. (Typically, the fault-tolerance of an algorithm or a translation is measured by comparing n, the total number of processors in the system, to t, the number of failures tolerated.) If the requirements of a particular translation between two types of failures are *necessary*, then this indicates that there is a certain "separation" between these two types failures. For example, Neiger and Toueg showed that any translation from crash to omission failures requires that a majority of processors remain correct ($n > 2t$); this indicated a fundamental difference between these two systems that has since been explored elsewhere [16].

The relationship between crash and arbitrary failures has not been fully established. While Coan provided a translation that spans the entire hierarchy, he did so in the context of *asynchronous* systems. His translations were not applicable to *synchronous* systems. Neiger and Toueg were able to compose their translations for synchronous system, yielding two translations from crash to arbitrary failures. For example, they composed a 2-round translation from crash to omission failures with a 2-round translation from omission to arbitrary failures (the latter requiring $n > 4t$), resulting in a 4-round translation from crash to arbitrary failures that required $n > 4t$. They also developed a 6-round translation that required only $n > 3t$. Because these translations were obtained by combining others, it seemed likely that developing *direct* translations from crash to arbitrary failures would be more efficient.

This paper explores direct translations from crash to arbitrary failures in synchronous systems. There were a number of indications that the following results could be obtained: a 2-round translation that requires $n > 4t$ and a 3-round translation that requires $n > 3t$. For one thing, Coan's translations for asynchronous systems had exactly these properties. Secondly, Neiger and Toueg had developed translations with these properties from omission to arbitrary failures. Since most problems can be

---

[1] All other translations referred to here, as well as those developed in this paper, are general.

solved in systems with crash and omission failures with algorithms of the same complexity, it seemed that translations from either of these failures models (to arbitrary failures) might also have the same complexity. A final indication came from examining certain solutions to *Distributed Consensus*. If crash or omission failures occur, there are very simple solutions that use 1-bit messages and run in $t + 1$ rounds. In the case of arbitrary failures, Berman and Garay [2,3] have developed the following solutions using 1- or 2-bit messages: an algorithm that requires $n > 4t$ and runs in $2t + 2$ rounds, and an algorithm that requires $n > 3t$ and runs in $3t + 3$ rounds. Again, we see a doubling of rounds if $n > 4t$ and a tripling if $n > 3t$.

This paper presents a complete characterization of translations from crash to arbitrary failures in synchronous systems. It considers translations that double, triple, and quadruple the number of rounds of communication used and, in each case, shows a tight lower bound on the number of processors needed to perform such a translation. In contrast to our original expectations, we achieve the following results:

- **2-round translations.** If $3t < n \leq 6t - 3$, no such translation is possible. We provide an efficient translation if $n > \max\{6t - 3, 3t\}$.

- **3-round translations.** If $3t < n \leq 4t - 2$, no such translation is possible. We provide an efficient translation if $n > \max\{4t - 2, 3t\}$.

- **4-round translations.** If $n \leq 3t$, no such translation is possible. We provide an efficient translation if $n > 3t$.

Note that there can be no translation if $n \leq 3t$.

The above results indicate that, in some cases with the same n and t, it is easier (takes fewer rounds) to translate from omission failures than to translate from crash failures. This shows that there is a fundamental difference between crash and omission failures with regard to their relationship to arbitrary failures.

2 Definitions, Assumptions, and Notation

This paper considers distributed systems in which computation proceeds in synchronous *rounds*. This section defines a formal model of such a system. This model is an adaptation of that used by Neiger and Toueg [15].

2.1 Distributed Systems

A *distributed system* is a set of processors $\mathcal{P} = \{p_1, \ldots, p_n\}$ (n is the number of processors in the system) joined by bidirectional communication links. Processors share no memory; they communicate only by passing messages along the communication links. For the sake of simplicity, the results in this paper assume that processors are fully connected. Work by Dolev and by Hadzilacos suggests that they may be extended to network topologies in which there is less (but sufficient) connectivity [7,10]. Each processor has a local *state*. Let $\mathcal{Q}$ be the set of states.

Processors communicate with each other in synchronous *rounds*. In each round a processor first sends messages, then receives messages, and then changes its state. Let $\mathcal{M}$ be the set of messages that may be sent in the system, let $\perp \notin \mathcal{M}$ be a value that indicates "no message," and let $\mathcal{M}' = \mathcal{M} \cup \{\perp\}$.[2]

2.2 Protocols

Processors run a *protocol* Π, which specifies the messages to be sent and the states through which to pass. A protocol consists of two functions, a *message function* and a *state-transition function*. The

---

[2] If a processor sends no message in a round, we say that it "sends" $\perp$, although no message is actually sent.

```
state = initial state;

for i = 1 to ∞ do
    message = μ_π(i, p, state);
    if message ≠ ⊥ then
        send message to all processors;
    foreach q ∈ P
        if received some m from q then
            rcvd[q] = m
        else
            rcvd[q] = ⊥;
    state = δ_π(i, p, rcvd)
```

Figure 1: Execution of protocol Π by processor p

message function is $\mu_\pi : \mathbf{Z} \times \mathcal{P} \times \mathcal{Q} \mapsto \mathcal{M}$ (where $\mathbf{Z}$ is the set of positive integers). If processor p begins round i in state s then Π specifies that it send $\mu_\pi(i, p, s)$ to all processors in that round. The state-transition function is $\delta_\pi : \mathbf{Z} \times \mathcal{P} \times (\mathcal{M}')^n \mapsto \mathcal{Q}$. If in round i processor p receives the messages $m_1, \ldots, m_n$ (m_j from processor p_j), then Π specifies that it change its state to $\delta_\pi(i, p, m_1, \ldots, m_n)$ after round i. Figure 1 illustrates the execution of a protocol Π. (Note that the following assumptions are made about protocols:

- they are "loquacious"; all processors send some message in every round;

- they restrict each processor to broadcast the same message to all (in any given round);

- their state transition function depends solely on the messages that a processor has just received and *not* its current state; and

- they never have processors "halt."

These assumptions are made only to simplify the exposition and do not restrict the applicability of the results.)

2.3 Histories

Histories are defined to describe the executions of a distributed system. Each history includes the following:

- the protocol being run by the processors,[3]

- the states through which the processors pass,

- the messages that the processors send, and

- the messages that the processors receive.

Let $\mathrm{Q}(i, p)$ be the state in which processor p begins round i. Let $\mathrm{s}(i, p, q)$ be the message that p sends to q in round i or $\perp$ if p sends no message to q. Let $\mathrm{R}(i, p, q)$ be the message that p receives from q in

---

[3] Given any history H we want to be able to identify incorrect behavior in H; this requires that H include the protocol Π that processors should be following.

round i or $\perp$ if p does not receive a message from q.[4] Let $\mathrm{R}(i,p)$ be an abbreviation for the sequence $\mathrm{R}(i,p,1),\ldots,\mathrm{R}(i,p,n)$. $\mathsf{H} = \langle \Pi, \mathrm{Q}, \mathrm{S}, \mathrm{R}\rangle$ is a *history of protocol* Π.

A *system* is identified with the set of all histories (of all protocols) in that system. Thus, a system can be specified by a set of histories. A system can also be defined by giving the properties that its histories must satisfy. If S is a system and $\mathsf{H} = \langle \Pi, \mathrm{Q}, \mathrm{S}, \mathrm{R}\rangle \in S$, then H is a *history of* Π *running in* S.

3 Correctness and Failures

Individual processors may exhibit *failures*, thereby deviating from *correctness*. They may do so by failing to send or receive messages correctly or by otherwise not following their protocol. This section formally defines crash and arbitrary failures. Neiger and Toueg provide formal definitions for omission failures [15].

3.1 Correctness

Protocol Π defines the actions that a correct processor takes when executing it. Consider a history $\mathsf{H} = \langle \Pi, \mathrm{Q}, \mathrm{S}, \mathrm{R}\rangle$. Processor p *sends correctly in round* i *of* H if

$$\forall q \in \mathcal{P}[\mathrm{S}(i,p,q) = \mu_\pi(i,p,\mathrm{Q}(i,p))].$$

Processor p *receives correctly in round* i *of* H if

$$\forall q \in \mathcal{P}[\mathrm{R}(i,p,q) = \mathrm{S}(i,q,p)].$$

Processor p *makes a correct state transition in round* i *of* H if

$$\mathrm{Q}(i+1,p) = \delta_\pi(i,p,\mathrm{R}(i,p)).$$

Processor p is *correct through round* i *of* H if it sends and receives correctly, and makes correct state transitions up to and including round i of H. Let

$$Correct(\mathsf{H},i) = \{p \in \mathcal{P} \mid p \text{ is correct through round } i \text{ of } \mathsf{H}\}.$$

Assume that all processors are initially correct, so $Correct(\mathsf{H},0) = \mathcal{P}$. Then let $Correct(\mathsf{H})$, the set of all processors *correct throughout history* H be $\bigcap_{i \in \mathbb{Z}} Correct(\mathsf{H},i)$. If a processor is not correct, it is *faulty*. Formally, $Faulty(\mathsf{H},i) = \mathcal{P} - Correct(\mathsf{H},i)$ and $Faulty(\mathsf{H}) = \mathcal{P} - Correct(\mathsf{H})$.

3.2 Crash Failures

A *crash failure* is the most benign type of failure that this paper considers [9]. A processor commits a *crash failure* by prematurely halting in some round. Formally, p commits a crash failure in round i_c of $\mathsf{H} = \langle \Pi, \mathrm{Q}, \mathrm{S}, \mathrm{R}\rangle$ if $i_c \in \mathbb{Z}$ is the least i such that $p \in Faulty(\mathsf{H},i)$ and

- either p crashes during sending:

  - it sends to each processor q either what the protocol specifies, or nothing at all:

    $$\forall q \in \mathcal{P}[\mathrm{S}(i_c,p,q) = \mu_\pi(i_c,p,\mathrm{Q}(i_c,p)) \lor \mathrm{S}(i_c,p,q) = \perp]; \text{ and}$$

  - it receives no messages in round i_c:

    $$\forall q \in \mathcal{P}[\mathrm{R}(i_c,p,q) = \perp];$$

---

[4]Because of failures, the states and messages specified by Π may be different from those indicated by Q and s, and $\mathrm{R}(i,p,q)$ need not equal $\mathrm{S}(i,q,p)$; see Section 3.

- or p crashes after sending:

  - it sends correctly in round i_c:

  $$\forall q \in \mathcal{P}[s(i_c, p, q) = \mu_\pi(i_c, p, \mathrm{Q}(i_c, p))]; \text{ and}$$

  - it receives from each processor q either what q sent or nothing at all:

  $$\forall q \in \mathcal{P}[\mathrm{R}(i_c, p, q) = s(i_c, q, p) \vee \mathrm{R}(i_c, p, q) = \perp].$$

In either case, p takes no action after the crash:

- it sends and receives no messages: $\forall i > i_c \, \forall q \in \mathcal{P}[s(i, p, q) = \mathrm{R}(i, p, q) = \perp]$, and

- it makes no state transitions: $\forall i > i_c[\mathrm{Q}(i, p) = \mathrm{Q}(i_c, p)]$.

The system $C(n, t)$ corresponds to the set of histories in which t processors are subject only to crash failures and all other processors are correct. That is, $\mathsf{H} \in C(n, t)$ if and only if $\mathcal{P}$ can be partitioned into sets C and F such that $C = Correct(\mathsf{H})$, $|F| \leq t$, and

$$\forall p \in F \, \exists i_c \in \mathbf{Z}[p \text{ commits a crash failure in round } i_c \text{ of } \mathsf{H}].$$

3.3 Arbitrary Failures

Crash failures considerably restrict the behavior of faulty processors. Different types of omission failures [8,14,17] place fewer restrictions on this behavior. In the worst case, faulty behavior may be completely arbitrary; processors may fail by sending incorrect messages and by making arbitrary state transitions [13]. Processor p *is subject to arbitrary failures* in history $\mathsf{H} = \langle \Pi, \mathrm{Q}, \mathrm{s}, \mathrm{R} \rangle$ if it may deviate from Π in any way. It may do one or more of the following:

- fail to send correctly: $\exists i \in \mathbf{Z} \, \exists q \in \mathcal{P}[s(i, p, q) \neq \mu_\pi(i, p, \mathrm{Q}(i, p))]$,

- fail to receive correctly: $\exists i \in \mathbf{Z} \, \exists q \in \mathcal{P}[\mathrm{R}(i, p, q) \neq s(i, q, p)]$, or

- make an incorrect state transition: $\exists i \in \mathbf{Z}[\mathrm{Q}(i + 1, p) \neq \delta_\pi(i, p, \mathrm{R}(i, p))]$.

The system $A(n, t)$ corresponds to the set of histories in which t processors are subject to arbitrary failures and all other processors are correct.

4 Translations between Systems with Failures

This section defines formally the concept of a translation from $C(n, t)$ to $A(n, t)$. The definition of translation used here is an adaptation of a more general definition used by Neiger and Toueg [15].

A *translation* is a function $\mathcal{T}_{ca}$ that converts protocol Π_c, designed to run correctly in $C(n, t)$, into a protocol $\Pi_a = \mathcal{T}_{ca}(\Pi_c)$ that runs correctly in $A(n, t)$. Π_a may use several rounds of communication to simulate each round of Π_c. If each round of Π_c is simulated by z rounds in Π_a, then round i is simulated by rounds $z \cdot (i - 1) + 1$ through $z \cdot i$ of Π_a; these z rounds are called a *phase*. The translation is said to have *round-complexity* z. One can think of Π_a as protocol Π_c running on an underlying layer of software that "hides" more severe failures from Π_c. The rounds of Π_a are part of this underlying software.

The state s of a processor executing a translated protocol $\Pi_a = \mathcal{T}_{ca}(\Pi_c)$ has two components, $s = \langle ss, cs \rangle$, called the *simulated state* and the *control state*, respectively. The simulated state ss corresponds to the state of a processor running Π_c. If a processor running Π_a is in state $s = \langle ss, cs \rangle$, then let $S_{sb}(s)$ denote the simulated state ss. Π_a updates the simulated state only at the end of the z rounds that make up each phase. For each translation defined, it will be clear what part of a processor's state is the simulated state.

Translation function $\mathcal{T}_{ca}$ *translates from* $C(n,t)$ *to* $A(n,t)$ *in* z *rounds* (or *is a z-round translation from* $C(n,t)$ *to* $A(n,t)$) if there is a corresponding *history simulation function* $\mathcal{H}_{ac}$ with the following property: given any protocol Π_c and any history H_a of $\Pi_a = \mathcal{T}_{ca}(\Pi_c)$ running in $A(n,t)$, $\mathcal{H}_{ac}$ maps H_a into a corresponding simulated history $H_c = \mathcal{H}_{ac}(H_a)$ of Π_c running in $C(n,t)$, where z rounds in H_a simulate each round of H_c. Formally, $\mathcal{H}_{ac}$ is such that for any protocol Π_c and any history $H_a = \langle \Pi_a, Q_a, S_a, R_a \rangle$ of $\Pi_a = \mathcal{T}_{ca}(\Pi_c)$ running in S, the following hold:

(a) $H_c = \mathcal{H}_{ac}(H_a) = \langle \Pi_c, Q_c, S_c, R_c \rangle$ is a history of Π_c running in $C(n,t)$,

(b) $Correct(H_a) \subseteq Correct(H_c)$, and

(c) $\forall i \in \mathbf{Z} \, \forall p \in Correct(H_a)[S_{ac}(Q_a(z \cdot (i-1)+1,p)) = Q_c(i,p)]$.

Condition (b) states that the translation preserves the correctness of processors. That is, any processor correct in H_a is also correct in the simulated history H_c. However, processors faulty in H_a may be correct in H_c. In fact, most translation techniques can mask minor failures, typically by using redundant communication. Condition (c) states that the states of *correct* processors at the beginning of round i of H_c are correctly simulated at the beginning of corresponding phase of H_a. Because the processors in H_a may fail arbitrarily, there is no way to ensure that the state of faulty processors are correctly simulated. If conditions (a)–(c) hold, then H_a *simulates* H_c. Neiger and Toueg showed that any problem solved by Π_c is "effectively solved" by Π_a. Since protocols are easier to design for systems with crash failures, translations simplify the task of designing fault-tolerant protocols. The designer can first derive (and prove correct) a protocol Π_c that tolerates only crash failures, a relatively simple task. Applying $\mathcal{T}_{ca}$ to Π_c automatically results in a protocol $\Pi_a = \mathcal{T}_{ca}(\Pi_c)$, which effectively solves any problem solved by Π_c and which tolerates more severe failures.

5 Impossibility Results

This section shows that, in certain cases, no translation of a specified round-complexity exists from $C(n,t)$ to $A(n,t)$. Neiger and Toueg [15] showed similar results for other failure models. For example, there can be no translation from crash to omission failures if $n \leq 2t$ because there is a problem that can be solved in a system with crash failures for any $n \geq t$ but which can be solved in a system with omission failures only if $n > 2t$.

We begin by noting a similar result by considering the classical problem of *Byzantine Agreement*. Hadzilacos [8] solved this problem in systems with crash failures for any $n \geq t$, but Lamport et al. [13] showed that it cannot be solved in a system with arbitrary failures if $n \leq 3t$. Together, these show that there can be no translation from $C(n,t)$ to $A(n,t)$ if $n \leq 3t$. (In Section 6.4, we give a translation from $C(n,t)$ to $A(n,t)$ that is correct as long as $n > 3t$.)

The balance of this section deals with considering cases when translations are possible (i.e., when $n > 3t$) and explores their round-complexity. We first observe that there can be no 1-round translation from $C(n,t)$ to $A(n,t)$ unless $t = 0$. Assume that $t > 0$ and consider any 1-round translation. Suppose that p is faulty and, in the very first round, it sends m_1 to q_1 and m_2 to q_2 (q_1 and q_2 are correct). Each q_i must simulate the receipt of the message sent to it because, for all it knows, p is correct. But then q_1 and q_2 receive different messages from p, and the simulated history is not one with only crash failures (in systems with crash failures, even faulty processors cannot send different messages to different processors in the same round). Any such translation must simulate each round by at least two so that such arbitrarily faulty behavior can be "masked."

Section 5.1 shows that a 2-round translation is impossible if $n \leq 6t - 3$ and Section 5.2 shows that a 3-round translation is impossible if $n \leq 4t - 2$. Both proofs consider executions of a very simple protocol Π_c: in the first round, a distinguished processor b (the broadcaster) sends its initial state (either 0 or 1) to all other processors. In round 2, all processors send the message they received (or f—for "faulty"—if they received $\perp$) to all others and, at the end of that round, set their state to be the vector of messages they received in round 2. After round 2, processors send no messages and stay in state $\perp$. Here is the specification of the protocol:

$$\mu_c(i, p, s) = \begin{cases} s & \text{if } i = 1 \text{ and } p = b \text{ or if } i = 2 \\ \perp & \text{otherwise} \end{cases}$$

$$\delta_c(i, p, r) = \begin{cases} r[b] & \text{if } i = 1 \text{ and } r[b] \neq \perp \\ f & \text{if } i = 1 \text{ and } r[b] = \perp \\ r & \text{if } i = 2 \\ \perp & \text{otherwise} \end{cases}$$

Any history of Π_c in a system with crash failures must have the following properties:

1. If b is correct and begins round 1 in state s, then all correct processors must be in state s at the end of round 1.

2. If any correct processor is in state f at the end of round 1, then the state s_q of each correct processor q at the end of round 2 is such that $s_q[b] = \perp$.

3. If some correct processor is in state s_0 at the end of round 2 such that $s_0[r_0] = 0$ (for any $r_0 \in \mathcal{P}$), then no correct processor can be in state s_1 at the end of round 2 such that $s_1[r_1] = 1$ (for any $r_1 \in \mathcal{P}$).

4. If correct processor p is in state s at the end of round 1, then the state s_q of each correct processor q at the end of round 2 is such that $s_q[p] = s$.

These facts will be central to the impossibility proofs below.

The protocol Π_c is neither contrived nor unnatural. Rather, it is the shortest protocol for which the desired impossibility results could be shown (it yields the simplest proofs). The communication used in the protocol is simply that of a processor broadcasting a message and the receiving processors relaying it. This communication structure is quite common.

Arguments by Coan [4] can be used to show that, if a translation exists, it can be assumed to be a *full-information translation*. Such a translation has each processor send, in each round, all the information it has (i.e., all the messages it received in the previous round). In the case of Π_c, such a translation has a processor send, in each round, all information that it has about b's initial state and the execution of Π_c.

Most of these results proven below are a result of the fact that, in many cases, the correct processors are uncertain as to the identity of the faulty processors. Because of that, the construction of all histories given below assume that all processors behave correctly unless otherwise noted.

5.1   Two-Round Translations

This section shows that there can be no 2-round translation from $C(n, t)$ to $A(n, t)$ if $3t < n \leq 6t - 3$.

Assume for a contradiction that some 2-round full-information translation T_2 exists from $C(n, t)$ to $A(n, t)$ that is correct with $3t < n \leq 6t - 3$ (note that this implies $t > 1$). Consider any execution of $\Pi_a = T_2(\Pi_c)$. Since T_2 is a full-information translation, b sends its initial state to all in the first round of phase 1. In the second, all correct processors echo that state to all others. Let v_p be the vector that p receives in the second round, where $v_p[q]$ is the state received from processor q.

Consider some set of processors G such that $b \in G$ and and $|G| = n - t$. If a processor $p \in G$ receives $m \in \{0, 1\}$ from b in the first round and then has $v[q] = m$ for all $q \in G$ at the end of the second, then p must set its simulated state to m at the end of phase 1. This is because, as far as p can tell, all processors in G (including b) are correct and b sent m in the first round (any faulty processors would be in $\mathcal{P} - G$) and, by property (1) above, p must set its state to m.

Consider a set G' such that $b \notin G'$ and $|G'| \geq n - 2t$. Suppose that, for some correct processor $p \in G'$, $v_p[q] = m$ for all $q \in G'$; then again p must set its simulated state to m at the end of phase 1. Consider the following history H_1. A set $F \subseteq \mathcal{P} - G'$ of t processors (including b) is faulty. In the first round, b sends 1 to t correct processors (including p) and 0 to all others. The t processors in F all echo 0 in the second round except to p, to whom they echo 1. Any correct processor that received 0 in the first round also received 0 from $n - t$ processors (including b) in the second and so, by the preceding paragraph, must set its state to m. Note, however, that p received 0 only from $(n-1) - t - (t-1) = n - 2t$ processors other than b. At this point, p cannot set its state to 1; if it did, properties (3) and (4) might not both hold. Thus, it must set its simulated state to either 0 or f. Assume that, in phase 2 of H_1, all processors behave correctly, with the faulty processors acting exactly like those that received 0 in the first round.

Suppose that p set its state to f at the end of phase 1. Then all correct processors q must be in a s_q at the end of phase 2, where $s_q[b] = \bot$ (this follows by property (2)). Consider now another history H_2, where the set of faulty processors contains exactly those processors to whom b sent 1 in H_1 (including p). All other processors (including b) are correct. In the first round, b sends 0 to all processors. The faulty processors all echo 1 to all processors in the second round. As noted above, all correct processors must set their simulated state to 0 at the end of phase 1. In phase 2, all faulty processors behave just as they did in H_1. By properties (1) and (4), every correct processor q must be in a state s_q with $s_q[b] = 0$ at then end of phase 2 of H_2. But processors correct in both histories (such a processor must exist because $n > 3t$) cannot distinguish the two histories, giving a contradiction if p is in state f at the end of round 1 of H_1. (Basically, p would appear to be a lone processor claiming b to be faulty.) Thus, p must set its state to 0 at the end of round 1 of H_1.

Consider now the following scenario. Let A_0 and A_1 be disjoint subsets of $\mathcal{P} - \{b\}$ such that $|A_0| = \lceil (n-1)/2 \rceil$ and $|A_1| = \lfloor (n-1)/2 \rfloor$. Note that $t > 1$ and $n > 3t$ imply that each set has at least $t + 1$ processors and $n \leq 6t - 3$ implies each has at least $n - 3t + 1$ processors (because $1 + 2(n - 3t + 1) = 2n - 6t + 3 \leq n$). Suppose that, at the end of the second round, some processor $p \in A_0$ has $v_p[q] = 0$ for all $q \in A_0 \cup \{b\}$ and $v_p[q] = 1$ for all $q \in A_1$. It is clear to p that b is faulty because it received at least $t + 1$ echoes of each value. Suppose now that, in phase 2, all processors but two behave as if they received the same vector as p in the second round. The two remaining processors, $q \in A_0$ and $r \in A_1$ send different messages. Processor q claims to have received 0 in the second phase from all processors in A_0 and from $t - 1$ processors in $A_1 - \{r\}$; thus, it received 0 from at least $n - 2t$ processors other than b and set its state to 0 at the end of phase 1. Processor r claims to have received 1 in the second phase from all processors in A_1 and from $t - 1$ processors in $A_0 - \{q\}$; thus, it received 1 from at least $n - 2t$ processors other than b and set its state to 1 at the end of phase 1. One of q and r must be correct and the other must be faulty, but there is no way for p to know which. Thus, at the end of phase 2, it will set $s[q]$ and $s[r]$ regardless of which is faulty and which is correct. But, by property (3), p must set $s[q] = 0$ and $s[r] \neq 1$ if q is correct and $s[q] \neq 0$ and $s[r] = 1$ if r is. This gives a contradiction, proving that the translation T_2 cannot exists. Note that this proof does not hold if $n > 6t - 3$, because one cannot partition $\mathcal{P}$ into two sets of $n - 3t + 1$ processors that do not include b.

5.2 Three-Round Translations

This section shows that there can be no 3-round translation from $C(n, t)$ to $A(n, t)$ if $3t < n \leq 4t - 2$.

Assume for a contradiction that some 3-round full-information translation T_3 exists from $C(n, t)$ to $A(n, t)$ that is correct with $3t < n \leq 4t - 2$ (note that this implies $t > 2$). Consider any execution of $\Pi_a = T_3(\Pi_c)$. Since T_3 is a full-information translation, b sends its initial state to all in the first round of phase 1. In the second, all correct processors echo that state to all others. At the end of the third round, each processor has received a vector of values from each other processor. Let $v_p[q]$ be the vector that p receives from q; let $v_p[q, r]$ be the component of this vector corresponding to processor r.

Consider some set of processors G such that $b \in G$ and and $|G| = n - t$. Suppose that, for some

	b	A_0	A_1	C_0	C_1	
b	0	1 $\cdots$ 1	0 $\cdots$ 0	0 $\cdots$ 0	1 $\cdots$ 1	1
	0	0 $\cdots$ 0	0 $\cdots$ 0	0 $\cdots$ 0	1 $\cdots$ 1	
A_0	$\vdots$	$\vdots$ $\quad$ $\vdots$	$\vdots$ $\quad$ $\vdots$	$\vdots$ $\quad$ $\vdots$	$\vdots$ $\quad$ $\vdots$	$t-1$
	0	0 $\cdots$ 0	0 $\cdots$ 0	0 $\cdots$ 0	1 $\cdots$ 1	
	1	1 $\cdots$ 1	1 $\cdots$ 1	0 $\cdots$ 0	1 $\cdots$ 1	
A_1	$\vdots$	$\vdots$ $\quad$ $\vdots$	$\vdots$ $\quad$ $\vdots$	$\vdots$ $\quad$ $\vdots$	$\vdots$ $\quad$ $\vdots$	$t-1$
	1	1 $\cdots$ 1	1 $\cdots$ 1	0 $\cdots$ 0	1 $\cdots$ 1	
	0	0 $\cdots$ 0	1 $\cdots$ 1	0 $\cdots$ 0	1 $\cdots$ 1	
C_0	$\vdots$	$\vdots$ $\quad$ $\vdots$	$\vdots$ $\quad$ $\vdots$	$\vdots$ $\quad$ $\vdots$	$\vdots$ $\quad$ $\vdots$	$\left\lceil \frac{n-2t+1}{2} \right\rceil$
	1	0 $\cdots$ 0	1 $\cdots$ 1	0 $\cdots$ 0	1 $\cdots$ 1	
	1	0 $\cdots$ 0	1 $\cdots$ 1	0 $\cdots$ 0	1 $\cdots$ 1	
C_1	$\vdots$	$\vdots$ $\quad$ $\vdots$	$\vdots$ $\quad$ $\vdots$	$\vdots$ $\quad$ $\vdots$	$\vdots$ $\quad$ $\vdots$	$\left\lfloor \frac{n-2t+1}{2} \right\rfloor$
	1	0 $\cdots$ 0	1 $\cdots$ 1	0 $\cdots$ 0	1 $\cdots$ 1	
	1	$t-1$	$t-1$	$\lceil (n-2t+1)/2 \rceil$	$\lfloor (n-2t+1)/2 \rfloor$	

Figure 2: Matrix received by processor p

correct processor $p \in G$, $v_p[q, r] = m$ for all $q, r \in G$. Then p must set its simulated state to m at the end of phase 1 because it is conceivable that the processors in $\mathcal{P} - G$ are faulty. This follows by property (1) above.

Consider a set G as defined above, and let $G' \subseteq G$ such that $b \notin G'$ and $|G'| = n - 2t$. Suppose that, for some correct processor $p \in G'$, $v_p[q, r] = m$ for all $q \in G'$ and $r \in G$. For all p knows, the processors in $G - G'$ are faulty and might have sent different messages to some other correct processor p'; they might have sent p' the same vector sent to p by processors in G'. Since p' would set its simulated state to m (see above), p must set its state to either m or f (otherwise, it might violate property (3) or (4)). Once again, we can argue that p must actually set its state to m, because it might not be able to convince the other correct processors to "crash" b in phase 2 (see the similar argument in Section 5.1 above). Thus, p must set its simulated state to m at the end of phase 1.

Consider now the following scenario. Let A_0, A_1, C_0, and C_1 be disjoint subsets of $\mathcal{P} - \{b\}$ such that $|A_0| = |A_1| = t - 1$, $|C_0| = \lceil (n - 2t + 1)/2 \rceil$, and $|C_1| = \lfloor (n - 2t + 1)/2 \rfloor$. Note that, since $t > 2$ and $n \leq 4t - 2$, $2 \leq |C_0|, |C_1| \leq t$. Suppose that, at the end of the third round, some correct processor $p \in C_0 \cup C_1$ receives vectors composing the matrix shown in Figure 2. It is clear to p that b is faulty: too many processors ($|A_0 \cup C_0 \cup C_1| > t$) give too much support to 0 for it to have correctly sent 1, and too many ($|A_1 \cup C_0 \cup C_1| > t$) give too much support to 1 for it to have correctly sent 0. In addition, it knows that either all of A_0 is faulty or all of A_1 is, but it does not know which. Suppose that, in phase 2, all processors but two behave correctly, indicating that they too had this matrix after round 1. The two remaining processors, $q \in A_0$ and $r \in A_1$ send different messages. Processor q claims to have set its state to 0 because it had the matrix in Figure 3 (with $G' = A_0 \cup A_1$ and $G = G' \cup C_0$); q is believable, because the processors in A_1 might be faulty. Processor r claims to have set its state to 1 because it had the matrix in Figure 4 (with $G' = A_0 \cup A_1$ and $G = G' \cup C_1$); r is believable, because the processors in A_0 might be faulty. Thus, one of q and r must be correct and the other must be faulty, but there is no way for p to know which. Thus, at the end of round 2, it will set $s[q]$ and $s[r]$ regardless of which is faulty and which is correct. But, p must set $s[q] = 0$ and $s[r] \neq 1$ if q is correct and $s[q] \neq 0$ and $s[r] = 1$ if r is. This gives a contradiction, proving that the translation $\mathcal{T}_3$ cannot exists. Note that this proof does not hold if $n > 4t - 2$, because one cannot partition $\mathcal{P}$ into the sets indicated.

	b	A_0		A_1		C_0		C_1		
b	0	1	$\cdots$ 1	0	$\cdots$ 0	0	$\cdots$ 0	1	$\cdots$ 1	1
	0	0	$\cdots$ 0	0	$\cdots$ 0	0	$\cdots$ 0	1	$\cdots$ 1	
A_0	$\vdots$	$\vdots$	$\vdots$	$\vdots$	$\vdots$	$\vdots$	$\vdots$	$\vdots$	$\vdots$	$t-1$
	0	0	$\cdots$ 0	0	$\cdots$ 0	0	$\cdots$ 0	1	$\cdots$ 1	
	0	0	$\cdots$ 0	0	$\cdots$ 0	0	$\cdots$ 0	1	$\cdots$ 1	
A_1	$\vdots$	$\vdots$	$\vdots$	$\vdots$	$\vdots$	$\vdots$	$\vdots$	$\vdots$	$\vdots$	$t-1$
	0	0	$\cdots$ 0	0	$\cdots$ 0	0	$\cdots$ 0	1	$\cdots$ 1	
	0	0	$\cdots$ 0	1	$\cdots$ 1	0	$\cdots$ 0	1	$\cdots$ 1	
C_0	$\vdots$	$\vdots$	$\vdots$	$\vdots$	$\vdots$	$\vdots$	$\vdots$	$\vdots$	$\vdots$	$\left\lceil \frac{n-2t+1}{2} \right\rceil$
	0	0	$\cdots$ 0	1	$\cdots$ 1	0	$\cdots$ 0	1	$\cdots$ 1	
	1	0	$\cdots$ 0	1	$\cdots$ 1	0	$\cdots$ 0	1	$\cdots$ 1	
C_1	$\vdots$	$\vdots$	$\vdots$	$\vdots$	$\vdots$	$\vdots$	$\vdots$	$\vdots$	$\vdots$	$\left\lfloor \frac{n-2t+1}{2} \right\rfloor$
	1	0	$\cdots$ 0	1	$\cdots$ 1	0	$\cdots$ 0	1	$\cdots$ 1	
	1	$t-1$		$t-1$		$\lceil (n-2t+1)/2 \rceil$		$\lfloor (n-2t+1)/2 \rfloor$		

Figure 3: Matrix received by processor $q \in A_0$, forcing it to receive 0

	b	A_0		A_1		C_0		C_1		
b	0	1	$\cdots$ 1	0	$\cdots$ 0	0	$\cdots$ 0	1	$\cdots$ 1	1
	1	1	$\cdots$ 1	1	$\cdots$ 1	0	$\cdots$ 0	1	$\cdots$ 1	
A_0	$\vdots$	$\vdots$	$\vdots$	$\vdots$	$\vdots$	$\vdots$	$\vdots$	$\vdots$	$\vdots$	$t-1$
	1	1	$\cdots$ 1	1	$\cdots$ 1	0	$\cdots$ 0	1	$\cdots$ 1	
	1	1	$\cdots$ 1	1	$\cdots$ 1	0	$\cdots$ 0	1	$\cdots$ 1	
A_1	$\vdots$	$\vdots$	$\vdots$	$\vdots$	$\vdots$	$\vdots$	$\vdots$	$\vdots$	$\vdots$	$t-1$
	1	1	$\cdots$ 1	1	$\cdots$ 1	0	$\cdots$ 0	1	$\cdots$ 1	
	0	0	$\cdots$ 0	1	$\cdots$ 1	0	$\cdots$ 0	1	$\cdots$ 1	
C_0	$\vdots$	$\vdots$	$\vdots$	$\vdots$	$\vdots$	$\vdots$	$\vdots$	$\vdots$	$\vdots$	$\left\lceil \frac{n-2t+1}{2} \right\rceil$
	0	0	$\cdots$ 0	1	$\cdots$ 1	0	$\cdots$ 0	1	$\cdots$ 1	
	1	0	$\cdots$ 0	1	$\cdots$ 1	0	$\cdots$ 0	1	$\cdots$ 1	
C_1	$\vdots$	$\vdots$	$\vdots$	$\vdots$	$\vdots$	$\vdots$	$\vdots$	$\vdots$	$\vdots$	$\left\lfloor \frac{n-2t+1}{2} \right\rfloor$
	1	0	$\cdots$ 0	1	$\cdots$ 1	0	$\cdots$ 0	1	$\cdots$ 1	
	1	$t-1$		$t-1$		$\lceil (n-2t+1)/2 \rceil$		$\lfloor (n-2t+1)/2 \rfloor$		

Figure 4: Matrix received by processor $r \in A_1$, forcing it to receive 1

5.3   General Impossibility Results

The results of the preceding two sections applied to a specific, two-round, protocol, Π_c. As mentioned earlier, this protocol captures the basic "broadcast and relay" structure that is seen in many distributed protocols; for example, any full-information protocol repeatedly uses this message structure. The two-round structure of Π_c simplifies the proof but does not limit our above results. An inductive argument could show that 2- or 3-round translations are impossible (in the appropriate cases) for a full-information protocol of two rounds or more Intuitively, the reason for this is that, by definition, a translation's round-complexity is fixed in advance; it cannot be reduced during execution in response to failures. Thus, a worst case round-complexity must be used.

If the definition of translation is changed so that round complexity can change during an execution, then, as failures occur and are detected, future simulation could be made more efficient. Section 7 briefly discusses such a translation which allows "shifting" among the translations developed in Section 6. The resulting translation is sensitive to failures as they occur and seeks to optimize the round-complexity of the translation during execution.

6   Translations

This section describes three translations from crash to arbitrary failures. Each has a different phase-complexity and requires that a different number of processors remain correct.

6.1   Common Properties

The purpose of these translations is to allow a protocol, written for a system with crash failures, to run correctly in a system with arbitrary failures. This is done by providing redundant communication so that the impact of arbitrary failures is minimized. The translations constrain the faulty processors so that they cannot "lie" about the messages they have received and so that they must send the same message to all processors. In addition, if a correct processor ever receives no message (i.e., $\perp$) from some faulty processor, then that faulty processor will appear to crash; no correct processor will receive any messages from it in later phases.

All three of these translations have essentially the same structure. One phase is composed of some number of rounds. In the first, a processor sends its message for that phase. In the later round(s), processors echo the messages sent in the first; the number of later rounds needed depends on the number of processors in the system. At the end of a phase, processors simulate the receipt of messages for that phase. In addition, each processor maintains some auxiliary information: the set of processors it considers faulty (the variable *faulty*), the set it considers to have crashed (*crashed*), and the messages from the previous phase that it considers to be "believable" (the array *believe*). (Although a processor may receive $\perp$ from a faulty processor, it may still be willing to believe that another processor received some non-$\perp$ message.) All three translations have the following properties:

1. Correctness. If a correct processor sends a message at the beginning of a phase, then all correct processors receive that message by the end of the phase.

2. Relay. If a correct processor receives a message at the end of a phase, then all correct processors consider that message believable by the end of the first round of the next phase.

3. Relayed Crashing. If a correct processor does not receive a message from processor p at the end of a phase, then all correct processors consider p faulty at that time and consider p crashed by the end of the first round of the next phase.

4. Consistency. If a correct processor considers a message from processor p believable (for a given phase), then no correct processor considers a different message from p believable for that phase and, if p is correct, then p indeed sent that message in that phase.

function $Valid(i, q) : \mathcal{M}'$;

> **if** $q \in$ *faulty* **then**
>   **return**($\bot$)
> **else if** $i = 1$ **then**
>   **if** received m from q and $m = \mu_{\pi_c}(1, q, s)$ for some $s \in \mathcal{Q}$ **then**
>   **return**(m)
>   **else**
>   **return**($\bot$)
> **else**
>   **if** received $[m, c]$ from q **and**
> $\forall r \in \mathcal{P}[((r \in c) \Rightarrow (r \in crashed)) \wedge ((r \notin c) \Rightarrow (believe[r] \neq \bot))]$ **and**
> $m = \mu_{\pi_c}(i, q, \delta_{\pi_c}(i-1, q, v))$ where
> $\forall r \in \mathcal{P}[((r \in c) \Rightarrow (v[r] = \bot)) \wedge ((r \notin c) \Rightarrow (v[r] = believe[r]))]$ **then**
>   **return**(m)
>   **else**
>   **return**($\bot$)

Figure 5: The function *Valid*

More specifically, in the first round of some phase i, a processor sends its phase i message to all; if $i > 1$, it also includes its set *crashed*. Upon receiving this message, other processors determine whether or not the message should be echoed. This is done by the function *Valid* (see Figure 5), which returns either the message originally sent (if it it is valid to be echoed) or $\bot$ (if it is not). The function *Valid* works as follows. The processor first checks to see if the sending processor is in its set *faulty* and echoes $\bot$ if it is. If $i = 1$, the processor checks that there is some state that would generate the message (using the message function μ_{π_c}). If $i > 1$, the processor checks the set *crashed* sent with the message. By Relayed Crashing, any processor in that set must be in its own set *crashed*; for any processor not in that set, it must have some non-$\bot$ believable message from the preceding round. If these tests are passed, the processor now has access, in its *believe* array, to all the messages that processor claims to have received in phase $i - 1$. It can then use the state transition function δ_{π_c} to determine the state in which the sending processor must have begun phase i. Finally, it checks if that state would indeed generate the message sent (using the message function μ_{π_c}).

In the second round of all three translations, each processor echoes all the messages it validated at the end of the first. Some translations then require one or two more rounds. At the end of the very last round, a processor determines what messages it can "receive" for that phase; how this determination is made depends on the translation. In all cases, messages are marked received by setting the array *rcvd*. If a processor does not receive a message from processor q, it adds q to its sets *faulty* and *crashed*. (By property (3) above, all correct processors will have $q \in$ *faulty* at that time and will echo no further messages for q; q will thus appear to crash.) Even if a processor does "receive" a message from q, it will add q to its set *faulty* if it can tell that q is faulty. In the first round of every phase (after the first), processors exchange their sets *faulty*. They then use the procedure *RelayFailures* (see Figure 6) to update their sets. If $t + 1$ processors send a processor a set *faulty* with $q \in$ *faulty*, it adds q to its own set *faulty*. If $n - t$ processors send it such a set, then the processor also adds q to its set *crashed*.

In the 2- and 3-round translations, each processor determines, at the end of the last round of a phase, the messages that it can "relay." This includes all messages the processor has "received" as well as certain other messages for which it received a sufficient number of echoes. In the first round

procedure *RelayFailures*;

    foreach $q \in \mathcal{P}$ **do**
        if received *faulty* with $q \in$ *faulty* from at least $t + 1$ processors **then**
            Add q to *faulty*;
            if received *faulty* with $q \in$ *faulty* from at least $n - t$ processors **then**
                Add q to *crashed*

Figure 6: The procedure *RelayFailures*

procedure *RelayValues*;

    foreach $q \in \mathcal{P}$ **do**
        if received *relay* with *relay*$[q] = m$ from at least $(n - 1) - t$ processors other than q **then**
            believe$[q] = m$
        else
            believe$[q] = \perp$

Figure 7: The procedure *RelayValues*

of the next phase, processors relay these messages. A processor then uses the procedure *RelayValues* (see Figure 7) to determine which of these relayed messages it considers "believable." If a message is relayed by at least $(n-1)-t$ processors other than its original sender, then it is considered believable. The 4-round translation determines believable messages in the fourth round of a phase and thus does not use the procedure *RelayValues*.

Recall that, for any translation $\mathcal{T}$ from $C(n,t)$ to $A(n,t)$, there must be a corresponding history simulation function that maps any history H_a of $\Pi_a = \mathcal{T}(\Pi_c)$ running in $A(n,t)$ to a history H_c of Π_c running in $C(n,t)$, such that correct processors and their states are preserved. Space considerations prevent a full presentation of the translation's history simulation function and a discussion, in the following sections, of its correctness. It is constructed so that any processor correct in H_a has its behavior mapped directly to H_c. The behavior of processors faulty in H_a is completely determined by the messages and echoes received by processors correct in H_a. It will be possible to show that any processor p that is faulty in H_c will fail by crashing. Specifically, it will crash in round $\lfloor j/z \rfloor$ of H_c, where j is the first round in which some processor correct in H_a places p in its *crashed*. If no processor ever places p in *crashed*, then it will turn out that p is correct in H_c. As desired, processors correct in H_a will be correct in H_c.

We now describe the three translations and discuss their correctness.

6.2 A Two-Round Translation

This section defines a translation $\mathcal{T}_2$ that translates protocols tolerant of crash failures into ones tolerant of arbitrary failures. This translation requires that $n > \max\{6t - 3, 3t\}$ (this implies $t > 1$). The translation of protocol Π_c, $\Pi_a = \mathcal{T}_2(\Pi_c)$, is given in Figure 8. (Although Π_a is defined only operationally in Figure 8, it can be defined formally in terms of the functions μ_{π_a} and δ_{π_a}; this definition is omitted for the sake of simplicity.) Phase i of Π_c is simulated by rounds $2i - 1$ and $2i$ of Π_a. In the first round, processors send their messages; in the second, valid messages are echoed. A processor "receives" a message at the end of round $2i$ if it receives echoes of the message from at least

/* Π_a is tolerant of arbitrary failures and requires $n > 6t - 3$ processors */

$state =$ initial state;
$faulty = \emptyset$;
$crashed = \emptyset$;

for $i = 1$ **to** ∞ **do**
    $message = \mu_{\pi_c}(i, p, state)$;   /* begin round $2i - 1$ */
    if $i = 1$ **then**
        send *message* to all processors
    else
        send [*message, crashed*] to all processors;
        send *relay* to all processors;
        send *faulty* to all processors;

    if $i > 1$ **then**
        RelayValues;
        RelayFailures;
    foreach $q \in \mathcal{P}$
        $echo[q] = Valid(i, q)$;

    send *echo* to all processors;   /* begin round $2i$ */

    foreach $q \in \mathcal{P}$
        if received *echo* with $echo[q] = m$ from at least $n - 2t$ processors other than q **then**
            $rcvd[q] = m$;
            if received *echo* with $echo[q] = m$
                from fewer than $(n - 1) - t$ processors other than q **then**
                Add q to *faulty*
        else
            $rcvd[q] = \perp$;
            Add q to *crashed*;
            Add q to *faulty*;
        if received *echo* with $echo[q] = m$ from at least $n - 3t + 1$ processors other than q **then**
            $relay[q] = m$
        else
            $relay[q] = \perp$;

    $state = \delta_{\pi_c}(i, p, rcvd)$

Figure 8: Protocol $\Pi_a = \mathcal{T}_2(\Pi_c)$ as executed by processor p

/* Π_a is tolerant of arbitrary failures and requires $n > 4t - 2$ processors */

$state =$ initial state;
$faulty = \emptyset$;
$crashed = \emptyset$;

for $i = 1$ **to** ∞ **do**
    $message = \mu_{\pi_c}(i, p, state)$;   /* begin round $3i - 2$ */
    if $i = 1$ **then**
        send $message$ to all processors
    else
        send $[message, crashed]$ to all processors;
        send $relay$ to all processors;
        send $faulty$ to all processors;

    if $i > 1$ **then**
        RelayValues;
        RelayFailures;
    foreach $q \in \mathcal{P}$
        $echo[q] = Valid(i, q)$;

    send $echo$ to all processors;   /* begin round $3i - 1$ */

    foreach $q \in \mathcal{P}$
        if received $echo$ with $echo[q] = m$ from at least $(n - 1) - t$ processors other than q **then**
            $forward[q] = m$
        else
            $forward[q] = \perp$;

Figure 9: Protocol $\Pi_a = \mathcal{T}_3(\Pi_c)$ as executed by processor p; part 1

$n - 2t$ processors (other than the original sender). This value is unique as long as $2t - 1 < n - 2t$, which is guaranteed since $n > 6t - 3$ and $t > 1$. A processor relays a message in round $2i + 1$ if, in round $2i$, it received echoes from at least $n - 3t + 1$ processors (other than the original sender). (The relayed message is unique if $3t - 2 > n - 3t + 1$, which is again guaranteed since $n > 6t - 3$.) Since $(2t - 1) + (t - 1) = 3t - 2 < n - 2t$, if some correct processor receives a message, then no other correct processor can receive a conflicting message; in fact, all correct processors will relay the received message.

$\mathcal{T}_2$ possesses the properties (1)–(4) described above. Correctness obviously holds. Relay holds because, if a processor receives a message, all relay it and thus believe it one phase later. Crashing is relayed because, if a processor receives $\perp$, it has gotten at most $n - 2t - 1$ non-$\perp$ echoes, meaning that no processor could have received $(n - 1) - t$ non-$\perp$ echoes (note that the sender must be faulty in this case, so only $t - 1$ echoing processors can fail) and thus all mark the sender as faulty. Consistency holds because *RelayValues* requires $(n - 1) - t$ relays; two processors cannot get different values from so many processors.

Together with the correctness of the history simulation function, these facts allow one to conclude that, if $n > 6t - 3$ and $t > 1$, $\mathcal{T}_2$ translates protocols from $C(n, t)$ to $A(n, t)$.

```
send forward to all processors;                                          /* begin round 3i */

foreach q ∈ P
    if received forward with forward[q] = m
        from at least n − 2t processors other than q then
            rcvd[q] = m;
        if received forward with forward[q] = m
            from fewer than (n − 1) − t processors other than q then
                Add q to faulty
    else
            rcvd[q] = ⊥;
            Add q to crashed;
            Add q to faulty;
    if received forward with forward[q] = m
        from at least n − 3t + 1 processors other than q then
            relay[q] = m
    else
            relay[q] = ⊥;

state = δ_{π_c}(i, p, rcvd)
```

Figure 10: Protocol $\Pi_a = T_3(\Pi_c)$ as executed by processor p; part 2

6.3 A Three-Round Translation

This section defines a translation T_3 that translates protocols tolerant of crash failures into ones tolerant of arbitrary failures. This translation requires that $n > \max\{4t - 2, 3t\}$ (this implies $t > 2$). The translation of Π_c, $\Pi_a = T_3(\Pi_c)$, is given in Figures 9 and 10. Phase i of Π_c is simulated by rounds $3i - 2$, $3i - 1$ and $3i$ of Π_a. In the first round, processors send their messages; in the second, valid messages are echoed. In the third round, a processor "forwards" (using the array $forward$) only those messages for which it received $(n − 1) − t$ echoes in the second round (from processors other than the sender). A processor p "receives" a message at the end of that round if at least $n − 2t$ processors (other than the original sender) forward it to p. It relays a message in first round of the next phase if the message is forwarded by $n − 3t + 1$ processors other than the original sender.

Using arguments given in the preceding section, it should be clear that, for any given sender, there can be only one value forwarded by the correct processors in the third round of a phase. Furthermore, any message "received" by a correct processor will be relayed by all in first round of the next phase. If the sender is correct, it is clear that only one value will be so relayed. Otherwise, the fact that $t − 1 < n − 3t + 1$ (which holds because $n > 4t − 2$) assures that only one value can be relayed by a given processor. Proofs of properties (1)–(4) are similar to the case of T_2, as is that of the overall correctness of T_3.

6.4 A Four-Round Translation

This section defines a translation T_4 that translates protocols tolerant of crash failures into ones tolerant of arbitrary failures. This translation requires that $n > 3t$. Recall that any translation requires $n > 3t$, so T_4 shows that, whenever a translation is possible, it can be done in four rounds. The translation of protocol Π_c, $\Pi_a = T_4(\Pi_c)$, is given in Figures 11 and 12. Round i of Π_c is simulated by rounds $4i − 3$, $4i − 2$, $4i − 1$ and $4i$ of Π_a. In the first round, processors send their messages; in

/* Π_a is tolerant of arbitrary failures and requires $n > 3t$ processors */

$state =$ initial state;
$faulty = \emptyset$;
$crashed = \emptyset$;

for $i = 1$ to ∞ do
    $message = \mu_{\pi_c}(i, p, state)$;   /* begin round $4i - 3$ */
    if $i = 1$ then
        send $message$ to all processors
    else
        send $[message, crashed]$ to all processors;
        send $faulty$ to all;

    if $i > 1$ then
        $RelayFailures$;
    foreach $q \in \mathcal{P}$
        $echo[q] = Valid(i, q)$;

    send $echo$ to all processors;   /* begin round $4i - 2$ */

    foreach $q \in \mathcal{P}$
        if received $echo$ with $echo[q] = m$ from at least $(n - 1) - t$ processors other than q then
            $forward[q] = m$
        else
            $forward[q] = \perp$;

Figure 11: Protocol $\Pi_a = \mathcal{T}_4(\Pi_c)$ as executed by processor p; part 1

the second, valid messages are echoed. In the third round, a processor forwards only those messages for which it received $(n - 1) - t$ echoes in the second round. In the fourth round, a processor sends a "force" message for any message that was forwarded to it by $2t + 1$ processors; it sends an "ok" message for any other message that was forwarded to it by $t + 1$ processors. A processor p "receives" a message at the end of a phase if at least $t + 1$ processors "force" it to p; if at least $t + 1$ processors send either "force" or "ok" messages for a message, then that message is marked as "believable."

It should be clear that, for any given sender, there can be only one value forwarded by the correct processors in the third round of a phase and that any message "received" by a correct processor will be believed by all in the fourth round of a phase. If the sender is correct, then only one value can be believed. Otherwise, the facts that a processor must receive $t + 1$ messages in phase $4i$ to believe a message and that correct processors cannot send disagreeing values in that round guarantee that property (4) holds. Proofs of the other properties are similar to the case of $\mathcal{T}_2$, as is that of the overall correctness of $\mathcal{T}_4$.

7 Conclusions

This paper has presented a complete characterization of protocol translations from systems with crash failures to those with arbitrary failures. It considers translations that double, triple, and quadruple send *forward* to all processors;   /* begin round $4i - 1$ */

foreach $q \in \mathcal{P}$
    if received *forward* with *forward*$[q] = m$ from at least $2t + 1$ processors **then**
        final$[q] = \langle m, force \rangle$
    else if received *forward* with *forward*$[q] = m$ from at least $t + 1$ processors **then**
        final$[q] = \langle m, ok \rangle$
    else
        final$[q] = \perp$;

send *final* to all processors;   /* begin round $4i$ */

foreach $q \in \mathcal{P}$
    if received *final* with *final*$[q] = \langle m, force \rangle$ from at least $t + 1$ processors **then**
        rcvd$[q] = m$;
        believe$[q] = m$;
        if received *final* with *final*$[q] = \langle m, force \rangle$
          from fewer than $(n - 1) - t$ processors other than q **then**
            Add q to *faulty*
    else
        rcvd$[q] = \perp$;
        Add q to *crashed*;
        Add q to *faulty*;
        if received *final* with *final*$[q] = \langle m, * \rangle$ from at least $t + 1$ processors **then**
          believe$[q] = m$
        else
          believe$[q] = \perp$;

$state = \delta_{\pi_c}(i, p, rcvd)$

Figure 12: Protocol $\Pi_a = \mathcal{T}_4(\Pi_c)$ as executed by processor p; part 2

the number of rounds of communication used and, in each case, shows a lower bound on the number of processors needed to perform such a translation. The lower bounds are shown to be tight because, in each case, translations are exhibited that match the proven lower bound. All these translations are efficient in that they generate protocols that do not require substantially more local computation than the original protocols. In all cases, if the largest message sent in original protocol is of size b, then the largest message sent in the translated protocol has size $O(bn)$.

As with all translations between systems with failures, these translations simplify the task of designing fault-tolerant protocols. The designer can work with the assumption that the only failures are those of crashing and then convert the protocol automatically to tolerate arbitrary failures. It is interesting to note that the number of processors required is, in general, more than previous results would suggest.

It is possible to link the three translations using a technique similar to the "shifting gears" developed by Bar-Noy et al. [1]. With the weakest assumption, $n > 3t$, processors can start by running the 4-round translation of Section 6.4. Processors can shift to the 3-round translation once d_1 failures have been detected, where $n - d_1 > 4(t - d_1) - 2$; they can shift to the 2-round translation once d_2 failures have been detected, where $n - d_2 > 6(t - d_1) - 3$. The details of this "shifting translation" are left to a future paper. Note that the resulting translation does not violate the impossibility results of Section 5. The definition of translation given in Section 4 assumes that the number of rounds used to simulate is constant over all executions (see Section 5.3).

It is clear that the notion of processor knowledge [11] is closely related to the work presented here, especially that in Section 5. For example, for a processor to refuse to receive a message from another, it must *know* that, by the end of the next round, all correct processors refuse messages from that other processor. We believe that the requirements of these and other translations can be expressed using a knowledge-based formalism that may make it possible to unify the proofs presented in Section 5 and provide framework within which to prove related results.

Acknowledgements

We would like to thank Vassos Hadzilacos, Sam Toueg, and the anonymous referees for suggestions that led to the improvement of this paper.

References

[1] Amotz Bar-Noy, Danny Dolev, Cynthia Dwork, and H. Raymond Strong. Shifting gears: Changing algorithms on the fly to expedite Byzantine agreement (preliminary report). In *Proceedings of the Sixth ACM Symposium on Principles of Distributed Computing*, pages 42–51, August 1987. Revised version received November 1988.

[2] Piotr Berman and Juan A. Garay. Asymptotically optimal consensus. In *Proceedings of the Sixteenth International Conference on Automata, Languages, and Programming*, volume 372 of *Lecture Notes on Computer Science*, pages 80–94. Springer-Verlag, 1989.

[3] Piotr Berman, Juan A. Garay, and Kenneth J. Perry. Towards optimal distributed consensus. In *Proceedings of the Thirtieth Symposium on Foundations of Computer Science*, pages 410–415. IEEE Computer Society Press, October 1989.

[4] Brian A. Coan. A communication-efficient canonical form for fault-tolerant distributed protocols. In *Proceedings of the Fifth ACM Symposium on Principles of Distributed Computing*, pages 63–72, August 1986. A revised version appears in Coan's Ph.D. dissertation [5].

[5] Brian A. Coan. *Achieving Consensus in Fault-Tolerant Distributed Computer Systems: Protocols, Lower Bounds, and Simulations.* Ph.D. dissertation, Massachusetts Institute of Technology, June 1987.

[6] Brian A. Coan. A compiler that increases the fault-tolerance of asynchronous protocols. *IEEE Transactions on Computers*, 37(12):1541–1553, December 1988.

[7] Danny Dolev. The Byzantine generals strike again. *Journal of Algorithms*, 3(1):14–30, 1982.

[8] Vassos Hadzilacos. Byzantine agreement under restricted types of failures (not telling the truth is different from telling lies). Technical Report 18-83, Department of Computer Science, Harvard University, 1983. A revised version appears in Hadzilacos's Ph.D. dissertation [9].

[9] Vassos Hadzilacos. *Issues of Fault Tolerance in Concurrent Computations.* Ph.D. dissertation, Harvard University, June 1984. Department of Computer Science Technical Report 11-84.

[10] Vassos Hadzilacos. Connectivity requirements for Byzantine agreement under restricted types of failures. *Distributed Computing*, 2(2):95–103, 1987.

[11] Joseph Y. Halpern and Yoram Moses. Knowledge and common knowledge in a distributed environment. *Journal of the ACM*, 37(3):549–587, July 1990.

[12] Joseph Y. Halpern and H. Raymond Strong, March 1986. Personal communication.

[13] Leslie Lamport, Robert Shostak, and Marshall Pease. The Byzantine generals problem. *ACM Transactions on Programming Languages and Systems*, 4(3):382–401, July 1982.

[14] Yoram Moses and Mark R. Tuttle. Programming simultaneous actions using common knowledge. *Algorithmica*, 3(1):121–169, 1988.

[15] Gil Neiger and Sam Toueg. Automatically increasing the fault-tolerance of distributed algorithms. *Journal of Algorithms*, 11(3):374–419, September 1990.

[16] Gil Neiger and Mark R. Tuttle. Common knowledge and consistent simultaneous coordination. In J. van Leeuwen and N. Santoro, editors, *Proceedings of the Fourth International Workshop on Distributed Algorithms*, volume 486 of *Lecture Notes on Computer Science*, pages 334–352. Springer-Verlag, September 1990. To appear in *Distributed Computing*.

[17] Kenneth J. Perry and Sam Toueg. Distributed agreement in the presence of processor and communication faults. *IEEE Transactions on Software Engineering*, 12(3):477–482, March 1986.

[18] Richard D. Schlichting and Fred B. Schneider. Fail-stop processors: an approach to designing fault-tolerant computing systems. *ACM Transactions on Computer Systems*, 1(3):222–238, August 1983.

[19] T. K. Srikanth and Sam Toueg. Simulating authenticated broadcasts to derive simple fault-tolerant algorithms. *Distributed Computing*, 2(2):80–94, 1987.

Efficient Distributed Consensus
with $n = (3 + \varepsilon)t$ Processors

(Extended Abstract)

Piotr Berman
Department of Computer Science
The Pennsylvania State University
University Park, PA 16802, USA

Juan A. Garay
IBM T.J. Watson Research Center
P.O. Box 704
Yorktown Heights, NY 10598, USA

Abstract

In a *Distributed Consensus* protocol n processors, of which at most t may be faulty, are given initial binary values; after exchanging messages all the processors that are correct must agree on one of the initial values. We measure the quality of a consensus protocol by the following parameters: the total number of processors n, the number of rounds of message exchange r, and the communication complexity, given by the maximal message size.

This paper presents a protocol that achieves Distributed Consensus with $n = (3 + \varepsilon)t$, $r = t + 1$, and polynomial message size for $\varepsilon \geq 1/\log t$, Thus, this is the first consensus protocol to use simultaneously nearly optimal number of processors ($n = 3t + 1$ is the known lower bound). optimal number of rounds, and short messages. Previous round-optimal protocols with polynomial communication required $n = \Omega(t^2)$, $n > 6t$ and $n > 4t$, respectively.

In order to obtain this result, we introduce new techniques for the abbreviation of messages in the classical full information consensus protocol. In particular, we introduce the *Dynamic Fault Masking* technique, called that way because the values that the correct processors use to mask the faults is computed on the fly during the execution of the protocol.

1   Introduction and Problem Statement

There are many situations in the management of distributed systems with one common characteristic: a collection of processors must coordinate a decision. This problem becomes non-trivial when some of the processors are faulty and cannot be relied upon to faithfully obey a protocol. It is reasonable to at least demand that, regardless of the behavior of the faulty processors, correct processors can reach an agreement consistent with the initial value of a correct processor. The *Distributed Consensus* problem (and its twin, *Byzantine Agreement*) provides an abstract setting in which methods for tolerating faults may be explored and perhaps influence practical designs. The problem can be formally stated as follows.

We are given a set of processors P of which some unknown subset T, $\#T \leq t$, are *faulty* and may exhibit arbitrary (*Byzantine*) behavior. Processors from $P - T$ are called *correct*. Every processor is given an initial value, 0 or 1. After the execution of the protocol, the final values of the correct processors have to satisfy the following two conditions, regardless of the behavior of the faulty processors:

- *Agreement*: they are all equal.

- *Validity*: they are all equal to the initial value, if the latter is unique.

We use the standard model of a synchronous network of processors numbered from 1 to n. The computation performed by the network evolves as a series of *rounds*, during which the processors send messages, receive them and perform local computations according to the protocol.[1]

There already exist many protocols for this problem because there are several quality parameters that can be optimized: the total number of processors n (as a function of t), the number of rounds of communication r, the computation time, and the communication complexity given by the maximal message size m. The known lower bounds for these parameters are $n = 3t + 1$ [LSP], $r = t + 1$ [DS] and $m = 1$, respectively. Typically, some parameters are optimized to the detriment of the others but there is no proof that all parameters cannot be made optimal simultaneously.

Probabilistic protocols are one way of circumventing the $t + 1$ lower bound on round complexity. In this setting, protocols that run for an expected constant number of rounds are possible. In particular, Feldman and Micali have designed a randomized consensus protocol that achieves optimal number of processors and polynomial message size [FM].

Table 1 compares the parameters of existing deterministic protocols with optimal number of rounds with the protocols presented in this paper. In particular, the original Byzantine Agreement protocol [LSP] (and its refinement, called EIG^2 by Bar-Noy *et al.* [BDDS]) achieves consensus using optimal number of processors but $\Omega(t^t)$ message size. In [W],

---

[1]A certain level of synchrony is necessary to obtain distributed consensus [FLP]. Thus every known protocol either assumes perfect synchrony, as here, or at least bounds the number of steps that a correct processor may take before receiving a reply from another correct processor. Interesting new results in this model were obtained recently by Attiya *et al.* [ADLS].

[2]*EIG* stands for *Exponential Information Gathering*.

Protocol	n	m
Lamport *et al.* [LSP], Bar-Noy *et al.* [BDDS]	$3t+1$	$O(t^t)$
Dolev, Reischuk and Strong [DRS]	$\Omega(t^2)$	$O(t^2)$
Bar-Noy and Dolev [BD]	$\Omega(t^2)$	1
Waarts [W]	$3t+1$	$O(t!)$
Moses and Waarts [MW]	$6t+1$	$O(t^4)$
Berman, Garay and Perry [BGP]	$4t+1$	$O(t^5)$
$ESDM^*$	$3t+1$	$O(1.5^t)$
$CVDM^*$	$(3+\varepsilon)t$	$O(t^5 2^{4/\varepsilon})$

* Protocols presented in this paper.

Table 1: Distributed Consensus in optimal number of rounds ($r = t + 1$).

Waarts describes a version of this protocol which allows for "early stopping" (see [DRS]), hereby reducing the communication overhead to $\Omega(t!)$. For many years it remained an open question whether a protocol with linear number of processors could be communication efficient; the issue was answered positively by Moses and Waarts [MW]. A later result by Berman *et al.* yielded better—yet not optimal—number of processors [BGP].

The main contribution of this paper is a new consensus protocol (*CVDM*, for *Cloture Votes with Dynamic Masking*) which brings the number of processors arbitrarily close to the optimum, i.e. it works for $n = (3 + \varepsilon)t$ processors, using $r = t + 1$ rounds of communication and maximal message size $m = O(2^{4/\varepsilon}t^5)$. This implies in particular that consensus with polynomial communication is possible for $\varepsilon \geq 1/\log t$, considerably narrowing the optimality gap.

For the sake of analysis, we present our result as the result of three consecutive transformations applied to the *EIG* protocol. The first transformation consists of the application of early stopping rules, which allow processors to conclude early that they do not need to receive certain classes of messages. The second transformation is the application of the new *Dynamic Fault Masking* technique, called that way because the value that the correct processors use to mask the discovered faults is computed on the fly during the execution of the protocol. This technique, in conjunction with the first transformation, reduces the number of times that the correct processors can be "seriously confused" by the faulty ones. The resulting consensus protocol (*ESDM*, for *Early Stopping with Dynamic Masking*) by itself has better message size than previous protocols with optimal number of rounds and number of processors $n < 4t$ (see Table 1).

The last transformation is the application of the *Cloture Votes* technique of Berman *et al.* [BGP]. The message size depends on the choice of a parameter called the "patience threshold," which monitors the degree of "corruption" that the faulty processors are responsible for. The analysis is novel, and allows us to invoke *Cloture Votes* in a much bolder manner than before.

The formalism of the theory of knowledge in distributed sytems [HM] improves both the presentation and analysis of our protocols. Essentially, the processors reach intermediate conclusions based on what they learn from the received messages. To describe this process we introduce a family of predicates to concisely express these conclusions, and rules to infer them. This makes the presentation of a variety of methods in a uniform manner possible, avoiding *ad hoc* procedures and data structures.

The remainder of this paper is organized as follows. Section 2 presents *EIG* and the first transformation based on the early stopping rules. Section 3 introduces the method of *Dynamic Fault Masking* and the resulting *ESDM* protocol. Section 4 describes the application of *Cloture Votes*, which yields our *CVDM* protocol and main result.

Because of space limitations, some of the proofs are omitted from this extended abstract. In this paper we assume that $n = 2p + t - 1$, where $p > t$ (equivalently, $n > 3t$ and $n - t$ is odd). We use $|\sigma|$ to denote the length of string σ.

2   EIG with Safe Rules for Early Stopping

In this section we present the first transformation to the round- and processor-optimal *EIG* protocol of Bar-Noy *et al.* [BDDS]. First we introduce the notion of *safe decisions* and the rules to obtain them. These rules by themselves allow us to re-derive a protocol first presented by Waarts [W] which omits certain messages of *EIG* and reduces its communication complexity to $m = O(t!)$.

We first present a short synopsis of *EIG*. Processors build trees of the following form: the set of nodes consists of the sequences from P^* without repetitions, of length at most $t + 1$; the root is λ (the empty sequence); the children of σ, $EXT(\sigma)$, are nodes of the form σj. In each tree, a node σ has a *value* denoted $Val(\sigma)$. Each correct processor initializes a one-node tree consisting of λ and sets $Val(\lambda)$ to its initial value. Then, for rounds $0, 1, \cdots, t$ each correct processor applies the following two primitives to every leaf σ of its tree:

- Send(σ): send $Val(\sigma)$ to all processors (provided i does not appear in σ).

- Receive(σ, j): upon receiving v when j executes Send(σ), create node σj and set $Val(\sigma j) = v$.

Once the tree is constructed, the processors use the rules given below to compute predicates $Dec(\sigma, v)$ and $PDec(\sigma, v)$ for each node σ and $v \in \{0, 1\}$ (We read $Dec(\sigma, v)$ ($PDec(\sigma, v)$) as "value v is the decision (a partial decision) for σ."):

$$PDec(\sigma, v) \equiv \begin{cases} Val(\sigma) = v & \text{if } \sigma \text{ is a leaf,} \\ \#\{\tau \in EXT(\sigma) : Dec(\tau, v)\} \geq p & \text{otherwise;} \end{cases}$$

$$Dec(\sigma, v) \equiv PDec(\sigma, v) \wedge \neg PDec(\sigma, \overline{v}).$$

The final value for the consensus problem is 1 iff $Dec(\lambda, 1)$. Note that $Dec(\sigma, v)$ implies $\neg Dec(\sigma, \overline{v})$, although it is possible that $\neg Dec(\sigma, 0) \wedge \neg Dec(\sigma, 1)$ holds.

Definition 2.1: A predicate $\pi(\sigma)$ is *common* iff π is true in the tree of every correct processor. A node σ is common iff either $\neg Dec(\sigma, 0) \land \neg Dec(\sigma, 1)$ is common, or for some v, $Dec(\sigma, v)$ is common.

The following two lemmas express the correctness of EIG in terms useful for our further discussion. Lemma 2.2 implies the Validity condition, while Lemma 2.3 implies Agreement.

Lemma 2.2: *Assume that either*

- $\sigma = \lambda$ *and the initial value of every correct processor is* v, *or*

- $\sigma = \tau i$, *in* i's *tree* $Val(\tau) = v$ *and* i *executes* $Send(\tau)$ *correctly.*

Then $Dec(\sigma, v)$ *is common.*

The proof follows directly from the definition by induction on $|\sigma|$; the base case is $|\sigma| = t$ and the induction parameter is decreasing.

Lemma 2.3: *The root of the EIG tree, λ, is common.*

Proof: (Sketch) By an induction similar to the one in the previous lemma, we prove that if a common node exists on every path from σ to a leaf, then σ is common. Note that every leaf has an ancestor of the form σi, where i is correct and, as a consequence of Lemma 2.2, node σi is common. ∎

In the protocols presented in this paper the values of predicates are actually computed differently and *not always consistently* with EIG. The following definition and lemmas provide justification for such deviations. Ideally, every processor would reach the same decision as in EIG for every node. However, we will show that it suffices to reach *safe* decisions for a well-defined class of nodes.

Definition 2.4: A predicate $\pi(\sigma)$ is *safe* if either

- $\pi(\sigma)$ is common, or

- $\sigma \notin T^*(P - T)^*$, or

- $\pi(\sigma) \equiv Dec(\sigma j, v)$ and the predicates $PDec(\sigma, v)$ and $PDec(\sigma j, v)$ are both common.

Note that the agreement value is 1 iff $Dec(\lambda, 1)$ is safe (i.e., iff $Dec(\lambda, 1)$ is common).

We now present *Early Stopping Rules* that can be used to predict the final values of nodes (and the consensus value) early. The first five rules deal with partial decisions.

1. If in the tree of a correct processor σ is created in round t and $Val(\sigma) = v$, then $PDec(\sigma, v)$ and $\neg PDec(\sigma, \overline{v})$ are safe.

2. If in the tree of a correct processor $\#\{\tau \in EXT(\sigma) : Val(\tau) = v\} \geq p + t - |\sigma|$, then $PDec(\sigma, v)$ is safe.

3. If in the tree of a correct processor $\#\{\tau \in EXT(\sigma) : Val(\tau) = v\} \geq p + 2(t - |\sigma|)$, then $PDec(\sigma, v)$ and $\neg PDec(\sigma, \overline{v})$ are safe.

4. If $\#\{\tau \in EXT(\sigma) : Dec(\tau, v)$ is safe $\} \geq p$, then $PDec(\sigma, v)$ is safe.

5. If $\#\{\tau \in EXT(\sigma) : PDec(\tau, \overline{v})$ is safe $\} \geq p + t - |\sigma|$, then $\neg PDec(\sigma, v)$ is safe.

The next set of rules concerns decisions.

6. If $PDec(\sigma, v)$ and $\neg PDec(\sigma, \overline{v})$ are safe, then $Dec(\sigma, v)$ is safe.

7. If $PDec(\sigma, 0)$ and $PDec(\sigma, 1)$ are safe, then $\neg Dec(\sigma, v)$ and $\neg Dec(\sigma, \overline{v})$ are safe.

8. If $PDec(\sigma, v)$ and $PDec(\sigma j, v)$ are safe, then $Dec(\sigma j, v)$ is safe.

We say that correct processor i *decides* on σ whenever i is able to apply rules (6-8) to a node σ; we say that i decides *promptly* on σ whenever it is able to apply rules (3) and (6) to σ.

The proof that the rules are sound is omitted from the abstract but follows from an easy induction and the following case analysis: If $\sigma \notin T^*(P - T)^*$ the claim is vacuous—any predicate is $\pi(\sigma)$ safe; if $\sigma \in T^*(P - T)^+$ then we can apply Lemma 2.2; lastly, if $\sigma \in T^*$, then at most $t - |\sigma|$ elements of $EXT(\sigma)$ store values received from faulty processors.

When discussing the conclusions that a correct processor may derive from the above Early Stopping Rules, we use notation from the theory of knowledge in distributed systems [HM]. We read $K_i^r \varphi$ as "after receiving round r's messages processor i knows fact φ." We will also use the "everybody knows" operator $E^r \varphi \equiv \bigwedge_{i \in P - T} K_i^r \varphi$. We will drop i and r when obvious from the context. It is assumed that the Early Stopping Rules are applied by all the correct processors; their only knowledge consists of the facts derived by applying these rules. We also assume that processors do not forget, i.e. $K_i^r \varphi$ implies $K_i^{r+1} \varphi$. A statement of the form $\neg K_i^r \varphi$ means that i cannot deduce φ by round r by applying the rules.

The next two lemmas, whose proofs are omitted from the abstract, are analagous to Lemmas 2.2 and 2.3 and show that the Early Stopping rules are sufficient to infer the consensus value.

Lemma 2.5: *Assume that processor i in round r executes $Send(\sigma)$ correctly, and in its tree $Val(\sigma) = v$. Then $E^{r+2}[Dec(\sigma i, v)$ is safe].*

Lemma 2.6: *Assume that after round r processor i has decided on an ancestor of every leaf node. Then i can also decide on λ.*

The next lemma shows that when a processor decides on a node early, it may also stop sending messages regarding that node early. The lemma shows that sending messages for two more rounds after a decision suffices.

Lemma 2.7: *Let $i \in P - T$ and $\sigma \in T^*(P - T)^*$. Then $K_i^r[Dec(\sigma, v)$ is safe] implies $E^{r+2}[Dec(\sigma, v)$ is safe]. Moreover, if i decides promptly on σ in round r, then $E^{r+1}[Dec(\sigma, v)$ is safe].*

This leads us to modify the *EIG* protocol to omit the sending of messages as follows: a processor does not executes Receive($\sigma\tau, j$), for any j, if it has decided on σ before the previous round, or if it has decided promptly on σ in the previous round. (The modification also works even if processors do not apply Early Stopping Rule (8).) This allows us to prove the following theorem, originally due to Waarts [W]:

Theorem 2.8: *The first modified version of EIG solves the Distributed Consensus problem using $n > 3t$ processors, $r = t + 1$ communication rounds and messages of size $m = O(t!)$.*

Remark 2.9: The modified version of the *EIG* protocol presented above satisfies the "early stopping" property, in the sense that it is able to terminate in a number of rounds proportional to f, the number of actual faults that occur during a run. However, the running time of this protocol is $r = min(t + 1, f + 3)$ while the known optimum is $r = min(t + 1, f + 2)$ [DRS].

In a recent development, using the Early Stopping Rules presented in this section together with a new *Safe Message Reconstruction* technique, the authors have designed an early-stopping consensus protocol that is optimal in both the number of processors and the number of rounds, thus settling a long-standing open question [BG]. ∎

3 Dynamic Fault Masking

In this section we introduce the *Dynamic Fault Masking* technique and apply it to the protocol of Theorem 2.8. The resulting protocol, *ESDM*—for *Early Stopping with Dynamic Masking*, is also round- and processor-optimal, but further reduces the message size to $m = O(1.5^t)$. This is possible because once a processor discovers that another processor is faulty (by the *Fault Discovery* Rule below), it may start *altering* all values reported to it by the faulty processor in order to maximize the applicability of the Early Stopping Rules. Dynamic Fault Masking provides the principle for such an alteration of values.

The validity of the following Fault Discovery Rule follows trivially from the analysis of the possible values reported by the correct processors about a given node:

9. If $\#\{k : Val(\sigma j k) \neq Val(\sigma j)$ or $K_i^r[k$ is faulty]$\} > t$, then $K_i^r[j$ is faulty].

From now on we assume that the Early Stopping Rules from the previous section and the above Fault Discovery Rule form the set of "legal" inference rules for all correct processors. The next lemma shows that the Fault Discovery Rule is sufficient to obtain the negative conclusions available from Lemma 2.5.

Lemma 3.1: *Assume that in round r processor i executes Receive($\sigma jk, l$). Then either i can decide on σj in that round, or $K_i^r[j$ is faulty].*

The Dynamic Fault Masking technique used by protocol *ESDM* increases the applicability of Early Stopping Rule (8). This rule prunes a node storing a value received from a faulty processor in a round in which p correct processors know that j is faulty (recall that we assume $n = 2p + t - 1$, where $p > t$). The traditional fault masking technique required that *all* correct processors know that j is faulty in order to achieve the same effect [C]. The following lemma provides the basis of the technique.

Lemma 3.2: (The principle of Dynamic Fault Masking) *Assume that for some v*

- $E^r[PDec(\lambda, v)$ is safe];

- $\sigma j \in T^*$;

- w equals v if $|\sigma|$ is even, $\bar{v}$ otherwise;

- p correct processors set $Val(\sigma j) = w$ in round r.

Then by round $r + 3$ all correct processors have decided on an ancestor of σj.

Proof: We utilize the fact that for every ancestor τ of σj, created in round $q \leq r$, $E^{q+3}[PDec(\tau, v_{|\tau|})$ is safe] holds for some $v_{|\tau|}$—the reason is that in round $q+3$ every correct processor decides on every node from $EXT(\sigma) \cap T^*(P - T)$; p out of these $2p - 1$ decisions have to be equal to some $v_{|\tau|}$.

The assumptions of the lemma assure that $v_0 = v$ and $v_{|\sigma j|} = w$. The way w is defined assures that the sequence of values $v_0, v_1, ..., v_{|\sigma j|}$ must contain a repetition. This repetition indicates that for some ancestor τk of σj, $E^{r+3}[PDec(\tau, v_{|\tau|})$ is safe] and $E^{r+3}[PDec(\tau k, v_{|\tau|})$ is safe] hold, hence $E^{r+3}[Dec(\tau k, v_{|\tau|})$ is safe] by rule (8). ∎

From now on we assume that whenever a correct processor knows that processor j is faulty, it replaces $Val(\sigma j)$ according to Lemma 3.2 in order to execute Send(σj).

We call a node σj created in round r *corrupted* if $E^{r+3}[Dec(\tau, v)$ is safe] holds for no v and no ancestor τ of σ. Obviously, if σ is corrupted, then all the processors in the sequence σ are faulty (Lemma 2.5). In order to achieve consensus with polynomial communication, the *Cloture Votes* technique of Berman et al. [BGP] requires that a faulty processor can create corrupted nodes—by executing Send's incorrectly—in one round only. Unfortunately, we do not know how to assure this when $n < 4t$. However, when $n = 2p + t - 1 = (3 + \varepsilon)t$, we will show that the situation can be *almost* as good: a faulty processor may create corrupted nodes in two consecutive rounds only. Moreover, the number of such rounds is $O(\varepsilon^{-1})$.

Definition 3.3: We say a round r is *bad* if there exist j, σ, τ such that σj is created in round r, τj created in round $r+1$, and both σj and τj are corrupted.

Let $F_i^r = \{j : K_i^r[j$ is faulty]$\}$, i.e. the set of faulty processors discovered by i after receiving round r's messages.

Lemma 3.4: (The price of a bad round) *If round r is bad, then*
$$\sum_{i \in P-T} \#(F_i^{r+2} - F_i^{r+1}) \; > \; p(p - t + |\sigma|).$$

Proof: Assume that round r is bad because of σ, τ and j. Because τj created in round $r+1$ is corrupted, j was not successfully masked in round $r+$; by Lemma 3.2 this means that less than p correct processors know after round $r+1$ that j is faulty, in other words
$$\#I \geq p, \text{ for } I = \{i \in P - T : \neg K_i^{r+1}[j \text{ is faulty}]\}.$$
Let $C = \{k : \sigma j k \in EXT(\sigma j]\}$. Because elements of I could not apply the Fault Discovery Rule in round $r+1$, it follows that
$$\#A_i \geq 2p - 1, \text{ for } A_i = C - \{k \in C : Val(\sigma j k) \neq Val(\sigma j)\} - F_i^{r+1}.$$
On the other hand, because σj is corrupted we know that $\neg K_i^{r+2}[Dec(\sigma j, Val(\sigma j))$ is safe] holds, for every $i \in I$. Thus no processor $i \in I$ could apply rule (5) in round $r+2$. Consequently,
$$\#B_i < p + t - |\sigma j|, \text{ for } B_i = \{k \in C : K_i^{r+2}[PDec(\sigma j k) \text{ is safe}]\}.$$
Thus for every $i \in I$ we have $\#(A_i - B_i) \geq p - t + |\sigma|$. Moreover, $A_i - B_i \subseteq F_i^{r+2} - F_i^{r+1}$, thus $\sum_{i \in P-T} \#(F_i^{r+2} - F_i^{r+1}) \geq \sum_{i \in I} \#(A_i - B_i) \geq p(p - t + |\sigma|$. ∎

Clearly, since the number of correct processors is less than $2p$, and each of them may know at most t faults, $\sum_{i \in P-T} \#F_i^r < 2pt$. Therefore we can conclude

Lemma 3.5: *In any run of EIG where correct processors apply Dynamic Fault Masking, the number of bad rounds is less than $2t/(p - t)$.*

Intuitively, the faulty processors may force the message size to increase much faster in bad rounds. In particular, the message size due to Dynamic Fault Masking is the square root of the size obtained with the traditional fault masking (see Theorem 3.6 below).

The *ESDM* protocol is shown in Figure 1. This protocol is run by every correct processor with parameter $Start = 0$, and the consensus value is stored as $Outcome[0]$; $ESDM$ is presented as a process since in the next section it will be repeatedly invoked by $CVDM$, in order to achieve this paper's main result. We note that the structure of the protocol is similar to the one presented in [BGP]; however the "real" computation is performed by the inference rules, which are different. $Tree$ is the tree maintained by the processor; Val maps nodes of $Tree$ into $\{0,1\}$; while $Level[0..2]$ consists of sets of nodes on the level of the tree currently being processed. $Level[0]$ contains the undecided nodes, while $Level[1]$ and $Level[2]$ are used to keep sending the messages on descendants of a node σ for two more rounds after σ is decided upon. Variable w stores the value currently used to mask the faults. Note that if $Level[0]$ becomes empty at the end of round r, processor i stops (early) communicating with other processors in round $r+2$.

```
process ESDM(Start, InitialValue);
begin
        Tree := {λ}; Val(λ) := InitialValue; w := 0;
        for k := 0 to 2 do Level[k] := {λ};
        for r := Start to t do begin
                for σ ∈ Level[2] do Send(σ);
                for k := 0 to 1 do
                        Level[k + 1] := ∪_{σ∈Level[k]} EXT(σ);
                for every σj ∈ Level[2] do
                        Rece..ve(σ, j); (* σj becomes a new node in Tree *)

                        (* dynamic fault masking *)
                w := w̄;
                if r = Start + 2 and K[PDec(λ, 1) is safe] then
                        w = 0;
                for every τj ∈ Level[2] s.t. K[j is faulty] do
                        Val(τj) := w;

                        (* early stopping *)
                for every σ ∈ Tree s.t. σ is decided on do
                        remove the descendants of σ from Level[0];
        Level[0] := ∪_{σ∈Level[0]} EXT(σ)
        end;
        if K[Dec(λ, 1) is safe] then
                Outcome[Start] := 1
        else
                Outcome[Start] := 0
end;
```

Figure 1: Protocol ESDM, code for processor i.

The main property of the *ESDM* protocol needed in the next section by protocol *CVDM* is given by Lemma 3.5. Nevertheless, when the total number of processors is moderate, *ESDM* by itself may be efficient enough, as its message complexity can be characterized as follows:

Theorem 3.6: *Protocol ESDM solves the Distributed Consensus problem using $n > 3t$ processors, $r = t + 1$ communication rounds and messages of size $m = O(1.5^t)$.*

The proof of the theorem also uses some lemmas not listed in this abstract, as they are not needed for the main result of the paper. We note that Remark 2.9 also applies here, i.e. *ESDM* can terminate in $r = min(t + 1, f + 3)$, where f is the actual number of faults that occur in a given run of the protocol.

4 Efficient Consensus with $n = (3 + \varepsilon)t$ Processors

In this section we show how to decrease the message size yet further by applying a modification of the *Cloture Votes* technique of [BGP]. The result is a consensus protocol (*CVDM*—for *Cloture Votes with Dynamic Masking*) that works for $n = 2p + t - 1 = (3 + \varepsilon)t$ processors, $r = t + 1$ rounds of communication and maximal message size $m = O(2^{4/\varepsilon} t^5)$. Thus, consensus with polynomial communication is achieved for $\varepsilon \geq 1/\log t$, considerably narrowing the optimality gap.

Cloture is a parliamentary procedure (also known as parliamentary guillotine) which allows unnecessary long debates to be curtailed. In the protocol, the unanimous will of the correct processors (akin to parliamentarian supermajority) may curtail the debate.[3] This is facilitated by the opening in each round of a new process (debate), which either ends quickly, with the conclusion "continue" or "terminate with the default value," or lasts through many rounds. Importantly, in either case the messages being sent are short.

A high-level sketch of *CVDM* is given in Figure 2. In Cloture Votes each correct processor opens in each round a new *ESDM* process from the previous section, while running an overhead *MAIN* process. Informally, we use $ESDM_k$ to refer to the version of *ESDM* started in round k. The purpose of parallel processes (debates) is the following. $ESDM_0$ should reconcile the original vote; when all correct processors are unanimous (Validity situation) this debate and all subsequent ones "die" right away (i.e. stop after three rounds). Otherwise, the message size in the original debate may grow. However, if it does grow beyond the threshold described in Lemma 4.1, then a correct processor forced to send such a message knows that

- the default consensus value will not defy the validity constraint;

- sufficiently many faults are universally known to assure consensus in the next cloture debate.

---

[3] The synopsis of Cloture Votes given here largely follows [BGP].

```
Round 0 of MAIN:
        start ESDM(0, InitialValue);
Round k:
                (* cloture *)
        if K[Dec(λ, 0) is safe] for some ESDM process then begin
                send only the messages required by that process
                (if many, the process with smallest Level[2]);
                output 0;
                halt
        end;
                (* start this round's debate *)
        if for an ongoing ESDM process #Level[2] ≥ S then
                start ESDM(k,0) (* vote for cloture *)
        else
                start ESDM(k,1);
At the end of round t:
        output min₀≤k≤t(Outcome[k]);
        halt.
```

Figure 2: Protocol $CVDM$, code for processor i.

Thus, once a correct processor is required by the protocol to send a very long message (exceeding limit S—the "patience" threshold), it will cast a vote for cloture. In every round a new debate starts—shall the deliberation end? Lemma 4.1 assures that these debates will terminate in agreement. Although the cloture debate itself may be extended, "filibuster speeches" exceeding length $n^2 S$ trigger an unanimous cloture vote ending all deliberations.

The success of *Cloture Votes* is based on two facts. One is that, as mentioned above, a very long message makes the cloture unanimous; this is due to the fact that the messages of correct processors differ in length by a factor smaller than n^2. The second is that if even only one correct processor votes for cloture, a sufficient number of faulty processors is known (and masked) to make the remaining number of rounds sufficient for achieving agreement (intuitively, each undetected fault may extend a debate by one round). Thus the following lemma, whose proof is based on Lemma 3.4, is sufficient to prove Theorem 4.2.

Let $F^r = \bigcap_{i \in P-T} F_i^r$, i.e., the set of processors universally known to be faulty at round r.

Lemma 4.1: (Sufficient Fault Detection) *Let $n = (3 + \varepsilon)t$. Assume that in a run of $CVDM$ a correct processor starts round $r \leq t$ with $\#Level[2] = l$. If $l > S = 2^{6+4/\varepsilon} t^3 n^2$, then $\#F^{r-3} \geq r$. Moreover, if $l > n^2 S$, then every correct processor started round $r - 2$ with $\#Level[2] > S$.*

Theorem 4.2: *Protocol CVDM solves the Distributed Consensus problem using $n = (3+\varepsilon)t$ processors, $r = t + 1$ communication rounds and messages of size $m = O(2^{4/\varepsilon} t^8)$.*

Remark 4.3: A factor of n^3 in message size can be saved by the simultaneous use of the pruning techniques presented above together with the message reconstruction technique described in [BG].

Same as *ESDM*, *CVDM* is also an early-stopping consensus protocol, i.e. it terminates in $r = min(t + 1, f + 3)$ rounds. ∎

5 Conclusions and Open Problems

In this paper we have presented the first protocol to achieve Distributed Consensus in optimal number rounds, using polynomial message size and nearly optimal number of processors ($n = (3 + \varepsilon)t$, where $\varepsilon \geq 1/\log t$).

It remains open whether polynomial time consensus with the optimal number of processors is achievable in the optimal number of rounds. A careful analysis of the techniques presented here for $n = 3t + 1$ leads to a $2^{O(\sqrt{t})}$ message size. This suggests that no simple proof of an exponential lower bound should be expected.

References

[ADLS] H. Attiya, C. Dwork, N. Lynch and L. Stockmeyer, "Bounds on the time to reach agreement in the presence of timing uncertainty," *Proc. 23rd STOC*, May 1991.

[BD] A. Bar-Noy and D. Dolev, "Families of Consensus Algorithms," *Proc. 3rd Aegean Workshop on Computing*, pp. 380-390, June/July 1988.

[BDDS] A. Bar-Noy, D. Dolev, C. Dwork and H.R. Strong, "Shifting gears: changing algorithms on the fly to expedite Byzantine Agreement," *Proc. 6th PODC*, pp. 42-51, August 1987.

[BG] P. Berman and J.A. Garay, "Optimal Early Stopping in Distributed Consensus," IBM Research Report *RC 16746*, December 1990. .

[BGP] P. Berman, J.A. Garay and K.J. Perry, "Towards Optimal Distributed Consensus," *Proc. 30th FOCS*, pp. 410-415, October/November 1989.

[C] B. Coan, "A communication-efficient canonical form for fault-tolerant distributed protocols," *Proc. 5th PODC*, pp. 63-72, August 1986.

[DRS] D. Dolev, R. Reischuk and H.R. Strong, "Eventual is Earlier than Immediate," *Proc. 23rd STOC*, 1982. Revised version appears in "Early Stopping in Byzantine Agreement," *JACM*, Vol. 37, No. 4 (1990), pp. 720-741.

[DS] D. Dolev and H.R. Strong, "Polynomial Algorithms for Multiple Processor Agreement," *Proc. 14th STOC*, pp. 401-407, May 1982.

[FLP] M. Fisher, N. Lynch and M. Paterson, "Impossibility of Distributed Consensus with one faulty process," *JACM*, Vol. 32, No. 2 (1985), pp. 374-382.

[FM] P. Feldman and S. Micali, "Optimal Algorithms for Byzantine Agreement," *Proc. 20th STOC*, pp. 148-161, May 1988.

[HM] J. Halpern and Y. Moses, "Knowledge and common knowledge in a distributed environment," *JACM*, Vol. 37, No. 3 (1990), pp. 549-587.

[LSP] L. Lamport, R.E. Shostak and M. Pease, "The Byzantine Generals Problem," *ACM ToPLaS*, Vol. 4, No. 3, pp. 382-401, July 1982.

[MW] Y. Moses and O. Waarts, "Coordinated Traversal: $(t+1)$-Round Byzantine Agreement in Polynomial Time," *Proc. 29th FOCS*, pp. 246-255, October 1988.

[W] O. Waarts, "Coordinated Traversal: Byzantine Agreement in polynomial time," M.Sc. Thesis, Weizmann Institute of Science, Rehovot, Israel, August 1988.

Randomized Consensus in Expected $O(n^2 \log n)$ Operations

Gabi Bracha   Ophir Rachman
Technion, Haifa, Israel [*†]

Abstract

We consider asynchronous shared memory distributed systems, and investigate coordination problems in this model. We provide a *wait-free randomized consensus* protocol that requires an expected $O(n^2 \log n)$ atomic operations.

1   Introduction

In the theory of distributed computation, the behavior of multi-processor environments is being studied. The various processors can be allocated in different sites, and can communicate through various communication media, such as massage passing networks, shared variables, and others. The behavior of the distributed system is studied with respect to the type of the communication media, and other characteristics of the system. For instance, the system may be synchronous, i.e., the processors have an access to a global clock, or to the contrary, the system may be totally asynchronous, i.e., the individual processors operate in rates that are independent on one another. Furthermore, the processors may be totally reliable, or may crash fail during their execution, (i.e., a processor may stop executing its code and never recover), or even worse, the processors may operate maliciously against the system.

All these characteristics and others specify the model of the distributed system. In spite of the large variety of models, in the heart of any distributed system is the ability to achieve coordination between the individual processors. The ability to coordinate gives an arbitrary set of scattered processors the power of a distributed system. Consequently,

---

[*]This research was partially supported by the U.S.-Israel bi-national science foundation grant 88-00-282.

[†]Both authors can be reached at: Department of Computer Science, Technion, Haifa 32000, Israel. (At e-mail: fimfam@csb.cs.Technion.ac.il).

coordination problems are one of the most investigated subjects in the field of distributed computation.

In this paper we consider coördination problems in the *shared memory* model, where the processors are totally asynchronous, and they communicate through shared *single-writer multi-reader* registers. We develop coordination protocols in this model that are highly resilient in the presence of crash faults. The correctness of the protocols is guaranteed even when all the processors crash fail except for one. Such protocols are called *wait-free* protocols since a processor can not wait for any event that depends on the activity of the other processors. A full and formal definition of the model can be found in several papers, [H 88], [L 86], [ALS 90], and others.

One of the most widely studied coordination problem in all distributed systems models and in particular in the shared memory model is the *consensus* problem. In a wait-free consensus protocol, each processor starts with a binary input value, (initially not available to other processors), and upon termination, decides on an output value. The protocol must satisfy the following conditions:

- **Consistency:** All the processors that terminate decide on the same value.

- **Validity:** The processors decide on a value that is an input value of some processor.

- **Termination:** Each processor terminates the protocol after a finite number of its own steps, regardless of the other processors' activity.

It has been shown in [CIL 89], [DDS 87], [LA 87], and others, that no deterministic protocol can satisfy the above conditions. However, it has been found that probabilistic protocols, that allow the processors to flip coins, can satisfy consistency and validity, and a probabilistic version of the termination condition:

- **Probabilistic termination:** Each processor terminates the protocol after an expected finite number of its own steps, regardless of the other processors' activity.

A probabilistic version of the consensus problem, namely, the *randomized consensus* problem, is defined by the probabilistic termination condition together with the previous consistency and validity conditions.

One of the first randomized consensus protocols was presented in [CIL 89]. This protocol assumed that the processors have an access to a global shared coin. A protocol that overcomes this strong assumption was introduced in [A 88]. However, the expected number of register operations in this protocol is exponential in the number of processors. The first randomized consensus protocol with expected polynomial number of operations was offered in [AH 90], and the first polynomial protocol that was also memory bounded was offered in [ADS 89]. In the latter two protocols, the expected number of operations is $O(n^4)$. So far, the most efficient protocols were introduced in [SSW 90], and in [BR 90]. The expected number of operations in these protocols is $O(n^3)$, and they use bounded memory. In this paper, we show how a modification in the protocol of [SSW 90] yields a significantly improved randomized consensus protocol with expected $O(n^2 \log n)$ operations.

The remainder of the paper is organized as follows: in section 2 we review the protocol of [SSW 90], and in section 3 we present the modified protocol.

2   The Protocol of [SSW 90]

The randomized consensus protocol of [SSW 90] is based on a primitive called a *shared coin*. A *shared coin* with agreement parameter λ, is a primitive that provides a value $\in \{0, 1\}$ to each processor, and that has the property that for each $b \in \{0, 1\}$ the probability that all the processors are provided the value b is $\geq \lambda$.

It has been shown in [SSW 90] that a shared coin with constant agreement parameter can be converted into a randomized consensus protocol with an expected constant overhead. Furthermore, a simply constructed shared coin primitive with constant agreement parameter was introduced, that requires $O(n^3)$ operations, and thus, yields a randomized consensus protocol with expected $O(n^3)$ operations.

```
0:    primitive shared-coin
1:    begin
2:        my_reg(total_flips, total_ones) := (0, 0)
3:        repeat
4:            coin := local flip
5:            my_reg := (my_reg.total_flips + 1, my_reg.total_ones + coin)
6:            read all the registers, and sum the fields total_flips and
              total_ones into local variables flips and ones, respectively
7:        until flips > n².
8:        if ones/flips ≥ 1/2
9:            then output 1
10:           else output 0
11: end
```

Figure 1: The original shared coin primitive

The shared coin of [SSW 90] is given in Figure 1, and can be simply explained as follows: each processor repeatedly flips a local coin, and updates its register to hold the total number of flips it has performed, and how many of those flips were 1. After each flip, the processor reads the registers of all the other processors, and counts the total number of coins and the total number of 1's that were flipped. If it counts at least n^2 total coins, it stops, and outputs the majority value (or 1 if tie). Otherwise, it continues to flip its local coin.

Since the processors are asynchronous, it is possible that different processors output a majority of different sets of coins. However, since each processor can flip its coin only once after the total reaches n^2, the total number of coins flipped can not exceed $(n^2 + n)$. Therefore, if there is a majority greater than n to either side (that is, if the majority value occurs at least n times more than the minority value among the flips,) then all the processors are guaranteed to compute the same majority. By the properties of the binomial distribution, this occurs with some constant probability.

The number of operations needed in the shared coin of Figure 1 is $O(n^3)$, since for each of the $O(n^2)$ flips, the set of n registers is read.

3 The Modified Protocol

In this section we show how to modify the shared coin primitive of Figure 1 to a shared coin primitive that preserves the constant agreement parameter, but requires only $O(n^2 \log n)$ operations. Thus, by the techniques of [SSW 90], we achieve a randomized consensus protocol with expected $O(n^2 \log n)$ operations.

```
0:   primitive shared-coin
1:   begin
2:        my_reg(total_flips, total_ones) := (0, 0)
3:        repeat
4:             for i = 1 to  n/log n  do
5:             begin
6:                  coin := local flip
7:                  my_reg := (my_reg.total_flips + 1, my_reg.total_ones + coin)
8:             end
9:             read all the registers, and sum the total_flips fields
                  into the local variable flips
10:       until flips > n².
11:       read all the registers, and sum the fields total_flips and
                  total_ones into the local variables flips and ones, respectively
12:       if ones/flips ≥ 1/2
13            then output 1
14:           else output 0
15: end
```

Figure 2: The modified shared coin primitive

The modified primitive appears in Figure 2. As in [SSW 90], each processor repeatedly flips a local coin, and updates its register to hold the total number of flips it has performed, and the total number of 1's. However, in the modified primitive, the processor reads the registers of the other processors only once in $k = \frac{n}{\log n}$ of its local flips. As before, if it sees at least n^2 coins, it stops, and outputs the majority value (or 1 if tie). Otherwise, it continues to flip its local coin.

As appears from Figure 2, we added a "cosmetic" modification in line 11. In that line, the processor reads all the registers once more after it has seen n^2 coins. We refer to this last read, that determines the output of the processor, as the *critical* read. As we will see, this "cosmetic" modification simplifies the analysis.

Let us refer to the first n^2 flips that were written into the shared registers as the *common* flips. It is clear from the construction of the primitive, that any processor starts the execution of its critical read after all of the common flips were written to the shared registers. Therefore, all the processors output a majority of a series of flips that includes the common flips. Furthermore, no processor will flip more than k flips that are not within the common flips. We will refer to these flips as the *extra* flips. Thus, each processor outputs the majority of the n^2 common flips, plus no more than nk extra flips.

Now, we prove some crucial properties of the primitive. We start with some simple probabilistic lemmas:

Lemma 1: Consider a series of N independent coin flips. For any $x \geq 1$, the probability that there is a majority greater than $x\sqrt{N}$ to either side, denoted by $\Psi(x)$, satisfies:

$$\frac{1}{2} \cdot \frac{1}{2x\sqrt{2\pi}} e^{-2x^2} < \Psi(x) < \frac{1}{2x\sqrt{2\pi}} e^{-2x^2}.$$

Proof: By the normal approximation of the binomial distribution.
□

Lemma 2: Consider the following probabilistic game between a player and a random coins generator. The game consists of rounds, where at each round the player asks the generator for a coin. According to the coin's value, the player can decide to store the coin in one of two bags: A (for Accepted coins), or W (for Withheld coins). The player is restricted to withholding in bag W only up to n coins at any given time, however, at any time, the player can remove coins from W to A (but not the other way around). Assuming the player plays no more than N rounds, the probability that in bag A there are coins with majority greater than $x\sqrt{N} + n$ to either side, (for $x \geq 1$,) is at most $N \cdot \Psi(x)$.

Proof: Consider a stronger player, that can foresee future coins results, and hence, can plan its strategy upon a known series of N independent coin flips. However, if in any prefix of the series, that is produced by the generator, there is no majority greater than $x\sqrt{N}$ for either side, then even this stronger player can not collect in bag A coins with majority greater than $x\sqrt{N} + n$.

By Lemma 1, for any series of independent coin flips of size $< N$, the probability of a majority greater than $x\sqrt{N}$ is at most $\Psi(x)$. Therefore, the probability that in any of the series's prefixes there is a majority greater than $x\sqrt{N}$, is at most $N \cdot \Psi(x)$.

$\square$

Lemma 3: Consider the following probabilistic game between a player and a set of n random coins generators. The game consists of rounds, where at each round the player chooses one generator, and asks it for a coin. According to the coin's value, the player can decide to store the coin in one of two bags: A (for Accepted coins), or W (for Withheld coins). The player is restricted to withholding in bag W no more than one coin from each generator at any given time, however, at any time, the player can remove coins from W to A (but not the other way around). Assuming the player plays no more than N rounds, the probability that in bag A there are coins with majority greater than $x\sqrt{N} + n$ to either side, (for $x \geq 1$,) is at most $N \cdot \Psi(x)$.

Proof: Since the random generators are independent, and since the player can withhold at most n coins at any given time, any strategy of the player in this game can be reduced to a strategy of a player in the game of Lemma 2. Therefore, the result of Lemma 2 still holds.

$\square$

Now, we prove that the shared coin primitive of Figure 2 preserves the constant agreement parameter.

Lemma 4: There is a constant probability that there is a majority greater than $3n$ in favor of 1 in the common flips.

Proof: Consider the first n^2 coins that are flipped in the primitive. These coins are not necessarily the common flips, however, at least $n^2 - n$ of them are within the common flips. Therefore, if in these coins there is a majority greater than $5n$ in favor of 1, then in the common flips there is a majority greater than $3n$ in favor of 1. Since the adversary (i.e., the scheduler) has no influence on the outcome of the first n^2 coin flips, we can consider them as a series of independent flips. By Lemma 1, the probability of a majority greater than $5n$ in a series of n^2 coin flips is a constant.

$\square$

Lemma 5: If there is a majority greater than $3n$ in favor of 1 in the common flips, then with probability greater than $1 - \frac{1}{n}$, all the processors will output 1.

Proof: Consider a processor p, that is executing its critical read. The flips that p collects consist of the n^2 common flips, plus up to nk extra flips. Since there is a majority greater than $2n$ in favor of 1 in the common flips, p will decide 1 unless it reads a majority greater than $2n$ in favor of 0 in the extra flips.

Assume that p is about to start its critical read. The power of the adversary in determining the extra flips that p reads is the following: the adversary can schedule any processor to flip its local coin, and then, either to let it write the coin's value into the shared memory, or to withhold it for some time from being written. Also, the adversary can schedule p to read the next register in its critical read. It is clear that if the adversary schedules p to read a register of some processor q, then all future schedules of q have no affect on the extra flips that p reads. Therefore, we can assume that after p reads the register of q, the adversary never schedules q again. Since this argument holds for any processor q, we can assume a scenario where the adversary schedules all the processors, except p, to flip or write coins, and then schedules p to read all the registers.

This scenario can be reduced to the game described in Lemma 3. The adversary is the player, the n processors are the set of n random coins generators, the coins written in the shared memory are the accepted coins, and $N = nk = \frac{n^2}{\log n}$. The rules of the game are satisfied, since the adversary can not schedule processors to flip more than $N = nk$ extra flips, and can not schedule a processor to flip a coin unless it writes the previous coin into the shared memory. Now, by the reduction to the game, the probability that p reads extra coins with majority greater than $3n$ in favor of 0, can not exceed the probability that in the game, the player accepts coins with majority greater than $(x\sqrt{N}+n)$ in favor of 0, where $x = 2\sqrt{\log n}$. By Lemma 3, this probability is less than $\frac{1}{n^2}$.

Summing, if the probability of a single processor to decide 0 is less than $\frac{1}{n^2}$, then the probability that any of the n processors will decide 0 is less than $\frac{1}{n}$.
$\square$

We can summarize the properties of the primitive in the following Theorem:

Theorem 1: The primitive of Figure 2 yields a shared coin with constant agreement parameter, that requires $O(n^2 \log n)$ operations.

Proof: Lemmas 4 and 5 guarantee that there is a constant probability that all the processors will decide 1 using the primitive. By symmetry of these lemmas, there is a constant probability that the processors will decide 0. Therefore, the primitive yields a shared coin with constant agreement parameter.

The total number of local coin flips is $O(n^2)$. The number of write operations is $O(n^2)$, and since the n registers are being read only once in $O(\frac{n}{\log n})$ flips, the number of read operations is $O(n^2 \log n)$.
$\square$

As was previously mentioned, the shared coin primitive of Figure 2 can be converted into a randomized consensus protocol with expected $O(n^2 \log n)$ operations.

Acknowledgments: Mostly to Prof. Benny Chor for his help and support. To Guy Even for helpful discussions. To Amir Ben-Dor for useful comments. To Rinat Rachman for her intelligent listening.

References

[A 88]   K. Abrahamson, On achieving consensus using a shared memory, *Proc. of the 7th ACM Symp. on Principles of Distributed Computing*, August 1988, pp. 291-302.

[A 90]   J. Aspnes, Time and space-efficient randomized consensus, *Proc. of the 9th ACM Symp. on Principles of Distributed Computing*, August 1990, pp. 325-331.

[AH 90]   J. Aspnes and M. Herlihy, Fast randomized consensus using shared memory, *Journal of algorithms*, September 1990, pp. 441-461.

[ADS 89]   H. Attiya, D. Dolev, and N. Shavit, Bounded polynomial randomized consensus, *Proc. of the 8th ACM Symp. on Principles of Distributed Computing*, August 1989, pp. 281-294.

[ALS 90]   H. Attiya, N. Lynch, and N. Shavit, Are Wait-Free Algorithms Fast? *Proc. of the 31st IEEE Symp. on Foundations of Computer Science*, 1990.

[BR 90]   G. Bracha and O. Rachman. Approximated Counters and Randomized Consensus. In *Technical Report #662, Computer Science Department, Technion, Haifa, Israel.* December 1990.

[CIL 89]   B. Chor, A. Israeli, and M. Li, On processor coordination using asynchronous hardware, *Proc. of the 6th ACM Symp. on Principles of Distributed Computing*, August 1987, pp. 86-97.

[DDS 87]   D. Dolev, C. Dwork, and L. Stockmeyer, On the minimal synchrony needed for distributed consensus, *Journal of the ACM, Vol. 34, No 1,* January 1987, pp. 77-97.

[H 88]   M. Herlihy, Wait Free Implementations of Concurrent Objects, *Proc. of the 7th ACM Symp. on Principles of Distributed Computing*, 1988, pp. 276-290.

[L 86]   L. Lamport, On interprocess communication, *Distributed Computing 1,2,* 1986, pp. 77-101.

[LA 87]   M. Loui and H. Abu-Amara, Memory requirements for agreement among unreliable asynchronous processes, *Advances in Computing Research, Vol. 4, JAI Press, Inc.,* 1987, 163-183.

[SSW 90]   M. Saks, N. Shavit and H. Woll, Optimal time randomized consensus-making resilient algorithms fast in practice, *Proc. 2nd ACM Symp. on Discrete Algorithms*, January 1991, pp. 351-362.

Using Adaptive Timeouts to Achieve At-Most-Once Message Delivery [†]

SOMA CHAUDHURI
Iowa State University
BRIAN A. COAN
Bellcore
JENNIFER L. WELCH
University of North Carolina

Abstract

We extend the algorithm of Liskov, Shrira, and Wroclawski for at-most-once message delivery so that it adapts dynamically to changes in message transmission time and degree of clock synchronization. The performance of their algorithm depends on its being supplied with a good estimate of the maximum message lifetime—the sum of the message delivery time and the difference in processor clock values between sender and recipient. We present two algorithms which are suitable for use in a system where the message lifetime is unknown or may change. Our extensions allow the automatic and continuous determination of a suitable value for the maximum lifetime. We prove that whenever the actual message lifetime is bounded, then our adaptive procedures converge to an accurate estimate of its true value. Our two algorithms make different assumptions about the behavior of the system and thus achieve different performance levels. Our formal statement of convergence is expressed in terms of the number of messages received, rather than time elapsed. We show that this formulation is necessary. Specifically, we prove that no method for estimating the lifetime can achieve convergence in a bounded amount of time.

1 Introduction

Reliable processor clocks can simplify the design of fault-tolerant distributed computer systems. They enable a processor to use the passage of time to make inferences about the rest of the system.

One of the most basic uses of a processor clock is the timeout, which allows a processor to deduce that a certain event will never happen based on the fact that a sufficient amount of time has passed on its own clock. For timeouts to be used, there is no need that the clocks of different processors be synchronized. All that is needed is that a processor's clock run at approximately the correct rate.

More general ways of using processor clocks entail that the clocks of different processors be synchronized to within some known fixed bound ϵ. Algorithms are known ([HSSD84, WL88, ST87] and many others) which do this with absolute certainty. Unfortunately they make assumptions which may be false in some networks. One such problematic assumption is the existence of a known, fixed upper bound on message delivery time. An alternative clock synchronization algorithm of Mills [Mil89] does not make these assumptions. It is a practical algorithm which presently is running on the Internet and operating very well. However, it is a probabilistic algorithm, and for any ϵ there is some small chance that processor clocks will deviate by more than ϵ.

[†]This work was done while the first author was at the University of North Carolina at Chapel Hill. The work of the first and third authors was supported in part by NSF grant CCR-9010730 and an IBM Faculty Development Award.

Liskov [Lis91] has argued that the Mills clock synchronization algorithm has two fundamental impacts on system design. First, because of its great efficiency, the Mills algorithm makes it completely feasible to write distributed algorithms that use approximately synchronized processor clocks. Second, because there is a (small) possibility that processor clocks will have a larger deviation than expected, it is important to avoid distributed algorithms whose correctness relies on synchronized clocks. Rather, the clocks should be used as a "hint," which is important for performance, but is not needed for correct operation.

An example of the type of algorithm advocated by Liskov is a message passing algorithm of Liskov, Shrira, and Wroclawski [LSW91] which makes use of an estimate of the maximum message lifetime—the sum of the maximum message delay and the maximum deviation of the sender's clock from the recipient's clock—to reduce the overhead of achieving at-most-once message delivery. The specific reduction in overhead is that the recipient is allowed to "forget" all state information about senders that have been inactive for sufficiently long. This ability to forget can make it much easier to restart the recipient after a failure and can reduce memory requirements at the recipient.

The [LSW91] algorithm must be supplied with an estimated upper bound on the message lifetime. Mistakes in making this estimation do not compromise correctness, but they do damage performance. The cost of having too small an estimated bound is that many messages can be lost (erroneously discarded as duplicates). The cost of having too large an estimated bound is that an unnecessarily large amount of memory is used at the server. Thus it is desirable to have a reasonably accurate estimate of the maximum message lifetime. In general, this may be difficult or impossible. For example, the message delay component of message lifetime may vary over time or with system load.

We believe that there are three reasons why it is useful for a distributed algorithm to have the ability to continuously adapt to the current prevailing message lifetime. First, it simplifies the job of configuring a system. There is no need to determine the lifetime and there is no need to install this parameter in the code for the algorithm. Second, in many systems the load, and hence the lifetime, varies in an approximately periodic way. For instance, days may be relatively busy and nights relatively idle. In such systems using an adaptive algorithm makes it possible to adjust the estimate of the lifetime accordingly. Third, it is reasonable to expect that over the long term there may well be a change in the message lifetime in a system. Again the benefit of an adaptive algorithm should be clear.

In this paper we present two extensions to the [LSW91] algorithm for at-most-once message delivery. Each algorithm uses a different method for dynamically determining an estimate of the message lifetime. Our idea of dynamically changing the estimate of the lifetime is inspired by prior work on adaptively changing timeout intervals [Jai86, Zha86]. We demonstrate that both our methods converge to an appropriate value whenever the actual message lifetime in the system is sufficiently well-behaved. Specifically, we demonstrate that our estimate is not too high by proving that it is no more than twice the true value, and we demonstrate that our estimate is not too low by proving that, as the length of an execution tends to infinity, the fraction of messages that are erroneously rejected because of the estimate tends to some suitable value. Our formal statement of the rate at which the message-lifetime estimate converges to the true value is expressed in terms of the number of messages received, rather than time elapsed. We conclude the paper by showing that this formulation is necessary. Specifically, we prove that no method for estimating the lifetime can achieve convergence in a bounded amount of time.

Our two algorithms differ in the following three respects: the requirements for the system to be considered well-behaved, the convergence properties that they ensure in well-behaved executions, and the extent to which the current behavior of the algorithm depends on history. We believe that the first algorithm is primarily of theoretical interest and the second is primarily of practical interest. For the first algorithm, an execution is considered well-behaved if there is some time after which there is an upper bound on the lifetime of any message. This algorithm ensures that in well-behaved executions, as the length of the execution tends to infinity, the fraction of messages that are erroneously rejected (due to the lifetime estimate) tends to 0. To achieve this convergence property, the behavior of this algorithm depends very heavily on history: the longer the execution, the more reluctant the server is to reduce its estimate of message lifetime. Our second algorithm eliminates this undesirable dependence on history and has a more lenient notion of well-behaved executions. Its convergence property is somewhat weaker, however. For the second algorithm an execution is considered well-behaved if there is some time after which there is an upper bound that holds on the lifetime of most messages (a more precise definition is in the body of the paper). This algorithm ensures that in well-behaved executions, as the length of the execution tends to infinity, the ratio of messages that are erroneously rejected (due to the lifetime estimate) to messages that are accepted tends to δ, for (almost) any fixed δ. We believe that its reduced sensitivity to history and its ability to ignore the occasional very late message contribute to the practical utility of this algorithm.

2   Problem Statement and Assumptions

Liskov, Shrira, and Wroclawski [LSW91] explain how an algorithm for at-most-once message delivery can be used as a part of a larger system that implements useful user-level functionality, for example, at-most-once remote procedure calls. The focus of this paper is not applications of at-most-once message delivery algorithms. Instead, we seek to improve the algorithm for at-most-once message delivery itself. Our specific interest in this regard is extending the algorithm to adapt to changes in the maximum message lifetime.

The system in which an at-most-once message delivery algorithm operates consists of one server process and a collection of client processes. These processes run on a collection of processors. They interact by sending messages over a communication network. The network can lose or duplicate messages, delivering zero, one, or many copies of each message that is sent.

Each client has a sequence of unique messages to be sent to the server. The objective of each client is to have as many of its messages as possible accepted (i.e., processed) by the server. Additionally, each client has the absolute requirement that none of its messages be accepted more than once and that none of its messages be accepted once a later message from that client has been accepted. The mechanism by which the client learns of the acceptance of a message is separate from the at-most-once message delivery algorithm: it is generally part of the higher-level application (e.g., remote procedure call) in which the message delivery algorithm is embedded.

The specific behavior of each client is to progress through its message sequence one message at a time, sending a single copy of the current message. Before proceeding on to send the next message in the sequence, the client waits until either it has learned of the acceptance of the current message or it has given up on the acceptance of the current message. A message on which the client has given up may or may not eventually be accepted by the server, but it will not be accepted out of order. Retransmission (of possibly lost messages) by clients is not permitted in our at-most-once message delivery protocol; although, this obviously useful functionality can be provide by a (slightly) higher level mechanism, which we sketch at the end of this section.

When the server receives a message from a client it either accepts or rejects it. Absolute requirements on the server are that it never accept a second copy of a message that it has already accepted (a *duplicate* message) and that it never accept a message that was sent earlier than the last accepted message from the same client (a *late* message). The server may on occasion reject messages that are not duplicates and are not late. To ensure that progress is made, it is desirable for the server to accept as many non-duplicate, in-order messages as practicable.

A straightforward solution to the at-most-once message delivery problem is for each client to append a sequence number to each message it sends, and for the server to keep track of the sequence number of the last message accepted from each client. The principal disadvantage of the straightforward solution is that it requires the server to keep information for each client that it has ever heard from. Liskov, Shrira, and Wroclawski [LSW91] use synchronized clocks to enable the server to forget about clients that have been inactive for a sufficiently long interval. Their technique works as follows. Every message sent by a client is tagged with the current value of the client's clock, the "timestamp." The server keeps a cache CT ("connection table") with an entry for each client. $CT[i]$ holds the timestamp of the most recent message from client i that has been accepted by the server. If the server does not hear from client i for a sufficiently long time, then the entry is replaced with *nil* ("garbage collected"). The server also keeps a variable *upper*, which is an upper bound on the largest timestamp of any entry that has ever been garbage collected from the connection table.

We now describe what happens when the server receives a message with timestamp t from client i. If t is greater than the (non-*nil*) entry in CT for i, then this message is not a duplicate and is not late and it is accepted. If t is less than or equal to the (non-*nil*) entry in CT for i, then this message is a duplicate or is late and it is rejected. If the entry in CT for i is *nil*, then t is compared against *upper*. If t is larger, then this message is not a duplicate and is not late, since *upper* is an upper bound on the largest garbage-collected entries in CT. The message is accepted, and t is stored in $CT[i]$. If t is less than or equal to *upper*, then this message is either a duplicate or a late-arriving non-duplicate, but to avoid accepting a duplicate, it is rejected.

Every so often the connection table is garbage collected based on a fixed estimate of the maximum message lifetime (the sum of the message delivery time and the uncertainty in the clock synchronization). For each entry in CT, the difference between that entry and the current clock value at the server is calculated. If this difference is greater than the fixed estimate of maximum message time, then the entry is garbage collected. When an entry is garbage collected, it is replaced by *nil* and *upper* is updated if necessary. The updating of *upper* is done to maintain the invariant that *upper* is at least as big as the largest entry that has ever been garbage collected from the connection table.

As discussed in [LSW91] the clients can be partitioned into different classes (e.g., local and remote) with different estimates of the maximum message lifetime maintained independently for each class. We will not discuss this point any further in this paper.

The [LSW91] algorithm uses a fixed value for the lifetime estimate. In this paper we develop a method for automatically and continually estimating lifetime. We use this method in conjunction with the basic framework of the [LSW91] algorithm. Throughout the remainder of this paper we use the term **standard form algorithm** to refer to an algorithm that uses the basic framework of the [LSW91] algorithm (e.g., a connection table, *upper*, and periodic garbage collection), but with estimates of the maximum lifetime that are computed as the execution proceeds.

We now give definitions of some terms that we will be using to discuss our algorithms. Throughout this paper, all times referred to are times on the server's clock (unless otherwise explicitly stated).

Definition *The* lifetime *of a message received by the server in an execution is the difference between the time on the server's clock when the server receives the message and the time on the client's clock when it sent the message.*

The lifetime is composed of the message delay plus the error in the clock synchronization. The server can calculate the lifetime of a message if the sender appends its current clock time to the message. In our algorithms, the client will always do this.

Definition *A message m is* late *in an execution if it arrives after a message that was sent by the same client after m was sent. A message is* lost *if it is not a duplicate, is not late, and, when it arrives at the server, it is rejected. A message is* lifetime-rejected *if it is rejected by comparing its timestamp with upper. A message is* CT-rejected *if it is rejected based on a comparison with a non-nil entry in CT.*

Note that no CT-rejected messages are lost, but that some lifetime-rejected messages may be lost.

Definition *Given a time t in an execution, let A_t be the number of messages accepted by the server up to time t, and let L_t be the number of messages lost by the server up to time t.*

Definition 1 *For time t and positive integer X, an execution is* (t, X, S, H)-bounded *provided that in every consecutive group of S messages arriving after time t, at most H messages have lifetime greater than X. If $H = 0$, then the parameter S gives no information, and we use the simpler notation* (t, X)-bounded.

Given fixed integers S and H and real number $\delta \geq 0$, we would like to guarantee the following conditions in any execution that is (t, X, S, H)-bounded for some t and X.

Condition 1 *No duplicate or late messages are accepted by the server.*

Condition 2 *There is some time $t' > t$ after which the server's estimate of the lifetime is always less than $2X$.*

Condition 3 *The limit of L_u/A_u is at most δ as u goes to infinity.*

The first condition is the basic safety condition for any at-most-once message delivery algorithm: specifically, it is that no duplicate or late messages ever be accepted by the server. Both of our algorithms satisfy this condition in any execution, bounded or not. The next two conditions ensure that the performance of our algorithm is adequate. The second condition requires that the algorithm not use an excessive amount of memory in the connection table or store old connection information for an excessively long period of time. The third condition requires that there be an adequate amount of progress (i.e., number of non-duplicate, non-late messages accepted). In any execution that is (t, X)-bounded, our first algorithm satisfies Conditions 2 and 3 (with $\delta = 0$). In any execution that is (t, X, S, H)-bounded, our second algorithm satisfies Conditions 2 and 3 (with δ any fixed value greater than or equal to $H/(S - H)$; see below).

In order to achieve these conditions, we must make three assumptions. The first assumption is needed because our conditions are inherently about infinite executions.

Assumption 1 *The server receives an infinite number of non-duplicate messages in any infinite execution.*

The second assumption is needed to prevent an adversary from swamping an algorithm with messages during some brief period when the bound on message lifetime is set too low. Without assuming an upper bound on the rate with which messages arrive at the server, the most we can guarantee is this: for the chosen δ, there is an infinite number of times u in the execution where $L_u/A_u \le \delta$.

Assumption 2 *There is an integer R such that for any times t_1 and t_2 in any execution, $t_1 < t_2$, the server receives at most $(t_2 - t_1) \cdot R$ messages in the interval $[t_1, t_2]$.*

Since, even when the execution is well-behaved, at least H messages (the ones making up the transient spike) can be lifetime-rejected out of every S messages received, we cannot hope to achieve our desired ends unless our desired ratio δ is not too small.

Assumption 3 $\delta \ge H/(S - H)$.

We return to the question of retransmission (of possibly lost messages) by clients. We describe a (slightly) higher level mechanism, HLM, which allows retransmission. We show how to implement the HLM on top of a standard form algorithm (SFA), which does not allow retransmission. The HLM adds one tag field to the header of each message.

In the HLM, each client has a sequence of unique messages to be sent to the server. The goal of the client is to have each of its messages accepted (i.e., processed) by the server exactly once and in order. To accomplish this goal in the presence of message loss by the underlying SFA, a client must be prepared to resend messages that may have been lost and the network must guarantee that starting at any time in an execution, if any processor sends sufficiently often, then at least one of the messages that it sent will be delivered. (A message that the HLP considers to be a retransmission is considered to be a brand new message by the SFA.) Because the client does not have perfect information about which messages have been lost, its resending behavior can result in duplicate copies of the same message being delivered in the SFA protocol layer at the server. The HLM provides the HLM layer at the server with a way to discard the duplicates.

In the HLM, the behavior of each client is to progress through its message sequence one message at a time, sending (using the SFA) multiple copies of the current message until it has been accepted by the HLM layer at the server. The HLM tag of the first copy of a message is the timestamp, t, that is used in the SFA. The HLM tags of all retransmissions of this message are also t, although the SFA assigns the retransmissions timestamps that differ from t. The HLM layer at the server uses the HLM tag on messages to discard duplicate messages that arise from retransmission at the HLM layer. A duplicate message is recognized at the server because it has a HLM tag that is identical to that of the last message accepted.

3 The Algorithms

In this section we describe two standard-form algorithms for at-most-once message delivery. Our algorithms use the basic framework of the [LSW91] algorithm, but they dynamically adjust the message lifetime estimate based on observations of the messages received.

The first algorithm is primarily of theoretical interest. It achieves a ratio of lost to accepted messages that approaches 0 (i.e., $\delta = 0$), but it requires that there be a time after which every message received, without exception, has a fixed upper bound on its lifetime. Also, the longer the system has been running, the longer it takes the estimated lifetime to be reduced once the observed lifetimes become small.

The second algorithm is probably of more practical interest. It only achieves a ratio of lost to accepted messages that approaches a constant $\delta = 1/p$ (subject to Assumption 3). However, it can tolerate transient spikes in the lifetimes of messages received, and the convergence behavior of the estimated lifetime does not depend on the entire history.

The times at which garbage collections can be done affect the method that can be used to tolerate server crashes, as discussed in [LSW91]. In the first algorithm, garbage collection can be done at arbitrary times, as long as there is an upper bound on the time between garbage collections. Thus either of the two methods presented in [LSW91] can be used. In the second algorithm, garbage collection must be done exactly when a certain number of messages have been received since the last garbage collection. Since there is no fixed upper bound on the elapsed time between two garbage collections, only one of the two crash recovery methods can be used with this algorithm.

3.1 Algorithm 1

Algorithm 1 satisfies the three correctness conditions with $H = 0$ and $\delta = 0$.

The algorithm is obtained by making some additions to the standard-form algorithm. Within each garbage collection interval, the server keeps track of the largest lifetime of any message received. At the end of each garbage collection interval, the server has the option of changing its current maximum lifetime estimate. The only possible values of the estimate are powers of 2. The reason for this restriction is explained below. If the maximum lifetime observed in this garbage collection interval is larger than the current estimate, then the current estimate is increased to be the smallest power of 2 at least as large as the maximum lifetime observed. If the maximum lifetime observed is smaller than the current estimate, then the current estimate is decreased to be the smallest power of 2 at least as large as the maximum lifetime observed, as long as the following condition holds. (Our algorithm ensures that the variable storing the maximum lifetime observed is never negative, by not considering any negative observed lifetimes.)

The general idea of the condition under which the estimate can be reduced is that the first time the estimate is decreased, the number of messages accepted must be larger than the number of messages lifetime-rejected, the second time the estimate is decreased, the number of messages accepted must be larger than twice the number of messages lifetime-rejected, and so forth. The implementation of this idea involves the variable *numacc*, holding the number of accepted messages, the variable *numrej*, holding the number of lifetime-rejected messages, and p, a counter that starts at 1 and is incremented every time the lifetime estimate is decreased. The estimate can be decreased only if *numacc* is larger than p times *numrej*. Whenever the estimate is decreased, *numrej* is reset to 0, and *numacc* is reduced by p times *numrej*. Thus we are saving the excess number of accepted messages to help us achieve the condition faster the next time. Once the lifetime estimate has been adjusted, it is used to garbage collect the connection table in the usual way.

The allowable values for the lifetime estimates are restricted to powers of 2 in order to minimize the number of step increases needed, and thus the number of messages lost, until the maximum estimate is achieved, once the execution is well-behaved. Suppose we allow the estimate to take on any value and consider this situation. All messages arriving in the first garbage collection interval have lifetime 1. In the second interval, the estimate is 1, but all messages have lifetime 2 and are lifetime-rejected. In the third interval, the estimate is bumped to 2, but all messages have lifetime 3 and are lifetime-rejected. Continuing this way until the upper bound of X on the observed lifetimes is reached will require X step increases, and the opportunity is there for messages to be lost the entire time. In contrast, if the estimate can only take on powers of 2, then at most $\log X$ step increases are required until the maximum is reached.

Below we describe the algorithm for the server, assuming n clients. Each client is assumed to timestamp each message it sends with the current value of its clock. The server has two transitions. The server's transition for receiving a message occurs whenever a message arrives. The server's transition for garbage collecting the connection table can occur at arbitrary times as long as there is an upper bound G on the amount of (server's clock) time elapsing between two consecutive garbage collections.

ALGORITHM 1:

State variables:

> *clock*: integer, current value of the clock
> $CT[1..n]$: array of integer, initially all entries are *nil* (say -1)
> *age*: powers of 2, initially 1
> *p*: integer, initially 1
> *upper*: integer, initially 0
> *numacc*: integer, initially 0
> *numrej*: integer, initially 0
> *maxage*: integer, initially 0

Transitions:

Receive message m from client i with timestamp t:

> $maxage := \max(maxage, clock - t)$
> if $((CT[i] \neq nil)$ and $(t \leq CT[i]))$ then
> > reject m
> else if $((CT[i] \neq nil)$ and $(t > CT[i]))$ or $((CT[i] = nil)$ and $(t > upper))$ then
> > accept m
> > $numacc := numacc + 1$
> > $CT[i] := t$
> else if $((CT[i] = nil)$ and $(t \leq upper))$ then
> > reject m
> > $numrej := numrej + 1$

Garbage collect CT:

> if $(maxage > age)$ then
> > $age := 2^j$, where $j = \min\{k : 2^k \geq maxage\}$
> else if $(numacc > p \cdot numrej)$ and $(age > maxage > 0)$ then
> > $age := 2^j$, where $j = \min\{k : 2^k \geq maxage\}$
> > $numacc := numacc - p \cdot numrej$
> > $numrej := 0$
> > $p := p + 1$
> $maxage := 0$
> for $i = 1$ to n do
> > if $CT[i] \leq clock - age$ then
> > > $upper := \max(upper, CT[i])$
> > > $CT[i] := nil$

Our analysis of Algorithm 1 consists of three theorems, one for each of the three correctness conditions.

Theorem 4 states that the algorithm never accepts duplicate or late messages. It is proved using Lemmas 1 through 3. The main idea is that the values taken on by *upper* are increasing, so the value of *upper* at any time is always at least as large as any previous value of a currently *nil* entry in the connection table, and thus the non-*nil* values taken on by any entry in the connection table are increasing.

Theorem 6 states that, once the execution is well-behaved, i.e., there is an upper bound on the maximum lifetime, the lifetime estimate will eventually converge to a value less than twice the upper bound. The proof of this theorem, as well as the next one, relies on Lemma 5, which gives an upper bound on how long it takes after the estimate stabilizes until no message is lifetime-rejected.

Finally, Theorem 8 states that in any well-behaved execution, the ratio of lost to accepted messages goes to 0. To prove this theorem, the execution is broken up into intervals, where a new interval begins whenever the lifetime estimate is decreased. Lemma 7 gives a constant upper bound on the number of messages lifetime-rejected in any interval (once the execution is well-behaved). The reason for the constant bound is that the variable age is increased at most a constant number of times in each interval, and messages can only be lost in a small space around the points when age is increased. The interesting case in the proof of Theorem 8 is when there is an infinite number of intervals. We want to show that for any q, there is a point in the execution after which the number of lost messages times q is no larger than the number of accepted messages. Consider a particular interval, say the i-th, that starts after the execution is well-behaved. At any time u in the i-th interval, the number of messages lost, L_u, is at most the sum of the number of messages lifetime-rejected in all the previous intervals, plus the constant bound from Lemma 7. On the other hand, the number of messages accepted, A_u, is at least the sum of the reduction in $numacc$ in all the previous intervals. By the way the algorithm works, in the j-th interval, for any j, the number of messages lifetime-rejected is equal to j times the reduction in $numacc$. Thus, for intervals before the q-th, the difference between q times the number lifetime-rejected and the reduction in $numacc$ is positive. However, for intervals after the q-th, the same difference is negative, and keeps increasing in magnitude. Eventually, for large enough i, the negative quantity overwhelms the positive, giving us the result.

Lemma 1 *In any execution, the sequence of values taken on by upper is increasing.*

Proof: The variable $upper$ is always set equal to the maximum of its old value and something else. ∎

Lemma 2 *For any execution, any i and any time t, suppose $CT[i]$ is equal to nil at time t. Let u be a non-nil value of $CT[i]$ at some time t' before t. Then, the value of upper at time t is at least u.*

Proof: Let t'' be the first time after t' when $CT[i]$ is set to nil. Clearly t'' exists and is less than t. Whenever $CT[i]$ is set equal to nil, $upper$ is set equal to the maximum of its old value and the old value of $CT[i]$. So, at time t'' when $CT[i]$ is set to nil, $upper$ is set to a value which is at least u. By Lemma 1, the values taken on by $upper$ are increasing. Therefore, at time t, the value of $upper$ is at least u. ∎

Lemma 3 *In any execution, for all i, the sequence of values taken on by $CT[i]$, excluding nil, is increasing.*

Proof: Consider any interval of time in which $CT[i]$ is not set equal to nil. Clearly the values taken on by $CT[i]$ form an increasing sequence. Consider an interval of time from t_1 to t_2 during which $CT[i]$ is nil. Let u_1 be the (non-nil) value of $CT[i]$ just prior to t_1 and u_2 be the (non-nil) value of $CT[i]$ just after t_2. By Lemma 2, the value of $upper$ at time t_2 is at least as large as u_1. By the code, the value of $upper$ at time t_2 is smaller than u_2. Thus $u_2 \geq u_1$. ∎

Theorem 4 *The algorithm never accepts the same message more than once and never accepts a late message in any execution.*

Proof: First consider the case of duplicates. Suppose two copies of the message (m, u, i) arrive at the server and the first copy is accepted. Then $CT[i]$ is set equal to u. Consider the time t when the second copy arrives. If $CT[i]$ is not nil at time t, then by Lemma 3, $CT[i] \geq u$, and thus the second copy is rejected. If $CT[i] = nil$ at time t, then by Lemma 2, $upper \geq u$, and thus the second copy is rejected. The case of late messages is straightforward. ∎

Lemma 5 *Let t_1 be any garbage collection time in any execution. Let B be the new value of age at time t_1. If t_2 is a garbage collection time such that $t_1 + B < t_2$, and age is non-increasing in $[t_1, t_2]$ (inclusive), then no message is lifetime-rejected in $[t_1 + B, t_2]$.*

Proof: Since *age* is non-increasing in $[t_1, t_2]$, this implies that B is an upper bound on the value of *age* in this interval. Also, no message with lifetime greater than the current value of the variable *age* is received in the time interval $[t_1, t_2]$. Consider a particular message which arrives at time t in the interval $[t_1 + B, t_2]$. Let s be the value of *age* at time t. Then, the timestamp of the message is at least $t - s$, since *age* is not increased in this interval. To show that this message is not lifetime-rejected, it is sufficient (by Lemma 1) to show that the value of *upper* at time t is less than $t - s$. Note that the value of *upper* at time t_1 is at most t_1.

If there is no change in the value of *upper* between times t_1 and t, the value of *upper* at time t is still at most t_1. Since $t > t_1 + B$ and $s \leq B$, it follows that $t - s > t_1$. Therefore, the value of *upper* at time t is less than $t - s$.

Suppose there is a change in the value of *upper* between times t_1 and t. (All changes to upper are made in garbage collection.) Let t' be the time of the last change to *upper* in the interval and let s' be the final value of *age* at time t'. Since $t > t'$ and $s' \geq s$, it follows that $t - s > t' - s$. Therefore, the value of *upper* at time t is less than $t - s'$. ∎

Theorem 6 *Consider any execution of the algorithm. Let t and X be such that the execution is (t, X)-bounded. Then there exists some real time $t' > t$ such that the value of the variable age at every time after t' is at most $2X$.*

Proof: Suppose the value of *age* is always less than $2X$ after time t. Then, we are done. Otherwise, let $t_1 > t$ be a point in time when the value of *age* is at least $2X$.

We show that at some point t'' after t_1, $numacc > p \cdot numrej$. Let s be the value of *age* at time t_1. Suppose the value of *age* decreases at some point after t_1. This means that $numacc > p \cdot numrej$ at that time. Otherwise, suppose *age* is never decreased. Since the execution is (t, X)-bounded, every message received after t_1 has lifetime at most X, and thus *age* is never increased. Then Lemma 5 applies to any interval starting at t_1, with $B = s$, and no message that arrives after $t_1 + s$ is lifetime-rejected. Therefore, at some point $t'' \geq t_1$, $numacc > p \cdot numrej$.

Now, at every garbage collection after t'', the server sets *age* to be the smallest power of 2 at least as large as the largest message lifetime it observed since the last garbage collection. By the (t, X)-bounded assumption, the largest lifetime observed is X, and thus *age* is always set to at most $2X$. So, starting at the time t' of the first garbage collection after t'', *age* is at most $2X$. ∎

Plotting time versus the value of *age* produces a step function. (See Figure 1.) Let u_i identify the point in time when *age* is decreased for the i-th time. At this time, *numrej* and *numacc* are reset and p is incremented. Let $u_0 = 0$. Define the i-th interval I_i to be $[u_{i-1}, u_i]$, $i \geq 1$. Consider the behavior of *age* in I_i. I_i consists of a sequence of one or more maximal **sub-intervals** which have the property that *age* is constant within each sub-interval. Each sub-interval consists of a sequence of consecutive garbage collection periods.

Let r_i be the number of messages lifetime-rejected in I_i. Notice that this is equal to the value of *numrej* at the end of I_i just before it is reset to 0. The variable p is set to the value i at the beginning of I_i. Let a_i be the amount subtracted from *numacc* at the end of I_i, namely $i \cdot r_i$.

Lemma 7 *Consider any execution that is (t, X)-bounded for some t and X. Then there exists a constant Z (depending on X) such that for all i with I_i beginning after t, $r_i \leq Z$.*

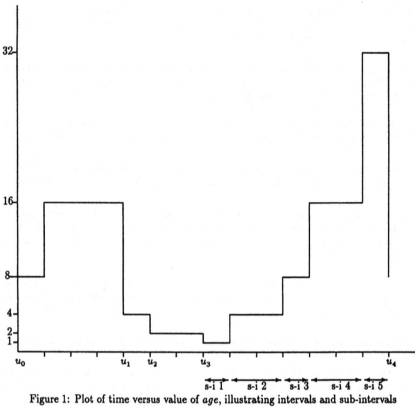

Figure 1: Plot of time versus value of *age*, illustrating intervals and sub-intervals

Proof: Consider any i with I_i beginning after t. I_i consists of one or more sub-intervals, where the value of *age* in each sub-interval is larger than the value in the preceding sub-interval. Each time *age* is increased, *age* is at least doubled, and the maximum value that *age* can reach is $2X$. Thus there are at most $\lceil \log X \rceil + 1$ sub-intervals in I_i.

Now, since *age* is constant in each sub-interval, the sub-intervals satisfy the condition of Lemma 5. Therefore, we can apply Lemma 5, with $B = 2X$, to each sub-interval (note that the maximum value of *age* in each sub-interval is at most $2X$). So, it follows that, in each sub-interval, messages are only lifetime-rejected in the last garbage collection period before *age* is increased, and in the first $2X$ time units after the beginning of the sub-interval. Therefore, the messages can be lifetime-rejected for at most $2X + G$ time in every sub-interval, where G is the maximum length of a garbage collection period. Thus the maximum number of messages that can be lifetime-rejected in I_i is $Z = (\lceil \log X \rceil + 1) \cdot (2X + G) \cdot R$, where R is the maximum rate at which the server receives messages. ∎

Theorem 8 *Consider any execution of the algorithm. Let t and X be such that the execution is (t, X)-bounded. Then the limit of L_u/A_u as u goes to infinity is 0.*

Proof: We must show that for any ϵ, there exists a time t_ϵ such that for all $u \geq t_\epsilon$, $L_u/A_u \leq \epsilon$.

Suppose there exists a time $t_f \geq t$ after which *age* never changes. By Lemma 5, after time $t_f + s_f$, where s_f is the value of *age* at t_f, no message is lifetime-rejected and every non-duplicate message is accepted. Let l be the number of messages lost up to time $t_f + s_f$. Fix ϵ. Let t_ϵ be the first time after $t_f + s_f$ by which at least a messages have been accepted, where $l/a \leq \epsilon$. By the assumptions, t_ϵ exists and for every time $u \geq t_\epsilon$, $L_u/A_u \leq \epsilon$.

Now suppose that *age* changes infinitely often after time t'. Clearly *age* is always at least 1. By Lemma 6, *age* never exceeds $2X$ after time t'. Thus, *age* increases infinitely often and decreases infinitely often.

We show that for all $\epsilon \geq 1/q$, where q is an integer, there exists t_ϵ such that $L_u \cdot q \leq A_u$ for all $u \geq t_\epsilon$.

Pick any u such that u is in I_i for some i and I_i begins after t. Then $L_u \leq r_1 + \cdots + r_{i-1} + r_i$ and $A_u \geq a_1 + \cdots + a_{i-1}$. By definition, $a_j = j \cdot r_j$ for all j. Thus

$$
\begin{aligned}
a_1 + \cdots + a_q &= 1 \cdot r_1 + \cdots + q \cdot r_q \\
&= q \cdot (r_1 + \cdots + r_q) - Q,
\end{aligned}
$$

where $Q = (q-1) \cdot r_1 + (q-2) \cdot r_2 + \cdots + r_{q-1}$.
Also,

$$
\begin{aligned}
a_{q+1} + \cdots + a_{i-1} &= (q+1) \cdot r_{q+1} + \cdots + (i-1) \cdot r_{i-1} \\
&= q \cdot (r_{q+1} + \cdots + r_{i-1}) + R_{i-1},
\end{aligned}
$$

where $R_{i-1} = r_{q+1} + 2 \cdot r_{q+2} + \cdots + (i-1-q) \cdot r_{i-1}$.
Combining these we get that

$$
a_1 + \cdots a_{i-1} = q \cdot (r_1 + \cdots r_{i-1}) - Q + R_{i-1}.
$$

Thus,

$$
\begin{aligned}
L_u \cdot q &\leq q \cdot (r_1 + \cdots + r_{i-1}) + q \cdot Z, \quad \text{since } r_i \leq Z \text{ by Lemma 7} \\
&= a_1 + \cdots a_{i-1} + Q - R_{i-1} + q \cdot Z \\
&\leq A_u + Q - R_{i-1} + q \cdot Z.
\end{aligned}
$$

We have two cases. If there exists m such that $r_j = 0$ for all $j > m$, then L_u is bounded above by some constant for all u. Therefore, $L_u \cdot q$ is bounded above by some constant for all u. Since A_u approaches infinity as u increases, there is some t_ϵ such that $L_u \cdot q \leq A_u$ for all $u \geq t_\epsilon$.

If, on the other hand, there is is an infinite number of non-zero r_j's, then R_{i-1} approaches infinity as i increases. Since Q and $q \cdot Z$ are constants, there is some i_0 such that for all $i \geq i_0$, $R_{i-1} \geq Q + q \cdot Z$. Let t_ϵ be the beginning of I_{i_0}. For all times $u \geq t_\epsilon$, $L_u \cdot q \leq A_u$. ∎

3.2 Algorithm 2

Algorithm 2 achieves the three correctness conditions for any fixed S and H and $\delta = 1/p$ for any $p \leq (S - H)/H$. The two constants S and H are embedded in the server's code.

As in Algorithm 1, the allowable values of the lifetime estimate are powers of 2. The server keeps a multiset of the lifetimes of all the messages received during the current garbage collection interval. When S messages have been received (and either accepted or lifetime-rejected), then the server tries to update the lifetime estimate and garbage collects the connection table. In determining whether to adjust the estimate, the server discards the H largest elements in the multiset of recorded lifetimes. The largest remaining element is compared against the current estimate. If it is larger, then the estimate is updated as in Algorithm 1. If it is smaller and the following condition is met, then the estimate is updated as in Algorithm 1.

The condition to be met in order to decrease the estimate is the same as the condition from Algorithm 1. However, the variable p has a fixed value throughout. As will be shown, this will allow the ratio of lost to accepted messages to converge to $1/p$, as long as p is at most $(S - H)/H$. No advantage regarding the ratio can be gained with a larger p since H out of every S messages might be lifetime-rejected simply due to transient spikes in the lifetimes.

Whenever the estimate is decreased, *numrej* is set to 0 and *numacc* is decreased by p times *numrej* plus 1. The plus 1 is important for achieving the desired ratio—it allows us to build up a reserve of extra accepted messages to counteract temporary surges in the number of lifetime-rejected messages.

ALGORITHM 2:

State variables:

    clock: integer, current value of the clock
    $CT[1..n]$: array of integer, initially all entries are *nil* (say -1)
    age: powers of 2, initially 1
    p: integer, initially some predetermined value at most $(S - H)/H$
    upper: integer, initially 0
    numacc: integer, initially 0
    numrej: integer, initially 0
    maxage: integer, initially 0
    ageMS: multiset of integer, initially empty

Transitions:

  Receive message m from client i with timestamp t:
       $ageMS := ageMS \cup \{clock - t\}$
       if $((CT[i] \neq nil)$ and $(t \leq CT[i]))$ then
           reject m
       else if $((CT[i] \neq nil)$ and $(t > CT[i]))$ or $((CT[i] = nil)$ and $(t > upper))$ then
           accept m
           $numacc := numacc + 1$
           $CT[i] := t$
       else if $((CT[i] = nil)$ and $(t \leq upper))$ then
           reject m
           $numrej := numrej + 1$

  Garbage collect CT: (do when *ageMS* has exactly S entries)
       $maxage := select(H + 1, ageMS)$
       if $maxage > age$ then
           $age := 2^j$, where $j = \min\{k : 2^k \geq maxage\}$
       else if $(numacc > p \cdot numrej)$ and $(age > maxage > 0)$ then
           $age := 2^j$, where $j = \min\{k : 2^k \geq maxage\}$
           $numacc := numacc - p \cdot numrej - 1$

$numrej := 0$
$ageMS :=$ empty
for $i = 1$ to n do
  if $CT[i] \leq clock - age$ then
    $upper := \max(upper, CT[i])$
    $CT[i] := nil$

The analysis of Algorithm 2 parallels that of Algorithm 1. It depends on the notion of a "normal" message, a message whose lifetime is not among the H highest lifetimes in the multiset when garbage collection occurs.

Theorem 9, stating that no duplicate or late messages are accepted, is proved exactly the same way as Theorem 4.

Lemma 10, analogously to Lemma 5, gives an upper bound on the time after the estimate stabilizes until no more normal messages are lifetime-rejected. Theorem 11 states that eventually the estimate converges, and is proved similarly to Theorem 6 using Lemma 10.

Lemma 12, analogously to Lemma 7, gives a constant upper bound on the number of normal messages that can be lifetime-rejected in any interval. (The constant is slightly different however.) It is used in the proof of Theorem 13, which states that the ratio of lost to accepted messages approaches $1/p$. The proof, as before, breaks the execution up into intervals. We want to show that after some time, the number of messages lost times p is at most the number of messages accepted. The number of messages lost up to some time in the i-th interval is at most the number of messages lifetime-rejected in all the previous intervals, plus the number of messages lost so far in the i-th interval. The number of messages accepted up to this time is at least the reduction in $numacc$ in all the previous intervals, plus the number of messages accepted so far in the i-th interval. As far as all the previous intervals are concerned, the numbers work out well, since by the code, the number of messages lifetime-rejected in each interval times p is at most the reduction in $numacc$ plus 1. Thus by the $i - 1$-st interval, we have a surplus of *at least* $i - 1$ accepted messages. We must now consider what has happened in the i-th interval up to the time of interest. Suppose the time is in the k-th garbage collection interval of the interval. Then the number of messages lost in the i-th interval so far is at most Z (the maximum number of normal messages lifetime-rejected) plus $k \cdot H$ (the number of abnormal messages received). But the number of messages accepted in the i-th interval so far is at least $(k-1)(S-H)$ (the total number of normal messages received) minus Z (the maximum number of normal messages lifetime-rejected). After multiplying the number of messages lost by p, the term $p \cdot k \cdot H$ cancels out the term $(k-1)(S-H)$ (because of the upper bound on p), leaving only a constant. Thus the $i - 1$ term, which grows without bound, eventually swamps all the constants.

Definition 2 *A message received during some garbage collection interval is* normal *if its lifetime is not among the H highest lifetimes in the multiset at the end of the garbage collection interval.*

Theorem 9 *The algorithm never accepts the same message more than once and never accepts a late message in any execution.*

Proof: The proof is exactly the same as for Algorithm 1. ∎

To show the desired liveness behavior of Algorithm 2 in (t, X, S, H)-bounded executions, we need a modified version of Lemma 5, which only refers to normal messages.

Lemma 10 *Let t_1 be any garbage collection time in any execution. Let B be the new value of age at time t_1. If t_2 is a garbage collection time such that $t_1 + B < t_2$, and age is non-increasing in $[t_1, t_2]$ (inclusive), then no normal message is lifetime-rejected in $[t_1 + B, t_2]$.*

Proof: The proof is essentially the same as before. ∎

The next theorem is the analog of Theorem 6.

Theorem 11 *Consider any execution of the algorithm. Let t and X be such that the execution is (t, X, S, H) bounded. Then there exists some real time $t' > t$ such that the value of the variable age at every time after t' is at most $2X$.*

Proof: The proof is essentially the same as before. ∎

To show that L_u/A_u converges to $\delta = 1/p$ in any (t, X, S, H)-bounded execution, we must modify Lemma 7 to refer only to normal messages. (The definitions of I_i, r_i, and a_i are the same as before.)

Lemma 12 *Consider any execution that is (t, X, S, H)-bounded for some t and X. Then there exists a constant Z (depending on X) such that for all i with I_i beginning after t, the number of normal messages lifetime-rejected in I_i is at most Z.*

Proof: The proof is essentially the same as before, except that Z is equal to $(\lceil \log X \rceil + 1)(2XR + S)$, since S is the maximum number of messages received in any garbage collection interval. ∎

Theorem 13 *Consider any execution of the algorithm. Let t and X be such that the execution is (t, X, S, H) bounded. Then the limit of L_u/A_u as u goes to infinity is $1/p$.*

Proof: We show that for the fixed p, there exists t_0 such that $L_u \cdot p \leq A_u$ for all $u \geq t_0$.

Pick any u, and let i be such that u is in I_i and I_i begins after time t. Let k be such that u occurs in the k-th garbage collection interval of I_i.

The number of messages lost in I_i up to time u is at most the number of messages lifetime-rejected in I_i, which is the number of normal messages lifetime-rejected plus the number of abnormal messages lifetime-rejected. The number of normal messages lifetime-rejected in I_i up to time u is at most Z, by Lemma 12. The number of abnormal messages lifetime-rejected in I_i up to time u is at most $k \cdot H$, the total number of abnormal messages received in the first k garbage collection intervals of I_i. Thus,

$$L_u \leq r_1 + \cdots + r_{i-1} + Z + k \cdot H.$$

The number of messages accepted in I_i up to time u is at least the number of normal messages accepted, which is at least the number of normal messages received minus Z, by Lemma 12. The number of normal messages received in I_i up to time u is at least $(k-1) \cdot (S-H)$. The reason is that in each of the $k-1$ preceding garbage collection intervals, $S - H$ normal messages are received, and in the worst case, no normal message have yet been received in the k-th garbage collection interval. Thus,

$$A_u \geq a_1 + \cdots + a_{i-1} + (k-1) \cdot (S-H) - Z.$$

By definition, $a_j = p \cdot r_j + 1$ for all j. Thus,

$$
\begin{aligned}
L_u \cdot p &\leq p \cdot r_1 + \cdots + p \cdot r_{i-1} + p \cdot Z + p \cdot k \cdot H \\
&= a_1 + \cdots + a_{i-1} - (i-1) + p \cdot Z + p \cdot k \cdot H \\
&\leq A_u - (k-1) \cdot (S-H) + Z - (i-1) + p \cdot Z + p \cdot k \cdot H \\
&< A_u - (i-1) + (p+1) \cdot Z - (k-1) \cdot (S-H) + k \cdot (S-H), \quad \text{since } p < (S-H)/H \\
&= A_u - (i-1) + (p+1) \cdot Z + S - H.
\end{aligned}
$$

Note that $i - 1$ goes to infinity, while $(p+1) \cdot Z + S - H$ is a constant. Thus there is some i_0 such that for all $i \geq i_0$, $i - 1 \geq (p+1) \cdot Z + S - H$. Thus for all times u starting with interval i_0, $L_u \cdot p \leq A_u$. ∎

4 Lower Bound

The proofs of Theorems 6 and 11 give us some information about how long it takes the estimate *age* to converge to less than twice the actual upper bound. In both cases, the time required for the estimate to converge depends on how long it takes to accept a certain number of messages. It might be more desirable for this amount of time to be a constant, but in this section we show that that is not possible. Specifically, we show that for any standard-form algorithm, there can be no bound, based on the previous history, between the time t after which the lifetime is at most X and the time t' after which the estimate is at most $2X$.

To state the next theorem, we need two more definitions. Let q_u be the state of the server at time u, for any u.

Let M be such that in any garbage collection interval in any execution, it is possible for the server to receive at least M messages. In general, each garbage collection interval may have a different upper bound on the number of messages that can be received; M is the minimum of all these upper bounds. (For Algorithm 1 as presented, since no lower bound is specified on the time between garbage collections, technically M is 0. However, if garbage collections were evenly spaced G apart, then M would be RG. For Algorithm 2, in every garbage collection interval exactly S messages are received; thus M is S.)

The theorem actually only holds for algorithms that try to achieve a ratio of lost to accepted messages that is smaller than M, but any reasonable algorithm would achieve a ratio much smaller than this.

Theorem 14 *Choose any standard-form algorithm satisfying the three conditions for $H = 0$ and any $\delta < M$. Let B be any function mapping states of the server to integers. Then there exists for some t, X, a (t, X)-bounded execution of the algorithm in which $t' - t > B(q_t)$, where t' is the time after which the estimate is always less than $2X$.*

Proof: Suppose, for contradiction, that in any (t, X)-bounded execution for any t and X, the time t' after which the estimate is at most $2X$ is such that $t' - t \leq B(q_t)$. We construct an infinite execution that violates Condition 3 as follows.

Let e_0 be any execution that is (t_0, X)-bounded for some t_0 and X and in which no messages are received in the interval $[t_0, t_0 + B(q_{t_0})]$.

Suppose execution e_{i-1}, $i > 0$, is defined so that it is (t_{i-1}, X)-bounded for some t_{i-1} and no messages arrive in $[t_{i-1}, t_{i-1} + B(q_{t_{i-1}})]$. Let t'_{i-1} be the time in that interval after which *age* is less than $2X$.

Let e_i be an execution that branches off from e_{i-1} at time t'_{i-1} in which M messages with lifetimes equal to $2X$ arrive in the next garbage collection interval and are lost. Furthermore, suppose that there is a subsequent time t_i at which a message with lifetime equal to X arrives, no messages arrive in the interval $[t_i, t_i + B(q_{t_i})]$, and e_i is (t_i, X)-bounded.

Let e be the limit of e_i as i approaches infinity. Note that e is $(t_0, 2X)$-bounded and in each interval $[t'_i, t'_{i+1}]$ one message is accepted and M messages are lost. Thus in e, L_u / A_u approaches M as u increases. Since $M > \delta$, we have a contradiction. ∎

A similar theorem can be shown even for a bound that does depend on the history after time t, as long as it is independent of the number of messages received after time t.

5 Acknowledgments

We thank Abel Weinrib, David Feldmeier, and the conference reviewers for helpful comments on earlier versions of this paper, and Liuba Shrira for clarifying for us the relationship between at-most-once message delivery and at-most-once remote procedure calls. The comments of Abel Weinrib on a very early draft inspired us to the development of our second, more practical algorithm.

References

[HSSD84] Joseph Halpern, Barbara Simons, Ray Strong, and Danny Dolev. Fault-Tolerant Clock Synchronization. In *Proceedings of the Third Annual ACM SIGACT-SIGOPS Symposium on Principles of Distributed Computing*, pages 89–102, August 1984.

[Jai86] Raj Jain. Divergence of Timeout Algorithms for Packet Retransmissions. In *Proceedings of the Fifth Annual International Phoenix Conference on Computers and Communications*, pages 174–179, March 1986.

[Lis91] Barbara Liskov. Practical Uses of Synchronized Clocks. In *Proceedings of the Tenth Annual ACM SIGACT-SIGOPS Symposium on Principles of Distributed Computing*, pages 1–9, August 1991. Invited talk at 1990 PODC Symposium.

[LSW91] Barbara Liskov, Liuba Shrira, and John Wroclawski. Efficient At-Most-Once Messages Based on Synchronized Clocks. *ACM Transactions on Computer Systems*, 9(2):125–142, 1991.

[Mil89] D. L. Mills. Internet Time Synchronization: the Network Time Protocol. Technical Report RFC 1129, Network Working Group, University of Delaware, October 1989.

[ST87] T. K. Srikanth and Sam Toueg. Optimal Clock Synchronization. *Journal of the ACM*, 34(3):626–645, 1987.

[WL88] Jennifer L. Welch and Nancy A. Lynch. A New Fault-Tolerant Algorithm for Clock Synchronization. *Information and Computation*, 77(1):1–36, 1988.

[Zha86] Lixia Zhang. Why TCP Timers Don't Work Well. In *Proceedings of ACM SIGCOMM Symposium*, pages 397–405, 1986.

Uniform Dynamic Self-Stabilizing Leader Election

(Extended Abstract)

Shlomi Dolev
Dept. of Computer Science
Technion — Israel

Amos Israeli
Dept. of Electrical Engineering
Technion — Israel

Shlomo Moran
Dept. of Computer Science
Technion — Israel

Abstract

A distributed system is *self-stabilizing* if it can be started in any *possible* global state. Once started the system regains its consistency by itself, without any kind of an outside intervention. The self-stabilization property makes the system tolerant to faults in which processors crash and then recover spontaneously in an arbitrary state. When the intermediate period in between one recovery and the next crash is long enough the system stabilizes. A distributed system is *uniform* if all processors with the same number of neighbors are identical. A distributed system is *dynamic* if it can tolerate addition or deletion of processors and links without reinitialization. In this work we present three dynamic, uniform, self-stabilizing protocols for leader election: The first protocol works on complete graphs. The second protocol works for systems with unbounded number of processor in which the size of the memory of a processor is unbounded. The third protocol works for systems whose communication graph has a bounded diameter; it uses a bounded amount of memory. We conclude this work by presenting a simple, uniform, self-stabilizing *ranking* protocol.

1 Introduction

A *distributed system* is a set of state machines, called *processors*. Each processor can communicate with some other processors, called its *neighbors*. Communication is held by use of shared variables. The system's *communication graph* is the graph formed by representing each processor as a node and by connecting each pair of neighbors by an edge. Each processor in the system is viewed as a RAM executing a program, the collection of these programs is a *protocol*.

A distributed system is *self-stabilizing* if it can be started in any *possible* global state. Once started the system regains its consistency by itself, without any kind of an outside intervention. The self-stabilization property makes the system tolerant to faults in which processors crash and then recover spontaneously in an arbitrary state. When the intermediate period in between one recovery and the next crash is long enough the system stabilizes. A distributed system is *uniform* if all processors with the same number of neighbors are identical. A system is *semi-uniform* if exactly a single processor is designated as a *special processor* and if all *normal* processors with the same number of neighbors, are identical. A system is *id-based* if every processor has a distinct identifier.

A distributed system is *dynamic* if it can tolerate addition or deletion of processors and links without reinitialization. Self-stabilizing protocols that tolerate dynamic topology changes are very useful. Following each dynamic change the system stabilizes and functions normally until the next change occurs. Semi-uniform systems are not dynamic since the special processor cannot be deleted. Id-based systems may be dynamic only if processor addition is a global procedure in which all existing id-s are scanned to allow a choice of a new distinct id for each added processor. In our protocols, on the other hand, the solutions for these problems are integrated into the protocols thus leader re-election and processor renumbering are executed routinely after each topology change as part of the system's normal operation.

The study of self-stabilizing systems started with the fundamental paper of Dijkstra, [Dij-74]. In [Dij-74] Dijkstra gave the basic definitions for self-stabilizing systems and presented three self-stabilizing, mutual-exclusion protocols for ring systems. A key observation made in [Dij-74] is that due to symmetry considerations there is no deterministic, uniform, self-stabilizing, mutual exclusion protocol for rings of composite size. Therefore Dijkstra has adopted the semi-uniform model. Most of the subsequent works use the semi-uniform model. Among them are [Kr-79, BGW-87, Bu-87, DIM-90, DIM-91]. Other works have taken a further step and assume the id-based model; those are the works of [La-86, AKY-90, AG-90].

The prime target aimed for, when using self-stabilizing protocols is fault tolerance. Our goal in this paper is to boost-up the system's fault tolerance by enabling the execution of any self-stabilizing semi-uniform or id-based protocol (including all aforementioned protocols) on dynamic systems. To do this we present uniform self-stabilizing protocols for *leader election* and *ranking*. Once these protocols are available our goal will be achieved by using *fair protocol composition*- a technique presented in [DIM-90]. All semi-uniform protocols can be executed on a uniform system when composed with the leader election protocol. All id-based protocols can be executed on a uniform system when composed with the ranking protocol. By arguments similar to those used in [Dij-74] it is not hard to see that no uniform, deterministic, self-stabilizing leader election protocols exists; the same holds for ranking. Therefore our protocols will be *randomized*. In proving the correctness of our protocols we assume a very weak atomicity - namely, we assume that tossing a coin is a single step, separated from the following write or read operation. Proving correctness in this model ensures also correctness under stronger assumptions. See e.g. [CIL-87], [Ab-88].

In this work we present three uniform, dynamic, self-stabilizing protocols for leader election. The complexity of our protocols is analyzed using round complexity (defined precisely in the next section). The first protocol works on complete graphs; its round complexity is $O(n \log n)$, where n is the number of processors in the system. The second protocol works for systems with any communication graph and any number of processors in this protocol the size of the memory of a processor is unbounded. Its round complexity is $O(\Delta D \log n)$, where Δ is the maximum degree of a processor in the systems communication graph and D is the *actual* diameter of the communication graph after the last topology change. Both protocols work without any assumption on the number of processors in the system.

The third protocol is a modification of the second protocol which assumes a bound on the number of processors and uses bounded amount of memory. Its round complexity is $O(\Delta D \log n)$. In a model in which each atomic step can generate up to $\log n$ random bits, the complexity of that protocol is $O(\Delta D)$. We conclude this work by presenting a simple uniform self-stabilizing *ranking* protocol. This protocol can use each of the leader choosing protocols as a subroutine and can be run on any systems which runs one of the leader choosing protocols. The ranking protocol issues distinct identities to the processors in the system. The issued identities are in the range 1 to n, where n is the number of the processors in the system after the last topology change.

Leader election is a fundamental task in distributed computing. Many well known deterministic id-based protocols for this task appear in the literature, e.g [Ga-78, Pe-82, Hu-84, KMZ-84, KKM-90]. Those protocols assume that the system starts from some predetermined initial state; processor id-s are used to break symmetry. Uniform randomized protocols for leader election are presented in

[IR-81, SS-89]; All those protocols are not self-stabilizing. A uniform, deterministic, self-stabilizing, mutual exclusion protocol for rings of prime size appear in [BP-88]. This protocol is not dynamic since it works only on prime rings. Randomized, uniform, self-stabilizing protocols for mutual exclusion in a general graph and for ring orientation appear in [IJ-90] and [IJ-90a] respectively.

Our unbounded protocol is the only protocol which solves the problem without any prior knowledge on the communication graph. Assuming some known bound on the graph's diameter solutions to the same problem have been suggested independently in [AKY-90], [AE-91] and in [V-91]. The complexity of all these solutions is inferior to the complexity of our bounded protocol.

The rest of this extended abstract is organized as follows: In Section 2 the formal model and requirements for uniform self-stabilizing protocols are presented. In Section 3 we present a new two player game which is helpful in proving correctness and analyzing the complexity of randomized protocols. Section 4 presents the protocol for complete graphs. Section 5 presents the protocols for general graphs. The ranking protocol is sketched in Section 6. For the sake of brevity, most of the proof in this abstract are omitted.

2 Model and Requirements

A *uniform distributed system* consist of n processors denoted by $P_1, P_2, \cdots, P_n$. Processors are *anonymous*, i.e. they do not have identities. The subscript $1, 2, \cdots, n$ are used for ease of notation only. Each processor can communicate with some other processors, called its *neighbors*. The system's *communication graph* is the graph formed by representing each processor as a node and by connecting each pair of neighbors by an edge. Each processor is a state machine and all processors with the same degree are *identical*. Communication between neighbors is carried out using *shared communication registers* (called *registers* throughout this work). Each register is serializable with respect to **read** and **write** operations. Each pair of neighbors is connected by a *link* composed of two registers. One neighbor writes in the first register and reads from the second, the other neighbor reads from the second register and writes to the first, hence each link supports two-way communication.

For ease of presentation we regard each processor as a RAM whose program is composed of *atomic steps*. An atomic step of a processor consists of an internal computation followed by one of the following actions: **read**, **write** or **coin toss**. The registers of a processor are the registers in which the processor writes. We assume that the state of a processor fully describes its internal state and the values written in all its registers. Denote by S_i the set of states of P_i. A *configuration*, $c \in (S_1 \times S_2 \times \cdots S_n)$, of the system is a vector of states of all processors.

Processor activity is managed by a scheduler. In any given configuration the scheduler activates a single processor which executes a single atomic step. To ensure correctness of our protocols, we regard the scheduler as an adversary. We let the scheduler choose *on line* the next activated processor using the system configuration as its input. An execution of the system is a finite or infinite sequence of configurations $E = (c_1, c_2, \cdots)$ such that for $i = 1, 2, \cdots$, c_i is reached from c_{i+1} by a single atomic step of a processor, $i = 1, 2, \cdots$. A *fair execution* is an execution in which every processor executes atomic steps infinitely often. A scheduler is *fair scheduler* if for any configuration c, an execution starting from c in which processors are activated by S is fair with probability 1.

In a distributed system each processor may execute atomic steps in any non constant rate. Various processors might be slow in various parts of the execution. The following definition of round complexity attempts to capture the rate of action of the slowest processor in any segment of the execution. Given an execution E we define the *first round* of E to be the minimal prefix of E, E', containing atomic steps of every processor in the system. Let E'' be the suffix of E for which $E = E' \circ E''$. The second round of E is the first round of E'', and so on. For any given execution,

E, the *round complexity* (which is sometimes called the execution time) of E to be the number of rounds in E.

We proceed by defining the self stabilization requirements for randomized distributed system. A behavior of a system is specified by a set of executions. Define a *task LE* to be a set of executions which are called *legitimate executions*. A configuration c is *safe* with respect to a task LE and a protocol PR if *any* fair execution of PR starting from c belongs to LE. Define now the randomized self stabilization requirement as:

Randomized Self Stabilization -
A protocol PR is randomized self stabilizing for a task LE if starting with any system configuration and considering any fair scheduler, the protocol reaches a safe configuration within an expected number of rounds which is bounded by some constant C. (The constant C may depend on n, the number of processors in the systems)

3   Scheduler-Luck Games

In this section we introduce a new method to analyze randomized distributed protocols. Towards this end we consider a two player game, called *sl-game*. An *sl-game* is determined by a protocol and by a set of *safe configurations*. In the context of self stabilization the set of safe configurations is determined by the protocol and some unspecified task, but this is not essential for using the method. The game is played as follows: The states of the game are system configurations; the two competitors are called *scheduler* (adversary) and *luck*, their opposing goals are to prevent the protocol from reaching a safe configuration and to help it reach a safe configuration, respectively. In each configuration the scheduler chooses the next activated processor. The only restriction on the schedulers behavior is the requirement for fairness. Each time the processor, activated by the scheduler, tosses a coin *luck* may (but does not have to) intervene and determine the result of the coin toss. Each execution of the game corresponds naturally to an execution of the protocol (but not vice versa). During the execution each of the players knows the configuration of the system and uses this knowledge to decide on its next step. The game terminates (if at all) when the system reaches a safe configuration. A similar, but more restricted, game appears in [Ab-88] for a different model and without any analysis.

In order to analyze the performance of our protocols, we say that an execution E takes r rounds if r is the number of rounds initiated in E.

Definition: Let SF and PR be a given set of safe configurations and a protocol, respectively. We say that *luck* has an *(f,r)-strategy* for SF and PR if for any initial configuration c and for every scheduler S, the *sl*-game (between S and *luck*) reaches a safe configuration in expected number of at most r rounds, and within at most f intervention of *luck*.

Lemma 1:   If *luck* has an (f,r)-strategy then the system is randomized self stabilizing within $r2^f$ expected number of rounds.

Proof: Let E be a given execution. Define the *first block* of E to be a prefix of E, B, satisfying one of the following:

1. **good block:** B is the minimal prefix of E which forms an *sl*-game which ends in a safe configuration (provided there is such a prefix), or

2. **bad block:** B is either (a) an infinite *sl*-game which does not reach a safe configuration, or (b) the minimal prefix of E which is not a prefix of any *sl*-game (note that in the latter

case a bad block must end with an atomic operation which contains a coin toss, where in the corresponding sl-game *luck* chooses to intervene, and the result of that coin toss is not the result determined by *luck*).

Every execution E must start with a good or a bad block. In either case, denote this block by B_1. In case that B_1 is a good block, then when it terminates the system is in safe configuration. Otherwise, B_1 is a bad block. If B_1 is finite, let E' be the suffix of E defined by $E = B_1 \circ E'$, and let B_2 be the first block of E'. Again, if B_2 is a good block then when it ends the system is in a safe configuration. Continuing this way, we associate with E a sequence $\mathcal{B} = (B_1, B_2, \cdots)$ of blocks, such that either $\mathcal{B}$ is a (possibly infinite) sequence of bad blocks, or $\mathcal{B}$ consist of l blocks, out of which the first $l - 1$ blocks are bad and the lth block is good.

Given a configuration c, a scheduler S and *luck* l, define $GT(c,S,l)$, the *sl-game tree* of c, S and l to be a directed tree that its nodes are configurations and its root is c. For each node c_i in a tree T, $PATH(T, c_i) = c, c_1, \cdots, c_i$ is the directed path from the root to c_i in T. c_{i+1} is a son of c_i in $GT(c,S,l)$ if immediately following the execution defined by $PATH(GT(c, S, l), c_i)$, the scheduler S activates a processor which executes an atomic step, a_i, that changes the system configuration from c_i to c_{i+1}. In case that during a_i coin is tossed and l does not intervene, c_i has two sons; otherwise c_i has at most one son. In addition, whenever $PATH(GT(c, S, l), c_i)$ contains a good block c_i has no sons. Each path $PATH(GT(c, S, l), c_i)$ is associated with the probability, p_i, to reach c_i in $GT(c,S,l)$. Let b be the number of nodes that have two sons in $PATH(GT(c, S, l), c_i)$, then $p_i = 2^{-b}$.

Towards the analyses of the expected number of rounds in an execution we define a set of *length functions*. A length function is denoted by $L(T, c)$ where T is a binary tree and c is a node in T. $L(T, c)$ has the following properties:
(1) If c_i is the father of c_j in T then $L(T, c_i) \leq L(T, c_j)$.
(2) If c_i and c_j have the same father in T then $L(T, c_i) = L(T, c_j)$.
In particular, the function $L_r(T, c)$ that counts the number of rounds in $PATH(GT(c, S, l), c_i)$ (that defines an execution) is a length function. Define the expected length of a leaf in T to be the sum $\Sigma L(T, c_i) p_i$ taken over the configurations that are leaves in T where p_i is the probability to reach c_i in T.

We now show that the expected length of the leaves in $GT(c,S,l)$, measured by $L_r(T, c)$, is equal to the expected number of rounds in the *sl-game* between S and l that starts with c. Note that there is 1-1 correspondence between the paths in $GT(c, S, l)$ and the executions of the *sl-game*. Moreover, the probability of some execution to occur is the probability defined for the last (if exist) node in the path that represent this execution in $GT(c, S, l)$. By the fact that the expected number of rounds in the *sl-game*, till a safe configuration is reached, is r, the probability that a *sl-game* contains infinitely many rounds is zero [1]. Thus, the probability that a safe configuration is reached following finite number of rounds is 1. Therefore, the sum of the probabilities of the leaves in $GT(c, S, l)$ is 1. Each leaf c_i in $GT(c,S,l)$ represent a good block that has probability $p_i > 0$ and the expected length of the leaves of $GT(c,S,l)$ represent the expected number of rounds in the first block of the *sl-game* of c, S and l.

Similarly to the definition of a *sl-game tree* $GT(c,S,l)$, define *blocks tree* $BT(c,S,l)$ to be a directed tree that its nodes are configurations and its root is c. $BT(c,S,l)$ contains all the nodes and links of $GT(c,S,l)$. In addition $BT(c,S,l)$ has an additional son for any configuration in which *luck* intervene. This, additional son represent the possibility that the coin toss result differs from the desired result of l. Note that there is 1-1 correspondence between the leaves in $BT(c,S,l)$ and the first (good or bad) blocks in the executions that start with c and in which S is the scheduler.

We use the trees $GT(c,S,l)$ and $BT(c,S,l)$ to prove the following claims. The claims hold for executions starting from arbitrary initial configuration c and every fair scheduler S.

---

[1] If this probability was $p > 0$ then the expected number of rounds till a safe configuration is reached would be infinite.

Claim 1: With probability at least 2^{-f}, the first block of an execution is a good block.

Proof of claim 1: Every leaf, c_i, of $GT(c, S, l)$ appears also in $BT(c, S, l)$ and $PATH(GT(c, S, l), c_i) = PATH(BT(c, S, l), c_i)$. For any such leaf c_i let b be the number of nodes with two sons in $PATH(GT(c, S, l), c_i)$. By the definition of the $(f, r) - strategy$ it holds that in $PATH(BT(c, S, l), c_i)$ the number of nodes with two sons is at most $b + f$. Let p_i be the probability of c_i in $GT(c, S, l)$. Then c_i has a probability of at least $2^{-f} p_i$ in $BT(c, S, l)$. The sum of the probabilities of the leaves of $GT(c, S, l)$ is 1, hence the sum of the probabilities of those leaves in $BT(c, S, l)$ is at least 2^{-f}. Thus, the probability that the first block in an execution is a good block is 2^{-f}. $\square$

Claim 2: The expected number of rounds in a (good or bad) block is at most r.

Proof of claim 2: The proof uses the following proposition:

Proposition: Let T and T' be two binary trees. Let σ and σ' be the expected length of a leaf in T and T', respectively. If T' is derived from T by addition of a new leaf as a son of a non-leaf node of T, then $\sigma \geq \sigma'$.

Proof of proposition: We now show that any addition of a leaf node, c_m, as a son of a non-leaf node in T may only decrease the expected length of a leaf. Let c_j be the non-leaf node which is father of c_m. Let $\{c_i \mid i \in I\}$ be the set of leaves that are descendants of c_j [2]. The only change in the calculation of the expected length of a leaf in T is due to the leaves $\{c_i \mid i \in I\}$ and to c_m, since the length and probability associated with any other leaf remains unchanged. We now compare the contribution of $\{c_i \mid i \in I\}$ and c_m in the calculation before the addition of c_m and following the addition of c_m. Before the addition of c_m the sum that $\{c_i \mid i \in I\}$ contributes to the expected length calculation is $\sigma_1 = \Sigma_{i \in I} L(T, c_i) p_i$. The addition of c_m changes the probabilities as follows:
(1) The probability associated with c_m is $\Sigma_{i \in I} (p_i / 2)$.
(2) The probability associated with each c_i $(i \in I)$ is reduced from p_i to $p_i / 2$.
Thus, the calculation now gives:
$\sigma_2 = \Sigma_{i \in I} L(T, c_i)(p_i / 2) + L(T, c_m)(\Sigma_{i \in I}(p_i / 2)) = \Sigma_{i \in I}((L(T, c_i) + L(T, c_m))/2) p_i$.
Since, $L(T, c_m) \leq L(T, c_i)$ for any $i \in I$ then $\sigma_1 \geq \sigma_2$. $\square$

We now complete the proof of claim 2. By the definition of the *sl-game* the expected length, measured by $L_r(T, c)$, of a leaf in $GT(c, S, l)$ is r rounds. $BT(c, S, l)$ is derived from $GT(c, S, l)$ by adding a leaf to any configuration in $GT(c, S, l)$ in which *luck* intervene. By the above proposition any addition of such a leaf may only decrease the expected number of rounds. The additional nodes may be ordered by lexicographic order $l_1, l_2, \cdots$. Let $T_1, T_2, \cdots$, be a (possible infinite) sequence of trees in which T_{i+1} is obtained by addition of l_i to T_i. By the above proposition it holds that the expected length of a leaf in T_i is great or equal to the expected length of a leaf in T_{i+1}. Thus, when the expected length of a leaf in $T_1 = GT(c, S, l)$ is r, the expected length of a leaf in $BT(c, S, l)$ is at most r. Therefor, the expected number of rounds in a (good or bad) block is at most r rounds. $\square$

Now we complete the proof of lemma 1. The existence of an (f, r)-strategy implies that for every possible initial state and for every fair scheduler S, an execution contains a good block with probability 1. We use Claim 1 to show further that the expected index of the first good block in an execution is at most 2^f. The above expected index may only increase if we assume that the probability that the first block in an execution to be a good block is *exactly* 2^{-f}. In this case, the probability that all the first i blocks in E are bad is $(1 - 2^{-f})^i$. Thus, the expected index of the first good block is at most $\sum_{i=1}^{\infty} i \times (1 - 2^{-f})^{i-1} \times 2^{-f} = 2^f$.

By Claim 2 and the fact that expectation of a sum is a sum of expectations, the expected number of rounds in an execution until the end of the first good block is at most $2^f \times r$. The proof is completed since when reaching the end of the first good block, the system is in a safe configuration. $\square$

---

[2] there are possibly infinitely many such leaves.

4 Leader Election in Complete Graphs

In this section we consider the special case of systems whose communication graph is complete. First we describe a simple protocol which is correct for the model which assumes that a coin toss is an internal processor action which cannot be separated from the following **read** or **write** operation. In this protocol each processor communicates with all its neighbors using a single writer multi reader binary register, called *leader*. This sums up to total $O(n)$ space complexity for a system of n processors.

Description of the first protocol: Starting the system with any possible combination of binary values of the *leader* variables of the processors, eventually the protocol fixes all the *leader* variables but one to hold 0. The single processor holding hold 1 in its *leader* variable is the leader. Each processor P_i repeatedly reads the *leader*'s of all its neighbors. Whenever P_i completes reading all these values, it checks whether one of the following conditions holds:

1. All *leader* variables (including the *leader* variable of P_i) were equal to 0 or

2. The *leader* variable of P_i holds 1 and there exist at least one more *leader* variable whose value was 1.

If one of the cases holds, P_i tosses a coin and writes the result (which is either 0 or 1) to its *leader* variable. Since P_i's program consists of at most $n + 1$ atomic steps, performed in a cyclic order (n reads and at most 1 write), it is not hard to see that if a coin-toss is not considered a separate atomic step then *luck* has an $(n, 3n)$-strategy for winning the game. Applying Lemma 1 on this game implies that the expected number of rounds guaranteed by the above strategy is $3n2^n$. In the full paper we show that indeed the scheduler has a strategy under which the expected time until this protocol reaches a safe configuration in bounded from below by a function exponential in n.

When assuming a model in which a coin toss is considered a separate atomic step, the protocol described above is not self-stabilizing. In this model the following strategy of the scheduler ensures that the protocol never stabilizes: Start the system in a configuration in which all *leader* variables hold 1. Let one processor notice that it has to toss a coin. If the coin toss result is 1 let this processor toss a coin again until the coin toss result is 0. Now stop the processor before it writes 0 in its *leader* variable and activate another processor in the same way. Once all processors are about to write 0 let them write. In a similar way the scheduler can force all processors to write 1 in their registers and so on and so forth. Thus, this strategy ensures that the system never stabilizes.

In order to overcome the above problem, we use a mechanism that guarantees some synchronization: the mechanism guarantees that between two successive coin tosses of a processor P_i, all other processors had read P_i's *leader* variable. For this we use two communication registers per link. We associate two "arrows" for each link of the communication link, each is *owned* by one of the two attached processors. The arrow associated with the link between P_i and P_j and is owned by P_i (P_j) is denoted by $arrow(i : j)$ ($arrow(j : i)$).

In all the next protocols we assume that each edge (P_i, P_j) of the communication graph is implemented by two communication registers, one of them, r_{ij}, is written by P_i and read by P_j and the other, r_{ji}, is written by P_j and read by P_i. Each time P_i reads r_{ji} it assigns the value it reads to a local variable ir_{ji}.

The implementation of $arrow(i : j)$ is by two binary fields of communication registers $r_{ij}.arrow(i : j)$ and $r_{ji}.arrow(i : j)$. Whenever $r_{ij}.arrow(i : j) = r_{ji}.arrow(i : j)$ then $arrow(i : j)$ is directed from P_j to P_i otherwise $arrow(i : j)$ is directed from P_i to P_j. A processor P_i turns the direction of an arrow (no matter if it is its own arrow, e.g. $arrow(i : j)$, or its neighbors arrow e.g. $arrow(j : i)$) only to be directed from P_i to its neighbor. Whenever a processor P_i tosses a coin it turns the direction of its own arrows (by executing the macro see_me_j) and waits for an *acknowledgment* from its neighbors i.e. turning those arrows back (till they execute ack_i). While P_i waits for the acknowledgment P_i continues to turn the arrows of its neighbors toward them (by executing ack_j).

In order to ensure that for every P_i and P_j P_i, executes ack_j only when P_j executed see_me_i and is waiting for acknowledgment we have to ensure that the arrows are stabilized. For this purpose we use a close field $r_{ij}.close$ for $arrow(i:j)$.

The protocol is presented by the code for P_i. This way of presentation is essential for proving correctness of a self stabilizing protocol since the protocol should be correct no matter what are the initial states of the processors in the system i.e. each processor might be executing any atomic step of the code. In writing the code for processor P_i we assume that it has d neighbors. Similarly to the previous protocol P_i has a binary *leader* variable that defines whether P_i is a leader or not.

```
0  for j := 1 to d do r_ij.close := 0;
1  do forever
2     for j := 1 to d do ack_j;
3     if (leader = 0 and {∀ j | ir_ji.leader = 0}) or (leader = 1 and {∃ j | ir_ji.leader = 1})
4     then
5        begin
6               coin:= choose(0,1); write(r_i.leader := coin);
7               for j := 1 to d do see_me_j;
8               while {∃ j | r_ij.arrow(i : j) ≠ ir_ji.arrow(i : j)} do
9                    for j := 1 to d do ack_j;
10       end
11 end

macro ack_j (* arrow(j : i) := (i,j) *)
12 ir_ji := read(r_ji)
13 if r_ij .arrow(j : i) ≠ ir_ji.arrow(j : i) then
14    begin
15       write r_ij.close := 1;
16       ir_ji := read(r_ji);
17       write (r_ij.arrow(j : i), r_ij.close) := (ir_ji.arrow(j : i), 0)
18    end

macro see_me_j (* arrow(i : j) := (i,j) *)
19 repeat
20        ir_ji := read(r_ji);
21 until ir_ji.close = 0;
22 write r_ij.arrow(i : j) := not ir_ji.arrow(i : j);
```

Figure 1: The Protocol for Complete Graph Systems

We define a *safe configuration* to be a configuration in which:
(1) For exactly one processor, say P_i, $leader_i = 1$ and $\{\forall j \mid ir_{ji}.leader = 0\}$.
(2) For every other processor, $P_j \neq P_i$, $leader_j = 0$ and $ir_{ij}.leader = 1$.
In such a configuration P_i is the only processor for which $leader_i = 1$ i.e. P_i is the only processor

that considers itself as a leader. Moreover, it is easy to see that in any configuration that follows a *safe configuration* P_i is the only leader.

The correctness proof of the protocol (which is omitted from this extended abstract) is by showing a wining strategy for *luck*.

Lemma 2: The system reaches a safe configuration in $O(n \log n)$ expected number of rounds.

5 Leader Election in General Graphs

5.1 Informal description of the unbounded protocol

In this protocol each system configuration c encodes a directed graph called the FSG (*father-son* relation graph) of c and denoted by $FSG(c)$. A safe configuration in this protocol is a configuration whose FSG is a single directed tree which contains all processors and which does not change anymore. The root of the FSG is the elected leader.

The protocol consists of two conceptual parts which are called *cycle elimination* and *tree elimination*. In the cycle elimination part all cycles in the FSG are removed. In the tree elimination part the number of directed trees in FSG is reduced until FSG consists of a single directed tree. Coin tosses are used in the tree elimination part in order to break symmetry between trees. Normal operation and completion of tree elimination depends crucially on completion of the first part which is undetectable by the nature of self-stabilizing protocols. On the other hand it is vitally important that the tree elimination part does not introduce new cycles.

FSG is defined by the processors using a relation between neighbors called *father-son* relation. Each processor can either be a *root* or can have one of its neighbors as its *father*. If P_i is the father of P_j in configuration c then there is a link from P_i to P_j in $FSG(c)$. Thus, in any configuration c there are at most n links in $FSG(c)$. Each tree of FSG is identified by a *tree-identifier*, abbreviated *tid*. A root is the only processor which changes the tree's *tid*; this is always done by extending the *tid* with a randomly chosen bit. Each non-root processor repeatedly copies its father's *tid*. Hence after an initial segment of every execution all processors in a tree T, hold a prefix of the *tid* of the root of T.

To achieve cycle elimination each processor computes its distance to the root of its subgraph. Whenever the processor realizes that this distance "grows" it cuts the link to its father and becomes a separate root. We prove that cycle elimination takes $O(\Delta)$ rounds, where Δ is the maximal degree of a processor in the communication graph. After this part is completed FSG is a collection of directed trees.

To achieve tree elimination each processor repeatedly scans its neighbors *tid*-s. Whenever a processor P_i discovers a neighbor P_j whose *tid* is larger than its own *tid*, P_i takes P_j to be its father. If previously P_i is a root, the number of trees is reduced by one. We prove that taking a new father never introduces new cycles in FSG.

To reduce the number of trees to one we first ensure that eventually there is unique *tid* in the system, and then that eventually there remains a unique tree in the system. Since the protocol should be correct for any initial configuration it has to be able to deal with a configuration in which the FSG contains two or more trees with the same *tid*. A root processor discovers that there are more roots with the (same) maximal *tid* by repeatedly recoloring of its tree. Each recoloring starts from the root which chooses the new color randomly. The root waits for each of its sons to confirm that all its sub-tree is recolored. Once all the tree is recolored the root chooses a new color once more and so on. A processor of a tree T detects the existence of another tree T' with the same *tid*, by observing that one of its neighbors is not colored neither by the old color of T, nor by its new color. In this case the processor "returns" this information to the root of T. Upon receipt of this information the root of T extends its tid by a random bit which is distributed again along the edges of T. At the same time T' may also extend its own tid. Since each extension is done randomly symmetry is eventually broken.

The code, that appears in Figure 2, is written for a processor P_i that has d neighbors ordered from 1 to d. Each processor, P_i, may read the variables $f, dis, tid, do\_color, ack, other\_trees$ of any of its neighbors. The variable f indicates which of the neighbors of a processor is its father. We assume that for every neighbors P_i and P_j, P_i repeatedly writes in the field $r_{ij}.order$ of its register the order of the link that connect it to P_j. The variables tid and dis indicates the tree identity for which P_i belongs and the distance of P_i from the $root$ of that tree, respectively. The tid is a string in $\{0,1\}^*$ that represent a binary fraction (i.e. 01 represent the value $1/4$). A processor with $dis = 0$ is a root. The variables do_color and ack are used to change repeatedly the color of the tree. do_color may contain six colors which are denoted by integers of values 1 to 6 and ack is a boolean variable. The variable $other\_trees$ is assigned by true whenever a processor notes that one of its neighbors has do_color that is different from both its own current do_color and the new do_color of its father.

The function $extend\_tid$ chooses a random bit and concatenate it to the tail of the current tid. The function $son(j)$ executed by P_i is true only when P_j is the son of P_i, i.e. $ir_{ji}.f = ir_{ji}.order$. The function $choose\_color$ chooses randomly between four (out of the possible six) colors that are not equal to the $previous\_color$ and to (the current) do_color.

Let tid_1 and tid_2 be two values of tid variables. The relation $tid_1 > tid_2$ indicates that: the fraction (represented in binary) in tid_1 is greater then the fraction in tid_2 or the two fraction values are equal and tid_1 is a longer string (i.e. tid_1 has suffix with one or more zeros that does not exist in tid_2). Define the relation $>$ over pair of processors variables (tid, dis) as follow: $((tid_1, dis_1) > (tid_2, dis_2))$ if $tid_1 > tid_2$ (as defined above) or when $tid_1 = tid_2$, then $dis_1 < dis_2$. In such a case we say that the pair (tid_1, dis_1) is $greater$ then (tid_2, dis_2).

Overview on the lines of the code:
The code consist of one do forever loop, the lines of this loop are described as follows:

Line 2 - P_i reads the registers of its neighbors.

Lines 3 to 5 - Using the read values, P_i calculate the maximal tid among the $tids$ of its neighbors. Then P_i finds the minimal dis of a neighbor among the neighbors that hold the maximal tid. At last P_i finds the index of the first neighbor that hold the above maximal tid and minimal dis.

Line 6 - If P_i finds that it has no neighbor with (tid, dis) that is greater than its own (tid, dis) then P_i becomes a root.

Lines 7 to 10 - P_i updates the value of its tid and dis according to the values it read.

Line 11 - Whenever P_i finds that the surrounding tid are equal P_i assumes that the cycles eliminating is over and participate in coloring its tree.

5.2 Correctness and complexity proofs

The proof is by the following lemmas. The proofs of those lemmas are omitted from this extended abstract.

Lemma 3: For any processor P_i and any execution the value of (tid_i, dis_i) can only be $increased$.

```
1 do forever
       (* Reading *)
2      for j:=1 to d do ir_ji := read(r_ji);
3      mtid := max(ir_ji.tid);
4      mdis:= min{ir_ji.dis | ir_ji.tid = mtid} + 1;
5      mf := {first j | ir_ji.tid = mtid, ir_ji.dis = (mdis−1)};
       (* Cycles eliminating *)
6      if (tid,dis) > (mtid,mdis) then write dis := 0 (* become a root *)
7      else if not (dis = 0 and tid ≥ mtid) then
8              begin (* become the son of the maximal neighbor *)
9                      write (f,tid,dis):= (mf,mtid,mdis)
10             end
11     if ∀ j (ir_ji.tid = tid) then one_tree_election;
12 end
```

Figure 2: Protocol for general graph

Lemma 4: Let $E' = (c_0, \cdots, c)$ be (a prefix of) an execution in which each processor executes a loop iteration. The following hold for any P_f and P_s such that P_f is the father of P_s in c :

1. Following the first read iteration in E', P_s executed the **write** operation in line 9 at least once.

2. In the configuration, c_i, that immediately follows the last **write** operation of P_s in line 9 during E' it holds that $(tid_f, dis_f) \geq (ir_{f_s}.tid, ir_{f_s}.dis) > (tid_s, dis_s)$.

Lemma 5: Let $E' = (c_0, \cdots, c)$ be an execution in which each processor executes a loop iteration. For any P_f and P_s such that P_f is the father of P_s in c, it holds in c that $(tid_f, dis_f) \geq (ir_{f_s}.tid, ir_{f_s}.dis) > (tid_s, dis_s)$.

Define a *forest* configuration to be a configuration, c, in which for any P_f and P_s such that P_f is the father of P_s in c, $(tid_f, dis_f) \geq (ir_{f_s}.tid, ir_{f_s}.dis) > (tid_s, dis_s)$. ¿From now on we consider only executions that start with a forest configuration. By the above lemmas a forest configuration is reached within $O(\Delta)$ rounds. The above lemmas show that every configuration that follows a forest configuration is also a forest configuration.

Lemma 6: Let c be a forest configuration. Then $FSG(c)$ is a forest of directed trees.

Lemma 7: For a configuration c, let $R(c)$ denote the set of the root processors in c. Then for any two successive forest configurations $c_i \rightarrow c_{i+1}$, $R(c_i) \supseteq R(c_{i+1})$.

In the following lemmas we show that when the system starts with forest configuration the number of roots decreases to 1 in $O(\Delta D \log n)$ expected number of rounds. Define $max\_tid(c)$ to be the maximal value of tid variable in c.

Lemma 8: There is a constant k_1 s.t. every execution that starts in a forest configuration c reaches in expected number of $k_1 \Delta D$ rounds a configuration c_i which is either a safe configuration or $max\_tid(c_i) > max\_tid(c)$.

```
13 Procedure one_tree_election
   (* Root *)
14 if (dis = 0) and { ∀ j | son(j) ⇒ ((ir_ji.do_color = do_color) and ack_ji)} then
15   begin
16       if { ∃ j | ir_ji.other_trees = true or ir_ji.do_color ≠ do_color }
             then write tid := extend_tid(tid);
17       write do_color := choose_color(previous_color, do_color);
18   end
19 else (* Not Root *)
20   if (dis ≠ 0) and (do_color ≠ ir_fi.do_color) then
21       begin
22           write other_trees := false;
23           if { ∃ j | ir_ji.do_color ∉ (do_color, ir_fi.do_color) } then write other_trees := true;
24           write (do_color, ack) := (ir_fi.do_color, false);
25       end
26   else if not ack and {∀ j | son(j) ⇒ ((ir_ji.do_color = do_color) and ack_ji)} then
27       begin
28           if {∃j | son(j) and ir_ji.other_trees = true} then other_trees := true;
29           write ack := true;
30       end
31 end
```

Figure 3: Procedure one_tree_election

Lemma 9: There is a constant k_2 s.t. every execution that starts in a forest configuration c reaches in $k_2 \Delta D$ rounds a configuration c_i in which the tid of every root processor is at least $max\_tid(c)$.

Corollary 10: There is a constant k_3 s.t. every execution that starts in a forest configuration c reaches in expected number of $k_3 \Delta D \log n$ rounds a configuration c_i which is either a safe configuration or the tid of every root processor in c_i is longer than its tid in c by at least $2 \log n$ bits.

The proof is completed by the following corollary:

Corollary 11: The system elects a leader in expected $O(\Delta D \log n)$ rounds.

5.3 The Bounded Protocol

In this section we give the main ideas of our bounded self stabilizing leader election protocol.

The bounded self stabilizing leader election protocol is achieved by combining the methods of the two previous protocols. The main modification of the unbounded protocol is the use of bounded tree identifiers. Each such identifier behaves as in the previous protocol until its length reaches some predetermined limit. Whenever an overflow occurs the (root) processor which is about to extend the tree identifier, beyond its capacity, *dismantles* its tree. Trees are dismantled from the leaves to the root. Each processor which belongs to a dismantled tree forms a separate tree which joins the tree elimination process once more.

The proof uses a strategy for *luck* which guarantees that some processor chooses a tree identifier which is larger than all other *tid*-s. Details and proofs are left to the full paper.

6 Ranking

In this section we sketch a self-stabilizing protocol which *ranks* the system's processors. First we present a self stabilizing ranking protocol that works on systems whose communication graph is a directed tree. This protocol assign each processor a rank which equals its DFS number. The time complexity of this protocol is $O(\Delta n)$ rounds, and its total space complexity is $O(\Delta n \log n)$. Then we use the technique of fair protocol composition of [DIM-90], and achieve a uniform self-stabilizing ranking protocol for general graphs by composition of this protocol with any of the protocols presented in the previous section. The time and space complexities of the combined protocol are the sum of the complexities of the two protocols.

References

[Ab-88]   K. Abrahamson, "On Achieving Consensus Using a Shared Memory", *Proceedings of the Seventh Annual ACM Symposium on Principles of Distributed Computing*, Toronto Canada, August 1988, pp. 291,302.

[AG-90]   A. Arora and M. Gouda: "Distributed Reset", to appear in *Proceedings of the Tenth Conference on Foundations of Software Technology and Theoretical Computer Science, Bangalore, India*, December 1990.

[AKY-90]   Y. Afek, S. Kutten and M. Yung, "Memory-Efficient Self-Stabilization on General Networks", it Proceedings of the 4th International Workshop on Distributed Algorithms, Bari Italy, September 1990.

[BGW-87]   G.M. Brown, M.G. Gouda, and C.L. Wu, "A Self-Stabilizing Token system", *Proc. of the Twentieth Annual Hawaii International Conference on System sciences* 1987, pp. 218-223.

[BP-88]   J.E. Burns and J. Pachl, "Uniform Self-Stabilizing Rings", *Aegean Workshop On Computing, 1988, Lecture notes in computer science* 319, pp. 391-400.

[Bu-87]   J.E. Burns, "Self-Stabilizing Rings without Demons", Technical Report GIT-ICS-87/36, Georgia Institute of Technology.

[CIL-87]   B. Chor, A. Israeli, and M. Li, "On Processor Coordination Using Asynchronous Hardware", *Proc. of the Sixth Annual ACM Symposium on Principles of Distributed Computation*, (1987), pp. 86-97.

[Dij-74]   E.W. Dijkstra, "Self-Stabilizing Systems in Spite of Distributed Control", Communications of the ACM 17,11 1974, pp. 643-644.

[DIM-90]   S. Dolev, A. Israeli and S. Moran, "Self Stabilization of Dynamic Systems", *Proc. of the Ninth Annual ACM Symposium on Principles of Distributed Computation*, Quebec City, August 1990, pp. 103-118.

[DIM-91]   S. Dolev, A. Israeli and S. Moran, "Resource Bounds for Self Stabilization Message Driven Protocols", *Proc. of the Tenth Annual ACM Symposium on Principles of Distributed Computation*, Montreal, August 1991, pp. 281-294.

[AE-91]   E. Angnostou and R. El-Yaniv More on the Power of Random Walks: Uniform Self-Stabilizing Algorithms

[Ga-78]   R. G. Gallagher, "Finding a leader in networks withe $O(E) + O(NlogN)$ messages", Internal Memo., M.I.T., Cambridge, Mass., 1978.

[GHS-83] R.G. Gallager, P.M. Humblet and P.M. Spira, "A distributed algorithm for minimum weight spanning trees", *ACM Trans. Program. Lang. Sys. 5* 1 (1983), pp. 66-77.

[Hu-84]   P. Humblet, "Selecting a leader in a clique in $O(n \log n$ messages. Inter. Memo., Laboratory for Information and Decision Systems, M.I.T, Cambridge, Mass., 1984.

[IJ-90]   A. Israeli and M. Jalfon, "Token Management Schemes and Random walks Yield Self Stabilizing Mutual Exclusion", *Proc. of the Ninth Annual ACM Symposium on Principles of Distributed Computation*, Quebec City, August 1990, pp. 119-132.

[IJ-90a] A. Israeli and M. Jalfon, "Self stabilizing Ring Orientation", Proceedings of the 4th International Workshop on Distributed Algorithms, Bari Italy, September 1990.

[IR-81]   A. Itai and M. Rodeh, "Probabilistic Methods for Breaking Symmetry in Distributed Networks", To appear in Information and Computation.

[KKM-90] E. Korach, S. Kutten and S. Moran, "A Modular Technique for the Design of Efficient Distributed Leader Finding Algorithms", *ACM Trans. Program. Lang. Sys. 12*, 1 (1990), 84-101.

[KMZ-84] E. Korach, S. Moran and S.Zaks, "Tight lower and upper bounds for some distributed algorithms for complete network of processors", *Proc. of the 3rd Annual ACM Symposium on Principles od Distributed Computing* (1984), pp. 199-207.

[KP-89]   S. Katz and K. J. Perry, "Self-stabilizing extensions for message-passing systems", *Proc. of the Ninth Annual ACM Symposium on Principles of Distributed Computation*, Quebec City, August 1990, pp. 91-101.

[Kr-79]   H.S.M. Kruijer, "Self-stabilization (in spite of distributed control) in tree-structured systems", Information Processing Letters 8,2 (1979), pp. 91-95.

[La-86]   L. Lamport, "The Mutual Exclusion Problem: Part II - Statement and Solutions", *Journal of the Association for Computing Machinery* , Vol. 33 No. 2 (1986), pp. 327-348.

[Pe-82]   G. L. Peterson, "An $O(n \log n)$ unidirectional algorithm for the circular extrema problem",

[SS-89]   B. Schieber and M. Snir "Calling Names on Nameless Networks", *Proceedings of the Eights Annual Symposium on Principles of Distributed Computing*, Edmonton, August 1989, pp. 319-328.

[V-91]   George Varghese, "Distributed Program Checking a Paradigm for Building Self-stabilizing Distributed Protocols", To appear in FOCS-91.

THE QUICKEST PATH PROBLEM IN DISTRIBUTED COMPUTING SYSTEMS

Yung-Chen Hung and Gen-Huey Chen

Department of Computer Science and Information Engineering,
National Taiwan University, Taipei, Taiwan
Fax: (886)-(2)-3628167

Abstract

Let $N=(V, A, C, L)$ be an input network with node set V, arc set A, positive arc capacity function C, and nonnegative arc lead time function L. The quickest path problem is to find a path in N to transmit a given amount of data such that the transmission time is minimized. In this paper, distributed algorithms are developed for the quickest path problem in an asynchronous communication network. For the one-source quickest path problem, we present two algorithms that require $O(rn^2)$ messages, $O(rn^2)$ time, and $O(rmn)$ messages, $O(rn)$ time, respectively, where $m=|A|$, $n=|V|$, and r is the number of distinct capacity values of N. For the all-pairs quickest path problem, we present an algorithm that requires $O(mn)$ messages and $O(m)$ time.

1. Introduction

An *asynchronous network* is a point-to-point (store-and-forward) communication network composed of processors and bidirectional non-interfering communication links. Each processor can be uniquely identified. Local memory is provided for each processor, but no common memory is shared by two or more processors. Hence, processors can communicate with each other only by exchanging messages over the communication links. Each processor processes messages received from its neighbours, performs local computation, and sends messages to its neighbours, all in negligible time. Messages can be transmitted independently in both directions on a communication link and arrive after an unpredictable but finite delay, without error and in *FIFO (First-In-First-Out)* order. When a message is received by a processor, it is inserted into a queue that is maintained by the processor to store unprocessed messages. We assume that each message is of fixed length.

When a distributed algorithm is executed in an asynchronous network, some processors may be nominated to perform special roles, executing special routines for this purpose. Nevertheless, it is assumed that at the start of the algorithm, each processor holds a copy of the entire code.

The performance of distributed algorithms can be measured according to two standards: message complexity and time complexity. (In some distributed algorithms, e.g., [2], [6], only the message complexity is considered in performance analysis, since in an asynchronous distributed environment the number of messages transmitted is significantly more important than the number of computation steps performed.) The *message complexity* of a distributed algorithm is the total number of messages transmitted during its execution. The *time complexity* is the maximum number of "time units" taken by the algorithm, assuming that the propagation delay of any communication link is at most one unit of time. Since processors in an asynchronous network are so far apart, the processing time for each message at a processor is small, in comparison with the message transmission time, and thus is negligible. In this paper, like most distributed algorithms, we only consider the message transmission time.

Let $N=(V, A, C, L)$ be a network with node set V, arc set A, positive arc capacity function C, and nonnegative arc lead time function L. The *quickest path problem*, which was originally proposed by Chen and Chin [4], is to find the quickest paths to transmit data in N. The quickest path problem is a variant of the shortest path problem. The *one-source quickest path problem* (*1_QSP* for short) is to find the quickest paths between one node and all other nodes. The *all-pairs quickest path problem* (*A_QSP* for short) is to find the quickest paths between every pair of nodes. Some centralized (single processor) algorithms have been proposed for the quickest path problem. For the 1_QSP, Chen and Chin [4] have presented an $O(m^2+mn\log n)$ time algorithm, where n and m are the numbers of nodes and arcs, respectively, in N. For the A_QSP, Hung and Chen [8] have presented an $O(mn^2)$ time algorithm.

In this paper, distributed algorithms are presented for the quickest path problem in asynchronous networks. For the 1_QSP, we present two algorithms that require $O(rn^2)$ messages, $O(rn^2)$ time, and $O(rmn)$ messages, $O(rn)$ time, respectively, where r is the number of distinct capacity values of N. For the A_QSP, we present an algorithm that requires $O(mn)$ messages and $O(m)$ time.

The rest of this paper is organized as follows. In the next section, we state the quickest path problem in detail. The distributed algorithms for 1_QSP and A_QSP are presented in Section 3 and Section 4, respectively. In Section 5, we conclude this paper with some final remarks.

2. The Quickest Path Problem

Let $N=(V, A, C, L)$ be a network, where $|V|=n$, $|A|=m$, $G=(V, A)$ is an undirected graph without self loops and without multiple arcs between any two nodes, and $C(u, v) \geq 0$ and $L(u, v) \geq 0$ are the capacity and the lead time, respectively, of an arc $(u, v) \in A$. More specifically, $C(u, v)$ denotes the maximal amount of data that can be transmitted between node u and node v per unit time, and $L(u, v)$ denotes the lead time required to send the data between node u and node v. If σ units of data are required to be transmitted between node u and node v through arc (u, v), then the required transmission time is computed as $L(u, v)+\sigma/C(u, v)$.

For example, if 8 units of data are required to be sent through an arc (u, v) with 2 units of capacity and 3 units of lead time, then no data reach node v at the first 3 time units. At time unit 4, 2 units of data are available at node v. Then, at each of subsequent time units, 2 units of data reach node v until the transmission is ended. Hence, the total transmission time is $3+8/2=7$ time units.

Suppose $P=(u_1, u_2, ..., u_k)$ is a path between node u_1 and node u_k. The lead time of P is defined as

$$L(P) = \sum_{i=1}^{k-1} L(u_i, u_{i+1}).$$

The capacity of P is defined as

$$C(P) = \min_{1 \leq i \leq k-1} C(u_i, u_{i+1}).$$

The total transmission time to send σ units of data between u_1 and u_k through path P is defined as

$$T(P, \sigma) = \text{the lead time of } P + \sigma/\text{the capacity of } P$$
$$= L(P) + \sigma/C(P).$$

The quickest path to send σ units of data between node s and node t, which is denoted by $QP(s, t, \sigma)$, is the path P satisfying

$$T(P, \sigma) = \min\{ T(P_i, \sigma) \mid \forall P_i=(u_{i1}, u_{i2}, ..., u_{ik}), \text{ where } s=u_{i1} \text{ and } t=u_{ik} \}.$$

Also, let $SP(s, t)$ denote the path between node s and node t in N that has the shortest lead time; namely, $SP(s, t)$ is the path P satisfying

$$L(P) = \min\{ L(P_i) \mid \forall P_i=(u_{i1}, u_{i2}, ..., u_{ik}), \text{ where } s=u_{i1} \text{ and } t=u_{ik} \}.$$

Throughout this paper, we assume that the distinct capacity values of N are $c_1>c_2> ... >c_r$. For each positive capacity c_i, we define $N(c_i)=(V, A_i, C, L)$ to be a subnetwork of N containing only those arcs with capacities larger than or equal to c_i. That is, $(u, v)\in A_i$ if and only if $(u, v)\in A$ and $C(u, v)\geq c_i$. Also, let $SP_i(s, t)$ denote the path between node s and node t in $N(c_i)$ that has the shortest lead time. We define the *shortest lead time path tree* of $N(c_i)$ rooted at s, which is denoted by $ST_i(s)$, as a spanning tree of $N(c_i)$ such that for each path P from root s to any other node v in $ST_i(s)$,

$$L(P) = L(SP_i(s, v)).$$

As an illustrative example, let us consider the network N of Figure 1(a), where $\sigma=12$ and each arc is associated with a pair of values: the lead time and the capacity. The path $P_1=(D, E, C)$ has $L(P_1)=L(D, E)+L(E, C)=5$, $C(P_1)=\min\{C(D, E), C(E, C)\}=2$, and $T(P_1, 12)=L(P_1)+12/C(P_1)=5+6=11$. The other paths between node D and node C are $P_2=(D, C), P_3=(D, B, C), P_4=(D, B, E, C), P_5=(D, A, B, C), P_6=(D, A, B, E, C)$, and $P_7=(D, E, B, C)$. Since $T(P_1, 12)=\min\{T(P_1, 12), T(P_2, 12), ..., T(P_7, 12)\}=\{11, 16, 23, 26, 30, 31, 36\}=11$, P_1 ($=QP(D, C, 12)$) is the quickest path between node D and node C. Also, since $L(P_2)=\min\{L(P_1), L(P_2), ..., L(P_7)\}=\{5, 4, 11, 22, 16, 27, 24\}=4$, P_2 ($=SP(D, C)$) is the shortest lead time path between node D and node C. The distinct capacity values of N are $c_1=6, c_2=4, c_3=3, c_4=2$, and $c_5=1$. Figure 1(b) shows the subnetwork $N(c_3)$ of N. Figure 1(c) shows the shortest lead time path tree $ST_3(A)$.

The most important concept of the quickest path problem is that the selection of the quickest paths depends on not only the characteristics of the network but also the amount of data to be transmitted. Taking these two factors into consideration will make the problem more reasonable. For example, when a certain amount of data are required to be transmitted between two nodes in a communication network, the baud rates of the transmission media between these two nodes can be treated as the capacities and the lengths of different paths between them can be treated as the lead times. For this example, if the amount of data is huge, the paths with larger capacities are preferred, and if the amount of data is quite small, the paths with smaller lead times are preferred. Thus, including the amount of data into the selection of the quickest paths is practically significant.

In the next two sections, the network N is embedded in an asynchronous network in the sense that there is a 1-1 correspondence between the nodes of N and the processors of the asynchronous network, and between the arcs of N and the communication links of the asynchronous network. At the beginning of our distributed algorithms, each processor knows only the identities of its neighbours, and the capacities and lead times of its incident arcs. When the algorithms terminate, a routing table for traversing the quickest paths is established within each processor.

3. Distributed Algorithms for 1_QSP

In this section, we first show that the quickest path problem can be solved by applying the shortest path algorithms.

Theorem 1. Let $P_1, P_2, ..., P_r$ denote the paths between node s and any other node t in $ST_1(s), ST_2(s), ..., ST_r(s)$, respectively, then $L(QP(s, t, \sigma)) + \sigma/C(QP(s, t, \sigma)) = \min_{1 \leq i \leq r}\{ L(P_i) + \sigma/C(P_i) \}$, where r is the number of distinct capacity values of N.

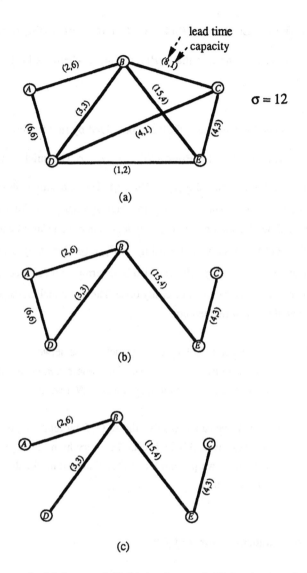

Figure 1. An example. (a) A network N. (b) A subnetwork $N(c_3)$, where $c_3 = 3$. (b) The shortest lead time path tree $ST_3(A)$.

Proof: By definition, $L(P_i) + \sigma/C(P_i) \geq L(QP(s, t, \sigma)) + \sigma/C(QP(s, t, \sigma))$, for $i \in \{1,2,...,r\}$. This theorem is proved by the fact that $QP(s, t, \sigma)=P_j$, where $j \in \{1,2,...,r\}$ and $C(QP(s, t, \sigma))=c_j$.

Q.E.D.

Based on the above theorem, the 1_QSP can be solved by first finding $ST_i(s)$, $i=1,2,...,r$, and then determining the minimum of $L(P_i)+\sigma/C(P_i)$, $i=1,2,...,r$. The quickest paths are those paths contained in the $ST_i(s)$ with minimum $L(P_i)+\sigma/C(P_i)$. The one-source shortest path algorithms proposed by Frederickson [5] and Lakshmanan, Thulasiraman, and Comeau [9] can find each $ST_i(s)$ (let the lead time of each arc with capacity less than c_i be infinity) and determin the value $L(P_i)$ (regard the lead times as weights) using $O(n^2)$, $O(mn)$ messages in $O(n^2)$, $O(n)$ time, respectively. Incidentally, the value $C(P_i)$ can also be determined if some additional operations are performed in their algorithms. This can be easily done. The additional messages and time are $O(n)$. Therefore, we have the following theorem.

Theorem 2. The 1_QSP can be solved in an asynchronous network N using $O(rn^2)$ messages in $O(rn^2)$ time, or using $O(rmn)$ messages in $O(rn)$ time, where r, n, m are the number of distinct capacity values, the number of nodes, the number of arcs in N, respectively.

According to Theorem 1, we can also solve the A_QSP by repeating the one-source shortest path algorithms nr times. However, this will result in a very high message complexity and time complexity. In the next section, we present a distributed algorithm for the A_QSP that requires only $O(mn)$ messages and $O(m)$ time.

4. A Distributed Algorithm for A_QSP

Since the amount of data to be transmitted is an important factor for the selection of the quickest paths, it is desired to find the quickest paths for all possible values of σ. Let $(0, \infty)$ be the range of σ. If σ is very small, then the shortest lead time path can be the quickest path. When σ increases, the shortest lead time path is no longer the quickest path, and another path becomes the new quickest path. Thus, we can divide the range of σ into intervals such that each of them has the same quickest path.

In the proposed algorithm, an $n \times n \times r$ matrix M is provided in each node to keep some information about the shortest lead time paths and intervals of σ. More specifically, for each node v, the entry $M_v(s, t, i)$, $1 \leq s \leq n$, $1 \leq t \leq n$, $1 \leq i \leq r$, records four values: $M_v(s, t, i).LT$, $M_v(s, t, i).NEXT$, $M_v(s, t, i).LEFT$, $M_v(s, t, i).RIGHT$. The value $M_v(s, t, i).LT$ represents the lead time of $SP_i(s, t)$. The value $M_v(s, t, i).NEXT$ represents the next node along $SP_i(s, t)$. If $SP_i(s, t)$ does not go through node v, then $M_v(s, t, i).NEXT$ is empty. When $M_v(s, t, i).LEFT \leq \sigma \leq M_v(s, t, i).RIGHT$, $SP_i(s, t)$ is the quickest path from node s to node t.

The algorithm consists of three stages. The first stage, referred to as *heap construction* stage, constructs a spanning tree of the network N, and then constructs a heap from the tree according to the capacities of arcs. The root node of the tree is as the algorithm leader. A *heap* is a tree, in which the value associated with each node is greater than or equal to the values associated with its children. The second stage, referred to as *shortest lead time path computation* stage, broadcasts all the arcs of N (including their end nodes and their capacities and lead times) over the tree, one by one in a nonincreasing order of their capacities. The broadcasting is initiated at the root node. Whenever a node v receives such a message (an arc and its relevant information), it updates the values $M_v(s, t, i).LT$ and $M_v(s, t, i).NEXT$. The third stage, referred to as *interval finding* stage, determines the intervals of σ for all pairs of nodes (i.e., determining the values $M_v(s, t, i).LEFT$ and $M_v(s, t, i).RIGHT$ for $s=1,...,n$, $t=1,...,n$, $i=1,...,r$.).

After the algorithm ends, the matrix M_v can serve as a routing table. That is, whenever the node v receives an amount σ of data that are required to be transmitted from a node s to another node t, it can determine the next node along the quickest path to forward the data by consulting the matrix M_v, using the values σ, s, t.

In the following, we describe the algorithm in more details. To simplify the description of the algorithm, the various acknowledgement messages will not be described.

Heap construction

The distributed algorithm proposed by Awerbach [1], which requires $O(m+n\log n)$ messages and $O(n)$ message transmission time, can be used to find a spanning tree of N. An arc is called a *tree arc* if it belongs to the tree, and a *nontree arc* otherwise.

In order to construct a heap from the tree, the root node broadcasts a message *init* over the tree. Initially, let the status of each arc be *enable*. When a nonleaf node receives a message *init* from its parent, it forwards the message to its children. Thus, the message *init* is propagated toward leaf nodes. When a leaf node v receives a message *init*, it returns a message $echo(v, w, L(v, w), C(v, w))$ to its parent, where (v, w) is a nontree arc that has the largest capacity. Then, the status of (v, w) is set to *disable* (node v will inform node w of this change). After a nonleaf node has received all the messages *echo* sent from its children, it selects the largest capacity arc (assume (x, y)) among the tree arcs joining it to its children, the arcs contained in the received messages, and the nontree arcs that are incident on it and are of status *enable*. Then, the nonleaf node sends a message $echo(x, y, L(x, y), C(x, y))$ to its parent. If the arc (x, y) is incident on the nonleaf node, then its status is changed to *disable*. Finally, this stage terminates when the root node has selected the largest capacity arc.

Note that in the above execution, each node sends the largest capacity arc that it holds to its parent. After this, the heap property may be violated. So, in order to maintain the heap, the node has to require its children to send the *echo* message again.

The message complexity and time complexity are analyzed as follows. Define the level of a tree arc (x, y) (assume that x is the parent of y) as the level of node x (see [7] for the definition of node level). Since there are one *init* message transmitted via each tree arc and at most i *echo* messages transmitted via each tree arc of level i, the total messages transmitted are $O(n^2)$, where n is the number of nodes in N. The time required to transmit the *init* message is proportional to the depth of the tree. Further, since on each path from the root node to a leaf node, the *echo* messages are transmitted in a pipelined manner, the time required to transmit the *echo* messages is also proportional to the depth of the tree. Thus, the total message transmission time is $O(n)$ in the worst case.

Shortest lead time path computation

In this stage, the root node broadcasts the message $arc(x, y, L(x, y), C(x, y))$ for each arc (x, y) in N, over the tree. These *arc* messages are broadcast in a nonincreasing order of $C(x, y)$'s. In the meantime, the *echo* messages are required to maintain the heap, as we have discussed in the heap construction stage. The total messages transmitted are $O(mn)$, where m is the number of arcs in N. Since the *arc* messages are sent from the root node in m consecutive time steps and the heap maintenance can be performed simultaneously, the total message transmission time is $O(m)$. In the rest of this subsection, we describe the necessary operations performed by a node v whenever it receives an *arc* message.

At first, the network N is regarded as empty (containing no arcs). Since new paths may be generated between any two nodes s and t when an arc (x, y) is added to N, the shortest lead time path between s and t may be altered. Thus, whenever a node v receives a message $arc(x, y, L(x, y), C(x, y))$, it must update the values $M_v(s, t, i).LT$, $M_v(t, s, i).LT$, $M_v(s, t, i).NEXT$ and $M_v(t, s, i).NEXT$ for every pair of nodes s and t, where $C(x, y)=c_i$ is assumed. (Recall that the node v receives *arc* messages in a nonincreasing order of $C(x, y)$'s.)

Although many new paths from node s to node t may be generated after adding an arc (x, y) to N, only two paths are needed to be considered. One contains the shortest lead time path from s to x (without passing the arc (x, y)), the arc (x, y), and the shortest lead time path from y to t (without passing (x, y)), and the other contains the shortest lead time path from s to y (without passing (x, y)), the arc (x, y), and the shortest lead time path from x to t (without passing (x, y)).

Initially, let $M_v(s, t, 0).LT = \infty$ and $M_v(s, t, 0).NEXT$ be empty. The updating of $M_v(s, t, i).LT$ and $M_v(s, t, i).NEXT$ is shown below. The values $M_v(t, s, i).LT$ and $M_v(t, s, i).NEXT$ can be updated similarly.

/* Suppose the arcs of N are broadcast in the sequence $(x_1, y_1), (x_2, y_2), ..., (x_m, y_m)$, where $C(x_1, y_1) \geq C(x_2, y_2) \geq ... \geq C(x_m, y_m)$. Also, assume that node v has just received a message $arc(x_k, y_k, L(x_k, y_k), C(x_k, y_k))$, where $C(x_k, y_k) = c_i$. */

$T_1 \leftarrow M_v(s, x_k, j).LT + L(x_k, y_k) + M_v(y_k, t, j).LT$;   /* $C(x_{k-1}, y_{k-1}) = c_j$ is assumed. */
$T_2 \leftarrow M_v(s, y_k, j).LT + L(x_k, y_k) + M_v(x_k, t, j).LT$;
$M_v(s, t, i).LT \leftarrow \min\{T_1, T_2, M_v(s, t, j).LT\}$;

if $T_1 = M_v(s, t, i).LT$
    then if $v = x_k$ then $M_v(s, t, i).NEXT \leftarrow y_k$
        else if $M_v(s, x_k, j).NEXT$ is not empty
            then $M_v(s, t, i).NEXT \leftarrow M_v(s, x_k, j).NEXT$
            else if $M_v(y_k, t, j).NEXT$ is not empty
                then $M_v(s, t, i).NEXT \leftarrow M_v(y_k, t, j).NEXT$
                else let $M_v(s, t, i).NEXT$ be empty;

if $T_2 = M_v(s, t, i).LT$ and $T_1 \neq T_2$
    then if $v = y_k$ then $M_v(s, t, i).NEXT \leftarrow x_k$
        else if $M_v(s, y_k, j).NEXT$ is not empty
            then $M_v(s, t, i).NEXT \leftarrow M_v(s, y_k, j).NEXT$
            else if $M_v(x_k, t, j).NEXT$ is not empty
                then $M_v(s, t, i).NEXT \leftarrow M_v(x_k, t, j).NEXT$
                else let $M_v(s, t, i).NEXT$ be empty;

Interval finding

In this stage, each node v determines the intervals of σ, which is the amount of data to be transmitted, for each pair of nodes s and t such that the quickest path between s and t is the same for each interval. No messages are transmitted in this stage.

Since each $SP_i(s, t)$, $1 \leq i \leq r$, has the shortest lead time among those paths from s to t whose capacities are larger than or equal to c_i, the transmission time of each path from s to t with capacity equal to c_i is greater than or equal to the transmission time of $SP_i(s, t)$ for all values of σ. Thus, only the paths $SP_1(s, t), SP_2(s, t), ..., SP_r(s, t)$ are necessary to be considered in constructing intervals of σ for nodes s and t.

Since the transmission time of each $SP_i(s, t)$ is a function of σ, we have

$$y = M_v(s, t, i).LT + x/c_i,$$

where $y=T(SP_i(s, t), \sigma)$ and $x=\sigma$. It is easy to see that the minimal transmission time for each value of σ is determined as $\min\{M_v(s, t, i).LT+x/c_i \mid i=1,...,r\}$. Equivalently, the lower portion of these r lines $y=M_v(s, t, i).LT+x/c_i$, $i=1,...,r$, are the minimal transmission time for all values of σ (see Figure 2), and the intersection points (s_1 and s_2 in Figure 2) along the lower portion separate the intervals.

Now the remaining problem is how to determine the intersection points along the lower portion. This can be done in linear time by adding the lines $y=M_v(s, t, i).LT+x/c_i$ to the plane one by one and in the sequence of $i=r, r-1, ..., 1$. The line $y=M_v(s, t, r).LT+x/c_r$ is first added to the plane. Then, let us consider the situation of adding a line $y=M_v(s, t, i).LT+x/c_i$, $1\le i<r$, to the plane. The newly added line intersects the current lower portion at one point z (since the slopes of the lines are positive and decreasing in their order), which can be determined by scanning the lower portion from right to left. The point z is then stored as a new intersection point, and the intersection points on the right of z are discarded (see Figure 3). Since each addition of a new line (except for the first line) produces exectly one intersection point, the total number of discarded points is not greater than $r-1$.

The values $M_v(s, t, i).LEFT$ and $M_v(s, t, i).RIGHT$ can be determined as follows. Initially, set $M_v(s, t, r).LEFT=0$ and $M_v(s, t, 1).RIGHT= \infty$. For each intersection point z, set $M_v(s, t, i).RIGHT=M_v(s, t, j).LEFT=z$, where $j<i$ and the two lines $y=M_v(s, t, i).LT+x/c_i$ and $y=M_v(s, t, j).LT+x/c_j$ intersect at z. Besides, set $M_v(s, t, k).LEFT=M_v(s, t, k).RIGHT= \infty$, for $j<k<i$.

The following theorem summarizes the main result of this paper.

Theorem 3. The A_QSP can be solved in an asynchronous network using $O(mn)$ messages, in $O(m)$ time.

It is easy to see that the matrix M_v can serve as a routing table for node v. Whenever node v receives σ units of data that are required to be transmitted from a node s to another node t, it can know how to route the data so as to minimize the total transmission time by looking up the matrix M_v. First, it determines an interval $(M_v(s, t, i).LEFT, M_v(s, t, i).RIGHT)$ such that $M_v(s, t, i).LEFT \le \sigma \le M_v(s, t, i).RIGHT$. Then, it can determine $M_v(s, t, i).NEXT$ as the next node to route the data, since the arc $(v, M_v(s, t, i).NEXT)$ belongs to the quickest path from s to t.

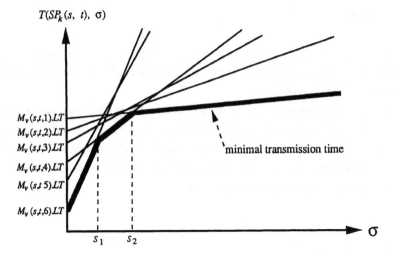

Figure 2. The minimal transmission time is the lower portion of $SP_i(s, t)$, $i=1,...,6$.

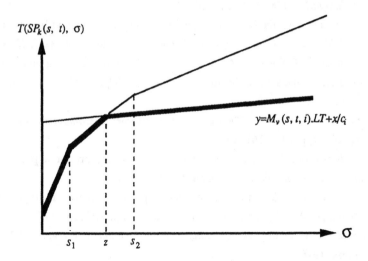

Figure 3. After a line $y=M_v(s, t, i).LT+x/c_i$ is added to the plane, a new intersection point z is generated and the intersection point s_2 is discarded.

5. Concluding Remarks

The quickest path problem is a variant of the shortest path problem. Unlike the shortest paths, the selection of the quickest paths depends on not only the characteristics of the network but also the amount of data to be transmitted.

In this paper, we have proposed distributed algorithms for the quickest path problem. For the 1_QSP, we proposed two distributed algorithms that require $O(rn^2)$ messages, $O(rn^2)$ time, and $O(rmn)$ messages, $O(rn)$ time, respectively, where m, n and r are the numbers of arcs, nodes and distinct capacity values, respectively, in the network N. For the A_QSP, we proposed a distributed algorithm that requires $O(mn)$ messages and $O(m)$ time.

References

[1] B. Awerbuch, "Optimal distributed algorithm for minimum weight spanning tree, counting, leader election and related problems," *Proceedings of 19th Annual ACM Symposium on Theory of Computing*, May 1987, pp. 230-240.

[2] B. Awerbuch and R. Gallager, "A new distributed algorithm to find breadth first search trees," *IEEE Transactions on Information Theory*, vol. 33, no. 3, pp. 315-322, May 1987.

[3] L. D. Bodin, B. L. Golden, A. A. Assad, and M. O. Ball, "Routing and scheduling of vehicles and crews: the state of the art," *Computers and Operations Research*, vol. 10, pp. 63-211, 1982.

[4] Y. L. Chen and Y. H. Chin, "The quickest path problem," *Computers and Operations Research*, vol. 17, pp. 153-161, 1989.

[5] G. N. Frederickson, "A distributed shortest path algorithm for a planar network," *Information and Computation*, vol. 86, pp. 140-159, 1990.

[6] M. L. Fredman and R. E. Tarjan, "Fibonacci heaps and their uses in improved network optimization algorithms," *Journal of the ACM*, vol. 34, pp. 596-615, 1987.

[7] E. Horowitz and S. Sahni, *Fundamentals of Data Structures*, Potomac MD: Computer Science Press, 1976.

[8] Y. C. Hung and G. H. Chen, "On the quickest path problem," *Proceedings of the International Conference on Computing and Information*, Ottawa, Canada, May 1991, to appear.

[9] K. B. Lakshmanan, K. Thulasiraman, and M. A. Comeau, "An efficient distributed protocol for finding shortest paths in networks with negative weights," *IEEE Transactions on Software engineering*, vol. 15, no. 5, May 1989.

The Communication Complexity of the Two List Problem

Alon Itai *

Dept. of Computer Science

Technion, Haifa 32000, Israel

email: itai@cs.technion.ac.il

Abstract

Let L and R be processors, each containing the n numbers $x_1, ..., x_n$ and $y_1, ..., y_n$ respectively, where each number consists of n bits. Their task is to determine whether there exists an i such that $x_i = y_i$. This problem requires $\Omega(n^2)$ bits for deterministic algorithms [7]. Here a simple $O(n)$ expected bit randomized (Las Vegas) algorithm is suggested. Its properties depend on the properties of universal hash functions, not on prime numbers or finite fields. Then the number of random bits is reduced by a general construction. Finally, practical algorithms using almost independent strings are presented.

1 Introduction

The *two list problem* is defined as follows: Let L and R be processors, each containing the n numbers $x_1, ..., x_n$ and $y_1, ..., y_n$ respectively, where each number consists of n bits. Their task is to determine whether there exists an i such that $x_i = y_i$. This problem has been investigated since it shows the power of nondeterminism and randomness over deterministic algorithms. Namely, Melhorn and Schmidt [7] have shown that any deterministic algorithm requires $\Omega(n^2)$ bits of communication while nondeterministically we can do with $\theta(n)$ bits. They also showed a randomized algorithm whose expected communication complexity is $O(n \log n)$ bits. Fürer reduced the expected bit complexity to $O(n)$. His method involves properties of finite fields and prime numbers. Fleischer et al. [5] have shown a $\log(n^2/\overline{C})$ lower bound for expected bit complexity $\Omega(n) \leq \overline{C} \leq O(n^2)$ and showed that expected bit complexity $\Omega(n \log n) \leq \overline{C} \leq O(n^2)$ can be achieved by using no more than $(1 + o(1)) \log(n^2/\overline{C})$ random bits.

Here we give a simpler randomized Las Vegas algorithm and show that its properties follow from a well-known construction – universal hash functions [3]. Moreover, we show how to reduce the number of random bits by a general method due to Newman [11]. We point out the difficulties with

---

*Supported by Technion V.P.R. – E. & J. Bishop Research Fund.

Newman's construction, suggest a better algorithm and show how methods for constructing almost independent random strings can be utilized to construct even more efficient algorithms.

The purpose of the paper is twofold, first to present very simple algorithms and second to demonstrate that general methods (universal hash functions, Newman's reduction and small sample spaces) may be used to devise efficient algorithms, the properties of which follow from general properties of randomness not from the properties of finite fields.

2 The model

We assume two processors L and R with unbounded computational power, which communicate over a single channel. The processors are asynchronous. Every message (bit) sent arrives, without error, after a finite but unpredictable delay. Our algorithms will ensure that the two processors do not attempt to use the channel simultaneously.

In this mode the only restricted resource is bits of communication. However, all our algorithms require $O(n)$ bit operations per communication bit. Also we will aim at "short" programs. I.e., some programs require large tables, the size of which depends on n. The more efficient algorithms do not need tables, and their storage requirements are $o(n)$.

Note that our model differs slightly from that of Fleischer et al [5] in that we pay for all acknowledgement bits.

One word about synchronous algorithms. In a synchronous environment, L can code its entire input $x_1, ..., x_n$ as one huge integer N and send a single bit at time N. R could then decode N and decide whether an appropriate i exists. The time of this algorithm is, of course, excessive. More reasonable tradeoffs between time and bit complexity are possible [12].

3 The basic algorithm

We shall make use of universal hash functions as defined in [3]: a class of functions H from A to B is a *universal$_2$ class of functions* iff for any $x \neq y \in A$

$$|\{h \in H \ : \ h(x) = h(y)\}| \leq |H|/|B| .$$

In this paper $B = \{0,1\}$. Thus, if $x \neq y$ and h is chosen at random from H then the probability that $h(x) = h(y)$ is $\frac{1}{2}$.

We now describe a distributed algorithm for the two list problem. We assume that $H = \{h_0, ..., h_{|H|-1}\}$ is a class of universal$_2$ hash functions.

The Algorithm

1. L generates m (to be specified later) random integers $r_1, ..., r_m$, $0 \leq r_i < |H|$ and sends them to R.

2. For $i := 1$ to n or until $x_i = y_i$ do (3)–(5) below:

3. For $j := 1$ to m or until R detects that $x_i \neq y_i$ do (4):

4. L computes $h_{r_j}(x_i)$ and sends it to R. R computes $h_{r_j}(y_i)$ and compares it to the value of $h_{r_j}(x_i)$ as received from L. If they are different then R detects that $x_i \neq y_i$ and notifies L; otherwise, R realizes that $h_{r_j}(x_i) = h_{r_j}(y_i)$, notifies L and the algorithm terminates.

5. If for $j = 1, \ldots, m$, $h_{r_j}(x_i) = h_{r_j}(y_i)$ then R sends y_i to L, where it can be compared to x_i.

Theorem 1: *The expected communication cost of the algorithm is bounded by $5n + 2m + (n-1)n2^{-m}$, plus the cost of sending the random integers $r_1, \ldots, r_m$.*

Proof: If $x \neq y$ and h_r is chosen at random $P(h_r(x) = h_r(y) \mid x \neq y) \leq 1/2$. Hence, the probability that $y_i \neq x_i$ is sent to L is 2^{-m}. Therefore, the expected cost of y_i being sent from R to L is $n2^{-m}$. The expected number of bits sent from L to R is $\sum_{i=1}^{m} 2^{-(i-1)} < 2$. With the acknowledgement bits this adds up to $4n$.

If $x_i = y_i$ the cost is $2m + n$.

The worst case occurs when $x_n = y_n$ and $x_i \neq y_i$ $(i < n)$. Thus, the total cost is $(n-1)n2^{-m} + 4(n-1) + 2m + n = 5n + 2m + (n-1)n2^{-m} - 4$. $\qquad\square$

Corollary 1: *For $m = 2 \log n$ the expected bit complexity is $5n + (2 \log n) \log |H| + O(n)$.*

A modification: The basic algorithm can be modified in a way that will prevent an adversary from forcing L to always check the entire input until finding and index i such that $x_i = y_i$. The modification is as follows: L decides at random whether to start checking the input from x_1 to x_n or from x_n to x_1. L must, of course, notify the direction to R (1 bit).

If there exists an i such that $x_i = y_i$ the expected number of trials until such an i is checked is bounded by $n/2$ (there can be more than one). Thus, the worst case of the modified algorithm occurs when there is no such i. In this case the expected communication complexity is $4n + 2m + (n-1)n2^{-m}$, plus the cost of sending the random integers.

4 Choosing the random functions

The cost of the previous algorithm was dominated by the cost of sending the hash functions. Our first choice is the following class of functions shown by Carter and Wegman to be universal$_2$: Let $h_r(x) = \sum_{i=0}^{n-1} x[i]r[i] \bmod 2 \stackrel{\text{def}}{=} \bigoplus_{i=0}^{n-1} x[i]r[i]$, where $x[0], \ldots, x[n-1]$ and $r[0], \ldots, r[n-1]$ are the bit representation of x and r respectively.

$$H_3 = \{h_r \ : \ 0 \leq r < 2^x\}.$$

Since $|H_3| = 2^n$ for $m = \log n$ the expected bit complexity of the algorithm is $n \log n + O(n)$.

5 Reducing the number of random bits

Using the result of Ilan Newman[11], we can reduce the number of random bits to $O(\log n)$.

A random algorithm A may be viewed as a set of r deterministic algorithms $A_1, \ldots, A_r$. Applying the random algorithm is equivalent to randomly choosing a deterministic algorithm and applying it. The average time of the randomized algorithm is the average time of all these algorithms.

Theorem 2: *[Newman] Let A be a randomized algorithm with probability bound ε and complexity c on a set X of 2^N inputs. Then there is a randomized algorithm A^* with error bound 2ε, complexity $O(c)$ that uses only $O(\log(N/\varepsilon))$ random bits.*

Proof: The assumption of A having ε error bound means that for each input $x \in X$ no more than ε fraction of the A_r's are incorrect.

Claim 1: *For any $\delta > 0$ there is a collection of $\ell = 2N/\varepsilon\delta^2$ algorithms $A_{i_1}, .., A_{i_\ell}$ out of $A_1, .., A_r$ such that for each input, no more than $(1+\delta)\varepsilon$ fraction of $A_{i_1}, .., A_{i_\ell}$ err.*

Proof: For an input $x \in X$ we say that an algorithm is x-bad if it is incorrect on x. Pick $\ell = 2N/(\varepsilon\delta^2)$ algorithms out of the original collection, independently and with equal probability. By the assumption on the error bound of A, for each input $x \in X$, the probability that the chosen algorithm is x-bad is bounded by ε. By Chernoff,

$$Prob(\text{there are more than } (1+\delta)\varepsilon\ell \ x-\text{bad algorithms})$$
$$\leq exp(-\delta^2 \varepsilon\ell/2)$$
$$= exp(-N).$$

Thus, $Prob(\exists x, \text{there are more than } (1+\delta)\varepsilon\ell \ x-\text{bad algorithms}) \leq 2^N \cdot exp(-N) < 1$. □

Now, take $\delta = 1$ and consider the randomized algorithm that picks at random, with uniform distribution, an algorithm from the new collection $A_{i_1}, .., A_{i_\ell}$, that is promised by the claim. It is guaranteed to have error bound of $(1+\delta)\varepsilon = 2\varepsilon$ and it uses only $\log \ell = O(\log(N/\varepsilon))$ random bits. □

The given proof is nonconstructive, but since for a given X there is a finite number of randomized algorithms with complexity c, it is possible to find the required set of algorithms by an exhaustive search. We now consider the algorithm of Section 3 as a common bit algorithm. As such it requires $4n + o(n)$ communication bits for sufficiently large n the average bit complexity is less than $5n$. Therefore, there is probability less than a quarter that it will send $20n$ bits. Modify the algorithm so that after sending $20n$ bits it stops returning the value ERROR indicating that an error occurred. Since the average communication complexity is less than $5n$, the probability for this to happen is at most $1/4$.

By the theorem, there exists an algorithm with error $\leq 2 \cdot (1/4) = 1/2$ which uses only $O\left(\log \frac{2n^2}{1/4}\right) = O(\log(8n^2)) = O(\log n)$ random bits.

To obtain a Las Vegas algorithm (one which doesn't make mistakes), we first notice that since each of the original algorithms always returned the correct result, the only possible error is if the value ERROR is returned. Thus, we modify the algorithm as follows: when ERROR is returned the algorithm is rerun. The probability of rerunning is at most a half and this transformation may only double the expected bit complexity.

The algorithm, as described, has infinite worst case running time. This also may be addressed by a standard technique: After sending n^3 bits, we run the $O(n^2)$ deterministic algorithm.

To summarize, we have a Las Vegas algorithm, with average bit complexity $10n$ and which uses $O(\log n)$ random bits.

6 A better algorithm

The previous algorithm has serious implementation problems: It must store all the ℓ random algorithms. I.e., it requires $\ell = 2n^2n/\epsilon$ "programs" each consisting of $2\log n$ n-bit integers. Thus, a total of $O(n^4 \log n)$ bits. Also, the time needed to construct the algorithm is excessive

We shall improve the memory requirements by decreasing the size of the sample space (the set H). The price we shall pay is that the class will no longer be universal, but it will satisfy the following weaker requirement: Let $\alpha \geq 1$, a class H of functions from A to B is α-universal, if $x \neq y$ then

$$|\{h \in H \ : \ h(x) = h(y)\}| \leq \alpha |H|/|B| \ .$$

In particular we shall use the following variant on H_3: Instead of choosing the r's randomly from $\{0,1\}^n$, we shall use only $2\sqrt{n}$ random bits for each r. Let σ, ρ be chosen randomly from $\{0,1\}^{\sqrt{n}}$. Consider the $\sqrt{n} \times \sqrt{n}$ matrix R, whose ℓ, k-th entry is $\rho[\ell] \cdot \sigma[k]$. The random vector r is the concatenation of the rows of R, i.e.,

$$r[k + (\ell - 1)\sqrt{n} \,] = \rho_\ell \sigma_k.$$

Obviously, r is not random, however, we can show

Lemma 1: *Let r be chosen as above. Then if $x \neq y$ then*

$$P(h_r(x) = h_r(y)) \leq \frac{3}{4}.$$

Proof: Let $z = x \oplus y$, (i.e., $z[k] = x[k] + y[k]$ mod 2).
$h_r(x) = h_r(y)$ if and only if $h_r(z) = 0$. Consider the matrix Z, such that the concatenation of its rows yields z, i.e., $Z[\ell, k] = z[k + (\ell - 1)\sqrt{n} \,]$.

To compute $h_r(z)$, we first perform the xor (exclusive or) on each row and then sum all the rows:

$$
\begin{aligned}
h_r(z) &= \bigoplus_m z[m]r[m] = \bigoplus_{\ell,k} z[k + (\ell-1)\sqrt{n} \,]r[k + (\ell-1)\sqrt{n} \,] \\
&= \bigoplus_\ell \bigoplus_k Z[\ell, k]\rho_\ell\sigma_k = \bigoplus_\ell \left(\bigoplus_k Z[\ell, k]\rho_\ell\sigma_k \right) \\
&= \bigoplus_\ell q_\ell \rho_\ell,
\end{aligned}
$$

where $q_\ell = \bigoplus_k Z[\ell, k]\sigma_k$.

If z is nonzero, there exists ℓ_0, k_0 such that $Z[\ell_0, k_0] = 1$. The class of functions $\{h_\sigma : \sigma \in \{0,1\}^{\sqrt{n}}\}$ is similar to H_3 except for the size of the domain. Hence it is also universal$_2$. With probability $\frac{1}{2}$, $q_{\ell_0} = \bigoplus_k Z[\ell_0, k]\sigma_k = h_\sigma(Z[\ell_0, \cdot]) = 1$. Thus, with probability $\geq \frac{1}{2}$, the vector $q = (q_1, \ldots, q_{\sqrt{n}})$ is not all zero. If $q \neq 0$ then by the same argument, with probability half, $h_r(z) = \bigoplus_\ell q_\ell \rho_\ell = h_\rho(q) = 1$. Therefore, if z is nonzero then $P(h_r(z) \neq 0) \geq \frac{1}{4}$. □

Note that if z contains only a single one, then $P(h_r(z) \neq 0) = \frac{1}{4}$.

Corollary 2: *If the number of random functions $m = 4 \log n$ then the expected communication complexity of the modified algorithm is $4n + o(n)$.*

7 More efficient algorithms

The heart of our construction is a space of almost random strings. Several previous authors [8, 1, 10, 2] also looked for small spaces of random strings. We shall use the construction of Alon et al. [2]. Their construction is somewhat more involved and requires more computation. However, to specify a function that with probability $\frac{1}{2}$, $x \neq y$ implies $h_r(x) \neq h_r(y)$, they need only $\log n + O(1)$ bits.

They have suggested several schemes. The simplest is that of a linear feedback shift register: Let $f(t) = 1 + f_1 t + \ldots + f_{k-1} t^{k-1} + t^k$ be an irreducible polynomial, and $s_0, \ldots, s_{k-1}$ a random bit string. The almost random string r is produced as follows:

$$r[i] = \begin{cases} s_i & \text{if } i < k \\ \bigoplus_{j=0}^{k-1} f_j \cdot r[i - k + j] & \text{otherwise.} \end{cases}$$

They proved that for any nonzero z

$$|P(h_r(z) = 0) - P(h_r(z) = 1)| \leq (n - 1)2^{-k}(1 + O(2^{-k/2})).$$

For $k = 3 + \log n$, this yields that for nonzero z,

$$P(h_r(z) = 0) \geq \frac{1}{4}.$$

To use this scheme L must choose $2k - 1$ random bits ($s_0, \ldots, s_{k-1}$ and $f_1, \ldots, f_{k-1}$). By choosing $m = 4 \log n$, we may achieve communication complexity $4n + O(\log^2 n)$, using only $2km = 8 \log^2 n + O(\log n)$ random bits. (The constant 8 can be reduced.)

8 Conclusions

We have presented several algorithms for the two list problem. The basic algorithm is very simple, and its analysis depends explicitly on the properties of universal hash functions. The schemes of previous solutions also yield universal hash functions. Thus, the intrinsic property is that of universal hash functions.

The number of random bits can be reduced either by a general technique, (which is spacewise inefficient) or by using "almost-random" sequences. Thus, again we show how to use a general method.

The performance of our algorithms is slightly inferior (a constant) to that of Fleischer et al. [5]. An interesting problem is to see the exact relationship between the various algorithms.

Acknowledgment

It is a pleasure to thank Noga Alon, Benny Chor, Roudolph Fleischer, and Avi Wigderson for for useful discussions and suggestions.

References

[1] "A fast and simple randomized parallel algorithm for the maximal independent set problem". Alon, N., L. Babai, A. Itai, J. Algorithms, 7, 567-583, (1986).

[2] Alon, N., O. Goldreich, J. Hastead and R. Peralta, "Simple construction of almost k-wise independent random variables". FOCS 90.

[3] Carter, J. L., and M. N. Wegman, "Universal classes of hash functions," *JCSS 18*, 143-154 (1979).

[4] Duris, P., and Z. Galil. "Two lower bounds in asynchronous distributed computation," 28 FOCS, 326-330, (1987).

[5] Fleischer, R., H. Jung, and K. Melhorn, "A time-randomness tradeoff for communication complexity", WDAG 4, Otranto, Italy, (1990). *Lecture Notes in Computer Science 486*, J. van Leewen and N. Santoro, (eds.), Springer-Verlag.

[6] Fürer, M., "The power of randomness for communication complexity," STOC 1987, 178-181.

[7] Melhorn, K., and E.M. Schmidt, "Las Vegas is better than determinism in VLSI and distributed computing," *STOC 1982*, 330-337.

[8] Luby, M., "A simple parallel algorithm for the maximal independent set problem", SIAM J. Computing, 15 (1986), 1036-1053.

[9] Melhorn, K., *Data structures and Algorithms 1: Sorting and Searching* EATCS Monographs on Theoretical Computer Science, Springer-Verlag, 1984.

[10] Naor, J., and M. Naor, "Small-bias probability spaces: Efficient construction and applications", 22nd ACM Symposium on the Theory of Computing, 213-223, (1990).

[11] Newman, I., "Communication complexity in the private random bits model is no harder than in the common bits model," unpublished manuscript, 1990.

[12] O'Reilly, U-M., and N. Santoro, "The expressiveness of silence", unpublished communication, (1991).

[13] Papadimitriou, C. H., and M. Sipser, "Communication complexity", Fourteenth ACM Symposium on the Theory of Computing, 196-200, (1982).

Distributed Algorithms for Updating Shortest Paths [*]
(EXTENDED ABSTRACT)

Giuseppe F. Italiano[†]
IBM Research Division
T. J. Watson Research Center
Yorktown Heights, NY 10598

Abstract

We present a distributed algorithm for updating all pairs shortest paths when an edge is either inserted or has its cost decreased. For unit edge costs (i.e., for maintaining all pairs minimum hops) our algorithm transmits a total of $O(V^3 \log V)$ messages during the insertion of $O(V^2)$ edges. This gives basically $O(V \log V)$ messages per change in the amortized sense. For integer edge costs in $[1, W]$, our algorithm transmits at most $O(WV^3 \log(VW))$ messages during $O(WV^2)$ edge insertions or edge cost decreases. Again, this gives $O(V \log(VW))$ messages per change in the amortized sense. The algorithm runs asynchronously, requires $O(V)$ time per change and $O(V)$ space per node.

This result favorably compares to using the best known distributed algorithm for computing all pairs shortest paths from scratch, and to other dynamic distributed algorithms previously proposed in the literature.

1 Introduction

1.1 Preliminaries and Terminology

We assume that the reader is familiar with the standard graph theoretical terminology as contained for example in [2, 14]. In particular, a *directed graph* $G = (V, E)$ (also referred to as a *digraph*) consists of a finite set V of *nodes* and a finite set E of edges such that each edge e has a *head* $h(e) \in V$ and a *tail* $t(e) \in V$. In the following, we use V and E to denote both the sets of nodes and edges and their cardinality. We consider the edge $e = (t(e), h(e))$ as directed from $t(e)$ to $h(e)$ and we say that e leaves $t(e)$ and enters $h(e)$. When edges are not directed, we say that the graph is *undirected*. Given a directed graph G, the undirected graph G^* obtained by considering each edge e of G as undirected is referred to as *the undirected graph corresponding to G*. In the following, when we speak about a graph we assume that the graph is directed, unless explicitly said otherwise.

A (directed) *path* $p = e_1, e_2, \ldots e_k$ is a sequence of edges such that $h(e_i) = t(e_{i+1})$ for $1 \le i \le k-1$. The path p is from node $t(p) = t(e_1)$ to node $h(p) = h(e_k)$ and contains the edges $e_1, e_2, \ldots, e_k$

---

[*]Work partially supported by the ESPRIT II Basic Research Actions Program of the EC under contract No. 3075 (Project ALCOM).

[†]This work was done while the author was at Columbia University.

and the nodes $t(e_1), t(e_2), \ldots t(e_k), h(e_k)$. The path is *simple* if all its nodes are distinct. The *length* of a path is the number of edges it contains. A path p from x to y is said to be of *minimum length* (or in short *minimum*) if there is no other path from x to y that contains a smaller number of edges. A (directed) *cycle* is a nonempty (directed) path from a node to itself. A node v is said to be *reachable* from a node u if there is a path p from u to v. If v is reachable from u, then u is said to be an *ancestor* of v and v a *descendant* of u. If $u \neq v$, u is a *proper ancestor* of v and v is a *proper descendant* of u. If there is an edge from u to v, then u and v are said to be *adjacent*. A digraph with no cycles is called a *directed acyclic graph* (in short *dag*). A *rooted tree* T is a dag satisfying the following properties: (i) there is only one node, referred to as the *root*, with no edges entering it; (ii) every node but the root has exactly one entering edge. A *leaf* of the tree is a node with no outgoing edges. If there is an edge (u, v) in T, u is said to be a *parent* of v and v a *child* of u. The *depth* of a node v in T is the length of the path from the root to v. Given a digraph $G = (V, E)$, a *spanning tree* is a rooted tree $T = (V, S)$ such that $S \subseteq E$.

Given a digraph $G = (V, E)$ and a node $x \in V$, a *tree of forward paths rooted at* x is a tree $T_F(x) = (X, S)$ rooted at x such that X is the set of descendants of x in G, and $S \subseteq E$. Similarly, a *tree of backward paths* rooted at x is a tree $T_B(x) = (X, S)$ rooted at x such that X is the set of ancestors of x in G, and $S \subseteq E^R$, where $E^R = \{(y, x) \mid (x, y) \in E\}$ is called the *set of reversed edges*. A tree of forward (backward) paths $T = (X, S)$ rooted at x is said to be of *minimum length* (or equivalently it is said to be a *tree of forward (backward) minimum paths*) if for each node $v \in X$, the depth of v in T coincides with the length of the minimum path from x to v. Notice that trees of minimum paths correspond to BFS-trees.

The above terminology can be extended to a graph G with edge costs. Denote by $c(i, j)$ the cost of edge (i, j). The cost of a path p can be defined as $c(p) = \sum_{(i,j) \in p} c(i, j)$. A path p from x to y is said to be the *shortest path* from x to y if there is no path from x to y whose cost is less than $c(p)$. A *tree of forward (backward) shortest paths* T is a tree of forward (backward) paths such that each path in T is a shortest path in G. We remark that trees of forward (backward) shortest paths rooted at a given node are not necessarily unique.

1.2 Network Model and Complexity Measures

In this paper we consider dynamic asynchronous communication networks. The network is represented by a directed graph $G = (V, E)$, whose nodes represent processors and whose edges represent links between processors. There is no global clock and no global memory is shared among the nodes. Communication is performed by exchanging messages in either direction over the edges. Each edge conforms to a FIFO discipline, i.e., messages always arrive (error-free) in the order sent and after an arbitrary but finite delay. For sake of simplicity, we assume that edge delays are normalized, i.e. the (normalized) time required to send a message over an edge is in the interval $[0, 1]$. When there is a change in the network concerning an edge (i, j), we assume that nodes i and j (but no other nodes) are able to detect the change.

The message complexity of a distributed algorithm is the total number of messages sent over the edges. We assume that each message contains at most $O(\log V)$ bits, where V is the number of nodes in the graph. The amortized message complexity is the total message complexity divided by the number of changes. The space complexity is the local space usage per node. The time complexity is the total (normalized) time elapsed from a change.

1.3 The Problems Considered

Problems on dynamic communication networks have been extensively studied [1, 4, 6, 7, 8, 9, 10, 11, 12], and many researchers have been focusing on the design of compilers that transform static algorithms into dynamic algorithms using distributed reset procedures (see for instance reference [1]).

One important problem in this area is the all pairs shortest path problem, where each node must maintain the shortest path to every other node. This is crucial for routing (such as in the ARPANET network [13]) and is also a basic subproblem in many other distributed network protocols. In this paper, we are concerned with updating all pairs shortest paths in directed graphs, where edges may change because of edge insertions and edge cost decreases.

Several solutions to this problem are known. One can recompute from scratch all pairs shortest paths after each change. However, this is quite expensive since it can require as much as $O(E^{1+\epsilon} \log W)$ messages, and $O(V^{1+\epsilon} \log W)$ time, just for recomputing the shortest paths from all the nodes to one distinguished node [5]. Here E is the current number of edges, W the largest edge cost, and $\epsilon > 0$. Better bounds can be achieved by using the algorithm of Ramarao and Venkatesan [12] that is able to update the all pairs shortest paths using $O(V^2)$ messages per change, $O(V)$ time per change, and $O(V)$ space per node. It is possible to further improve the message complexity at the expense of the local space required per node. Indeed, following for instance the general approach of Awerbuch et al. [6], the message complexity can be reduced from $O(V^2)$ to $\Theta(V)$, while the space per node increases from $O(V)$ to $O(E) = O(V^2)$.

It is therefore quite natural to ask whether there is any trade-off between the message complexity and space required at each node. We answer this question by presenting a distributed algorithm for updating all pairs shortest paths during edge cost decreases and edge insertions when edge costs are integers in a fixed range, say $[1, W]$. Our algorithm requires a total of $O(WV^3 \log(VW))$ messages over any sequence of (at most $O(WV^2)$) edge insertions and edge-cost decreases (any edge can be inserted at most once and have its cost decreased at most $O(W)$ times). This gives an $O(V \log(VW))$ amortized message complexity per change. Furthermore, each change can be dealt with in $O(V)$ time and with $O(V)$ space per node. The results are summarized in Table 1.

Author(s)	Communication	Space	Time
[6]	$\Theta(V)$	$O(V^2)$	$O(V)$
[12]	$O(V^2)$	$O(V)$	$O(V)$
This paper	$O(V \log(VW))$	$O(V)$	$O(V)$

Table 1

The remainder of the paper consists of four section. Section 2 reviews the sequential algorithm of Ausiello et al. [3] for dynamic shortest paths. Section 3 describes our distributed algorithm, that is inspired by the sequential algorithm in [3]. In Section 4 we analyze the message complexity of our algorithm, while Section 5 lists some concluding remarks.

2 The Sequential Algorithm

We now describe the sequential algorithm given in [3] using a somewhat different terminology. To each node v of G, the tree of forward and backward shortest paths rooted at v is associated. We refer to these trees as $T_F(v)$ and $T_B(v)$, respectively. Furthermore, to each node v an array D_v is associated, defined as follows. For each node u, $D_v[u]$ stores the length of the shortest path from v to u. $D_v[v] = 0$ and when there is no path between two nodes the corresponding entry of the array contains a special value $(+\infty)$.

We now show how to update the shortest paths when a new edge (i, j) is inserted. Let $d(x, y)$ denote the length of the shortest path from node x to node y. Let $d_{new}(x, y)$ denote the length of the shortest path after the insertion of an edge (i, j) and let $d_{old}(x, y)$ denote the length of the shortest path before the insertion of (i, j), then the following property must be true:

$$d_{new}(x, y) = \min\{d_{old}(x, y), d_{old}(x, i) + 1 + d_{old}(j, y)\}, \forall x, y \in V. \tag{1}$$

Equation (1) states that the insertion of an edge (i, j) can introduce new shorter paths only from ancestors x of node i (for which $d(x, i) < +\infty$) to descendants y of node j (for which $d(j, y) < +\infty$). In this case, the trees of forward shortest paths might have to be updated for nodes in $T_B(i)$ and the trees of backward shortest paths might have to be updated for nodes in $T_F(j)$. We now sketch how to update trees of forward shortest paths because of the insertion of an edge (i, j) (a similar argument can be applied to the trees of backward shortest paths). In the update of the trees of forward shortest paths, a crucial role is played by the two trees $T_B(i)$ and $T_F(j)$: indeed, in order to guarantee a correct update of the forward trees, it is sufficient that for each node x contained in the backward tree $T_B(i)$, its forward tree $T_F(x)$ is updated using $T_F(j)$.

The update algorithm can be described as follows. We start from node i and update both the array D_i and the tree $T_F(i)$ with the help of the tree $T_F(j)$. The details of this update will be given later on. At the end of the update of $T_F(i)$, let $T \subseteq T_F(j)$ be the tree such that $v \in T$ if and only if the distance from i to v has decreased after the insertion of edge (i, j). Notice that T is the subtree of $T_F(i)$ (after the update) rooted at node j. This tree T is then sent to the children of i in $T_B(i)$ which will perform the updates of their forward trees of minimum paths in a recursive fashion. In other words, each node x in $T_B(i)$ starting from node i receives a tree T that is a (possibly pruned) copy of $T_F(j)$. After receiving T, node x updates $T_F(x)$ with the help of T. At the same time, x performs a further pruning of T and passes it to its children in $T_B(i)$.

The update of $T_F(x)$ for each node $x \in T_B(i)$ is carried out as follows. Node x visits the tree T it receives, starting from the root of T. The subsequent update of $T_F(x)$ and the pruning of T are subject to the following two rules.

(i) When a node y in T for which $d(x, i) + 1 + d(j, y) < d(x, y)$ is reached, then a shorter path from x to y has been found. In this case, node y is inserted into $T_F(x)$ either as a child of i if $y = j$ or as a child of the same parent it has in $T_F(j)$ otherwise. By inserting a node y into a tree, we mean that any previous occurrence of y in the tree is implicitly deleted. At the end of this step, all the children of y in T are examined with the same algorithm in a recursive fashion.

(ii) When a node y in T for which $d(x, i) + 1 + d(j, y) \geq d(x, y)$ is reached, then no shorter path from x to y was introduced by the insertion of the edge (i, j). In this case no update is performed in $T_F(x)$ and the tree T is pruned by cutting the subtree rooted at y. In fact, for each node y' in such a subtree the new arc (i, j) does not decrease the distance from x to y'.

After $T_F(x)$ has been updated, the pruned copy of T is passed to the children of x in $T_B(i)$.

We notice that the heart of the algorithm consists of using a *bidirectional pruning* in the update. Indeed, assume we are updating the tree $T_F(x)$ with the help of the tree T, and let y be an encountered node of T that satisfies the condition of rule (ii). As noticed above, in this case no shorter path from node x to node y was introduced by the insertion of the edge (i, j). The implementation of rule (ii) implies two facts. First, the nodes in the subtree of T rooted at node y are not looked up for updating $T_F(x)$ (*forward pruning*). Second, the node y (and consequently its subtree) is pruned from T and will not be passed to the children of x in the backward tree $T_B(i)$ (*backward pruning*).

Ausiello et al. [3] showed that this sequential algorithm requires a total of $O(V^3 \log V)$ time on any sequence of (at most $O(V^2)$) edge insertions. When edge costs are integers in a small range, say $[1, W]$, then the same algorithm achieves a $O(WV^3 \log(VW))$ total time bound on any sequence of (at most $O(WV^2)$) edge insertions and edge-cost decreases. As shown in [3], these bounds improve almost one order of magnitude on previously known algorithms and are only a logarithmic factor away from the best possible bounds.

3   The Distributed Algorithm

We now describe a distributed algorithm that exploits some of the ideas of the previous sequential algorithm (recall that the main idea below the sequential algorithm was to apply a bidirectional pruning during the insertion of an edge (i, j)). For our distributed algorithm, we maintain the following data structures. To each node v, the tree of forward shortest paths $T_F(v)$ rooted at v and the array D_v are associated. As said before, we denote by G^* the undirected graph corresponding to the directed graph G. Throughout the algorithm, we maintain a spanning forest F^* of G^*. We notice that, differently from the sequential algorithm, we do not maintain explicitly the trees of backward shortest paths $T_B(x)$. Rather, each time there is a change in an edge (i, j), the tree $T_B(i)$ is computed on the fly.

Our distributed algorithm works in three phases. When there is a change involving an edge (i, j), the two nodes i and j are able to detect the change. In the first phase, node i starts a broadcast of the change that uses only edges of F^*. Then $T_B(i)$, a tree of backward shortest paths rooted at node i, is computed. This is called the *backward tree computation phase*. In the second phase, referred to as the *actual update phase*, the update of the trees of forward shortest paths $T_F(x)$, for all the nodes x ancestor of i, is carried out using the computed tree $T_B(i)$. Finally, the last phase consists of updating the forest F^*. This is called the *forest update phase*. During the three phases, all the nodes in $T_B(i)$ can be in three different states: *ready*, *active* or *done*. Informally, a node x is *ready* if the tree $T_F(x)$ has to be (possibly) updated because of the change in edge (i, j), x is active if it is in the middle of updating its tree $T_F(x)$, while it becomes *done* when $T_F(x)$ has been fully updated. Each node of $T_B(i)$ starts as *ready* at the beginning of the second phase and finishes the second phase as *done*. Possible transitions between these states are either from *ready* to *active* and to *done*, or directly from *ready* to *done*. We now describe the three phases in more detail.

3.1 The backward tree computation phase

The first phase consists actually of three subtasks. First, the change in G is broadcast to all the interested nodes. Next, a tree of backward shortest paths is actually computed. Finally, the nodes get ready to start the subsequent update phases.

During the initial broadcast, messages are transmitted only along edges of F^*. The message is $START(i,j,c(i,j))$ and contains information about the edge (i,j) that has changed, and its new cost $c(i,j)$. When node i detects the change in edge (i,j) (remember that (i,j) is an edge directed from i to j), it starts the broadcast by sending a $START(i,j,c(i,j))$ message to all its neighbors in F^*. A node x, when receiving the $START$ message from its neighbor u, forward the same $START$ message to all its neighbors except u. By the end of this phase, all nodes in the same connected component of G^* as i have received the $START$ message. By definition of G^*, this set of nodes includes all the ancestors of i in G. Furthermore, each node x, upon receiving the $START$ message can easily verify whether it is an ancestor of i by simply checking if $D_x[i]$ is different from $+\infty$. Consequently, at the end of the broadcast, all the ancestors of i have been informed of the change in edge (i,j). Clearly, this phase requires at most $O(V)$ messages and $O(V)$ time.

We now describe how the computation of $T_B(i)$ takes place. As said before, upon receiving a $START$ message, each node x can easily verify whether it as an ancestor of i. If it is, then i must appear in the tree $T_F(x)$. In this case, x elects as its *backward parent* the second node (i.e., the closest node to x) in the path from x to i in $T_F(x)$. This is done by notifying this node using an $ELECT$ message. When a node u receives an $ELECT$ message from node x, u knows that it has been chosen by x as a backward parent. As a consequence of the definition of the trees $T_F(x)$, all the backward parents define a tree of backward shortest paths $T_B(i)$. As a special case, we say that node i chooses node j as its backward parent. Note that the computation of $T_B(i)$ can be accomplished within the bound of $O(V)$ messages and $O(V)$ time.

Before starting the second phase, the nodes get prepared to the actual update of their data structures. They do so by sending back $READY$ messages through the forest F^*. This is initiated by the leaves in F^* that received a $START$ message. After a leaf u has received a $START$ message from a node v, and after u has finished its internal computation (i.e., u has checked whether it is an ancestor of i in G, and possibly computed its backward parent), u puts itself in the *ready* state, and then sends a $READY$ message back to v on edge (u,v). In general, let $u \neq i$ be a non-leaf node in F^* that received a $START$ message from a node $v \neq j$. As soon u receives a $READY$ message from all the nodes it sent a $START$ message, and as soon as u has finished its internal computation, it puts itself in the *ready* state and sends a $READY$ message back to v. When i receives a $READY$ message from all the nodes it sent a $START$ message to, it puts itself in the *ready* state. Then i sends a $WAKEUP$ message to j over edge (i,j). As soon as j gets the $WAKEUP$ message, it puts itself in the *done* state and starts the second phase. We recall that a node x is *ready* if the tree $T_F(x)$ has to be (possibly) updated because of the change in edge (i,j), x is *active* if it is in the middle of updating $T_F(x)$, while x becomes *done* when $T_F(x)$ has been fully updated. At this point (i.e., immediately before the second phase starts) all the nodes of $T_B(i)$ are *ready*, no one is *active*, and the only *done* node is j (indeed the change in (i,j) does not cause any update in $T_F(j)$). Once again, all this requires $O(V)$ messages and $O(V)$ time.

3.2 The actual update phase

Similarly to the sequential algorithm described in Section 2, throughout the update algorithm a node x of $T_B(i)$ receives from its backward parent a pruned copy of the tree $T_F(j)$. The communication uses only the edges of $T_B(i) \cup \{(i,j)\}$.

We now describe the protocol ruling this exchange. Let x be any node in $T_B(i) - \{i\}$ and let u be its backward parent (we will describe $x = i$ as a special case later). Notice that since $x \neq i$, we have that $u \neq j$. The pruned copy of $T_F(j)$ is passed from u to x on edge (u,x) according to the following handshake protocol, where we denote u as the *master* and x as the *slave*. Recall that if u is the backward parent of x then x must have sent to u an $ELECT$ message during the previous phase. As soon as the master u becomes *done*, it starts sending to all the nodes that have elected u as a backward parent (i.e., the slaves) pieces of the subtree of the updated $T_F(u)$ rooted at j, starting from j and in a pre-order fashion. Node u does so by sending messages of the form $OFFER(y, p(y), D_u[y])$, where y is a node in the subtree of $T_F(u)$ rooted at j, $p(y)$ is the parent of y in $T_F(y)$, and $D_u[y]$ is the (updated) length of the shortest path from u to y. As soon as a slave x receives an $OFFER(y, p(y), D_u[y])$ message over the edge (x,u), it compares $D_x[u] + D_u[y]$ with $D_x[y]$, giving rise to the following two cases.

If $D_x[u] + D_u[y] < D_x[y]$, then the change in edge (i,j) introduces a shorter path from x to y. If x was *ready*, then it puts itself in the *active* state. Furthermore x performs the resulting updates in its data structures: it sets $D_x[y]$ to $D_x[u] + D_u[y]$, and inserts y in $T_F(x)$ as a child of node $p(y)$. Finally, x sends u an $ACCEPT(y)$ message.

If $D_x[u] + D_u[y] \geq D_x[y]$, then the change in edge (i,j) does not introduce any shorter paths from x to y. In this case, as for the sequential algorithm, there cannot be any shorter path from any ancestor of x in $T_B(i)$ to any descendant of y in $T_F(u)$. Consequently, x performs no other internal computation and sends u a $REFUSE(y)$ message.

Upon receiving an $ACCEPT(y)$ message from node x, node u computes the node v in the subtree of $T_F(u)$ rooted at j that immediately follows y in the pre-order numbering. On the other hand, upon receiving a $REFUSE(y)$ message, u computes the next node v (again according to pre-order numbering) that is in the subtree of $T_F(u)$ rooted at j but such that it is not in the subtree of $T_F(u)$ rooted at y (in other words it skips all the descendants of y in its pre-order visit of $T_F(u)$). If such a node v is found, then u starts the handshake protocol again with an $OFFER(v, p(v), D_u[v])$ message to x. Otherwise, no such node v exists; this implies that u has finished passing to x the pruned copy of $T_F(j)$ and therefore $T_F(x)$ has been fully updated. In this case, u sends x a $HALT$ message.

As soon as x receives a $HALT$ message from its backward parent, it does the following. If x was in the *active* state, x puts itself in the *done* state and starts the handshake protocol (as a master) with its children in $T_B(i)$ (i.e., the nodes that have chosen x as a backward parent) as slaves. Otherwise (x was either *ready* or *done*), no updates were needed on $T_F(x)$. In this case, x forwards the $HALT$ message to all the nodes that have x as a backward parent.

This explains the case $x \neq i$. We now describe how this protocol is started, i.e., when $x = i$ and $u = j$. The only difference in this case lies in the type of $OFFER$ messages sent over the edge (i,j). Indeed, to make the update in i possible, j sends i messages of the form $OFFER(y, p(y), c(i,j) + D_j[y])$ (rather than of the form $OFFER(y, p(y), D_j[y])$). This will guarantee that i performs the correct comparison of $D_i[y]$ versus $D_i[i] + c(i,j) + D_j[y] = c(i,j) + D_j[y]$. When $y = j$ (i.e., for the first $OFFER$ message sent by j) there is the problem that j has no parent in $T_F(j)$ (i.e., $p(j)$

is undefined). We recall that the use of $p(y)$ is that node i has to make y child of $p(y)$ in $T_F(i)$, provided that i finds out that its distance to y has decreased. Consequently, it is enough that node j starts the handshake protocol by sending an $OFFER(j, i, c(i,j))$ message over edge (i,j).

3.3 The forest update phase

The last phase updates the forest F^* after the change in edge (i,j). Indeed, if the change involving edge (i,j) is an insertion, and i and j are in two different trees of the forest F^*, this causes the two trees to be merged into one. This can be done as follows. As soon as node i becomes *ready*, it starts a spanning tree computation in $F^* \cup \{(i,j)\}$. Namely, he sends to all its neighbors in $F^* \cup \{(i,j)\}$ an *INVITATION* message. When a node $x \neq i$ receives an *INVITATION* message for the first time, say from a node u, it replies to u with a *JOINING* message and sends the *INVITATION* message to all the other nodes that are its neighbors in $F^* \cup \{(i,j)\}$ but u. From now on, x ignores any other *INVITATION* received, i.e., it performs no computation. Similarly, i ignores any *INVITATION* received. At the end of this phase, we have that either F^* did not change (nodes i and j were already in the same tree of F^*) or F^* is augmented with edge (i,j) (nodes i and j were in two different trees of F^*). Since $F^* \cup \{(i,j)\}$ has V nodes and edges, the above algorithm updates F^* in $O(V)$ time and using $2V$ messages. We remark that the forest update phase is started as soon as i becomes *ready*, i.e. after the backward tree computation phase, and can be carried out simultaneously with the actual update phase. Indeed, during the actual update phase a node x may have to update the tree $T_F(x)$ and the array D_x, while during the last phase a node x may have to update only its neighbors in F^*. Consequently, there are no conflicts in the data structures used and the two phases can be carried out at the same time.

4 Message Complexity of the Algorithm

The correctness of the algorithm hinges on the following theorem, whose lengthy proof has been omitted.

Theorem 4.1 *Let x be any node in $T_B(i)$. When x has received a HALT message from its backward parent, then $T_F(x)$ has been correctly updated.*

We now turn to the message and time complexity of our algorithm. As said before, the backward tree computation phase and the forest update phase require a total of $O(V)$ messages and $O(V)$ time. We have still to bound the message and time complexity of the actual update phase.

Let σ be a change (i.e., either an insertion or an edge cost decrease) involving edge (i,j). During the update, we have four kinds of messages: $OFFER$, $ACCEPT$, $REFUSE$, and $HALT$. Denote by O_σ the total number of $OFFER$ messages, by A_σ the total number of $ACCEPT$ messages, by R_σ the total number of $REFUSE$ messages, and by H_σ the total number of $HALT$ messages during the change σ. Since each $OFFER$ message is answered with either an $ACCEPT$ or a $REFUSE$ message, then $O_\sigma = A_\sigma + R_\sigma$. Furthermore, since each node in $T_B(i)$ has only one backward parent, it receives at most one $HALT$ message. Consequently, $H_\sigma \leq V$. This implies that the total number of messages transmitted during an actual update phase for change σ is bounded above by $V + 2(A_\sigma + R_\sigma)$. The total number of messages transmitted during any sequence S of (at most $O(WV^2)$) edge insertions or edge cost decreases will therefore be $\sum_{\sigma \in S}(V + 2(A_\sigma + R_\sigma)) = O(WV^3) + O(\sum_{\sigma \in S}(A_\sigma + R_\sigma))$.

To bound this quantity, we only need to bound $\sum_{\sigma \in S}(A_\sigma + R_\sigma)$. We do this by applying a similar analysis of the sequential algorithm of [3]. Before this, we need a little new terminology.

We define a pair $\langle x, y \rangle$ to be *good during* σ if and only if the change σ causes the distance from x to y to decrease. Notice that this happens if and only if $x \in T_B(i)$, during the update algorithm node x receives an $OFFER(y, *, *)$ message from its backward parent, and it answers with an $ACCEPT(y)$ message. Otherwise, we say that a pair $\langle x, y \rangle$ is *bad during* σ if and only if $x \in T_B(i)$, during the update algorithm x receives an $OFFER(y, *, *)$ message from its backward parent, but it answers with a $REFUSE(y)$ message. Since there is exactly one $ACCEPT$ message for each good pair and one $REFUSE$ message for each bad pair, we have the total number of good pairs during σ is A_σ and the total number of bad pairs during σ is R_σ.

The following lemma is a consequence of the bidirectional pruning used by the algorithm.

Lemma 4.1 *Let $\langle x, y \rangle$ be any bad pair during change σ. Denote by π_x the path from x to i in $T_B(i)$ and by π_y the path from j to y in $T_F(j)$. For any choice of node $u \neq x$ in π_x and node $v \neq y$ in π_y, $\langle u, y \rangle$ and $\langle x, v \rangle$ are good pairs during σ.*

Proof: We proceed by contradiction. Consider first a bad pair $\langle x, y \rangle$ and assume that there exists a node $u \neq x$ in π_x such that also the pair $\langle u, y \rangle$ is bad. This means that the distance from u to y did not decrease during change σ and therefore y did not change its position in $T_F(u)$. Consequently, y is not in the subtree of $T_F(u)$ rooted at j after the update. This implies that neither u nor all the nodes of π_x between x and u will never send an $OFFER(y, *, *)$ message, which contradicts the assumption that $\langle x, y \rangle$ is a bad pair.

To prove the second part of the lemma, consider a bad pair $\langle x, y \rangle$ and assume that there exists a node $v \neq y$ in π_y such that also the pair $\langle x, v \rangle$ is bad. This means that x sent a $REFUSE(v)$ message to its backward parent. Upon receiving this $REFUSE(v)$ message, according to the protocol the backward parent of x skips all the descendants of v in its updated tree $T_F(v)$. As a result, x will never receive an $OFFER(y, *, *)$ message and therefore $\langle x, y \rangle$ cannot be a bad pair, again a contradiction. $\square$

Theorem 4.2 *Let S be any sequence of changes, and let A_σ and R_σ be respectively the number of ACCEPT and REFUSE messages transmitted during change σ. Then*

$$\sum_{\sigma \in S} A_\sigma \leq W(V-1)V^2. \tag{2}$$

$$\sum_{\sigma \in S} R_\sigma = O(WV^3 \log(VW)). \tag{3}$$

Proof: As said before, A_σ and R_σ equal respectively the number of good and bad pairs during change σ.

We start by bounding the total number of good pairs. During change σ, for each good pair $\langle x, y \rangle$ we have that the entry $D_x[y]$ is either changed from $+\infty$ to a positive integer value $\ell \leq (V-1)W$ or decreased at least by 1. Since each entry of an array D can hold only the values $1, 2, \ldots (V-1)W, +\infty$ and there are at most V^2 different such entries, the total number of good pairs is bounded above by $\sum_{\sigma \in S} A_\sigma \leq (V-1)WV^2$, which proves equation 2.

In order to bound the total number of bad pairs, we use an amortization argument based upon the credit technique by Tarjan (see reference [14] for a definition of amortized computational complexity). Our credit policy is the following:

During a change σ that either inserts or decrease the cost of edge (i, j), for each good pair $\langle x, y \rangle$ during σ give $4 \left\lceil \frac{VW}{d(x,y)} \right\rceil$ credits both to x and y, where $d(x, y) \geq 1$ is the distance from x to y after change σ. Furthermore, give 1 credit to the change σ itself.

Denote by Φ_σ the number of credits given during change σ. Since we give credits to good pairs, their distance strictly decreases. Therefore, the total number of credits given during any sequence S of changes is $\sum_{\sigma \in S} \Phi_\sigma \leq \sum_{\sigma \in S} \left(1 + O \left(\sum_{x \in V} \sum_{y \in V} \sum_{d=1}^{(V-1)W} \left\lceil \frac{VW}{d} \right\rceil \right) \right) = O(WV^2 + WV^3 \log(VW)) = O(WV^3 \log(VW))$.

To complete the proof, we have to show that there are more credits after all the changes than bad pairs. We actually show something stronger, namely that there are more credits Φ_σ than bad pairs during each change σ.

Consider any bad pair $\langle x, y \rangle$. Denote by π_x the path in $T_B(i)$ from x to i and by π_y the path in $T_F(j)$ from j to y. Let $\ell_x \geq 0$ and $\ell_y \geq 0$ be, respectively, the total number of edges of π_x and π_y. We have now two cases, depending on whether $\ell_x + \ell_y = 0$ or $\ell_x + \ell_y > 0$.

If $\ell_x + \ell_y = 0$ then $x = i$ and $y = j$. Since by assumption $\langle x, y \rangle$ was a bad pair, this implies that the change in the edge (i, j) did not decrease any shortest path in G. By Lemma 4.1, $\langle i, j \rangle$ is the only possible bad pair during change σ. Since $\Phi_\sigma \geq 1$, then Φ_σ is greater than the number of bad pairs during σ.

Assume now $\ell_x + \ell_y > 0$. Because of Lemma 4.1, for any node $u \neq x$ in π_x and $v \neq y$ in π_y, the pairs $\langle u, y \rangle$ and $\langle x, v \rangle$ are good pairs. This implies that x was given $4 \lceil \frac{VW}{d(x,v)} \rceil$ credits because of v and y was given $4 \lceil \frac{VW}{d(u,y)} \rceil$ credits because of u. As a result, the total number of credits given to node x because of nodes in π_y is bounded below by $4 \sum_{h=\ell_x+1}^{\ell_x+\ell_y} \left\lceil \frac{VW}{hW} \right\rceil \geq 4V \log_e \left(\frac{\ell_x + \ell_y + 1}{\ell_x + 1} \right)$. Similarly, the total number of credits given to node y because of nodes in π_x is bounded below by $4 \sum_{h=\ell_y+1}^{\ell_x+\ell_y} \left\lceil \frac{VW}{hW} \right\rceil \geq 4V \log_e \left(\frac{\ell_x + \ell_y + 1}{\ell_y + 1} \right)$. As a consequence, the number of credits given to both x and y during change σ is at least $4V \log_e \frac{(\ell_x + \ell_y + 1)^2}{(\ell_x + 1)(\ell_y + 1)}$. Since $\ell_x + \ell_y > 0$, we have that $\frac{(\ell_x + \ell_y + 1)^2}{(\ell_x + 1)(\ell_y + 1)} \geq 2$. Then the credits given to x and y are at least $4V \log_e 2 > 2V$. In this case the credits given to both x and y are more than all all the bad pairs of the form $\langle x, * \rangle$ and $\langle *, y \rangle$ which are at most $2V$. Since this argument can be repeated for any bad pair, it proves that if $\ell_x + \ell_y > 0$ once again Φ_σ is greater than the number of bad pairs during σ. This completes the proof of (3). $\square$

As a consequence of Theorem 4.2, we have that the total number of messages transmitted during any sequence of (at most $O(WV^2)$) changes is $O(WV^3 \log(VW))$. It can be shown that the time required by the algorithm is $O(V \log(VW))$ amortized per change. A better implementation of the algorithm (omitted here) is able to achieve $O(V)$ time per change while preserving the same bounds on the message complexity. The main idea behind this improved algorithm is to allow each node to transmit messages to its neighbors in $T_B(i)$ before putting itself in the *done* state.

5 Conclusions and Open Problems

We have described a distributed algorithm which is able to update a solution to the all pairs shortest paths problem during a sequence of changes (either edge insertions or edge cost decreases). When the edge costs are integers in the range $[1, W]$ our algorithm requires a total of $O(WV^3 \log(VW))$ messages for any sequence of (at most $O(WV^2)$) changes. This gives basically an $O(V \log(VW))$ amortized message complexity per change. In the case of unit edge costs (i.e., $W = 1$), the above bound becomes $O(V \log V)$ per change. The algorithm runs asynchronously, requires $O(V)$ time per change and $O(V)$ space per node.

We are currently investigating the case where changes can happen simultaneously in different points of the network. Furthermore, it would be interesting to study the trade-off between messages, time and space complexity also for other dynamic changes such as edge deletions (i.e., link failures) and edge cost increases. This seems more difficult than the case of edge cost decreases.

Acknowledgments : I am grateful to Giorgio Ausiello, Alberto Marchetti-Spaccamela and Umberto Nanni for many useful discussions.

References

[1] Y. Afek, B. Awerbuch, and E. Gafni. Applying static network protocols to dynamic networks. In *Proc. 28th Annual IEEE Symp. on Foundations of Computer Science*, pages 358–370, 1987.

[2] A. V. Aho, J. E. Hopcroft, and J. D. Ullman. *The Design and Analysis of Computer Algorithms*. Addison-Wesley, Reading, MA, 1974.

[3] G. Ausiello, G. F. Italiano, A. Marchetti Spaccamela, and U. Nanni. Incremental algorithms for minimal length paths. In *Proc. 1st Annual ACM-SIAM Symp. on Discrete Algorithms*, pages 12–21, 1990. A journal version to appear in *J. Algorithms*.

[4] B. Awerbuch. On the effects of feedback in dynamic network protocols. In *Proc. 29th Annual IEEE Symp. on Foundations of Computer Science*, pages 231–245, 1988.

[5] B. Awerbuch. Distributed shortest paths algorithms. In *Proc. 21st Annual ACM Symp. on Theory of Computing*, pages 490–500, 1989.

[6] B. Awerbuch, I. Cidon, and S. Kutten. Communication-optimal maintainance of replicated information. In *Proc. 31st Annual IEEE Symp. on Foundations of Computer Science*, pages 492–502, 1990.

[7] B. Awerbuch and M. Sipser. Dynamic networks are as fast as static networks. In *Proc. 29th Annual IEEE Symp. on Foundations of Computer Science*, pages 206–219, 1988.

[8] E. Gafni. Topology resynchronization: a new paradigm for fault tolerance in distributed algorithms. In *Proc. of the Amsterdam Workshop on Distributed Algorithms*, 1987.

[9] R. G. Gallager. A shortest path routing algorithm with automatic resynch. Technical report, MIT, Lab. for Information and Decision Systems, 1976.

[10] R. G. Gallager. An optimal routing algorithm using distributed computation. *IEEE Trans. on Commun.*, COM-25:73–85, 1977.

[11] J. Jaffe and F. Moss. A responsive distributed routing protocol. *IEEE Trans. on Commun.*, COM-30:1758–1762, 1982.

[12] K. V. S. Ramarao and S. Venkatesan. On finding and updating shortest paths. In *Proc. 24th Annual Allerton Conf. on Communication, Control and Computing*, pages 1079–1988, 1986.

[13] A. S. Tanenbaum. *Computer networks*. Prentice-Hall, Englewood Cliffs, NJ, 1981.

[14] R. E. Tarjan. *Data structures and network algorithms*, volume 44. CBMS-NSF Regional Conference Series in Applied Mathematics, SIAM, 1983.

Minimal Shared Information for Concurrent Reading and Writing

Prasad Jayanti[*]   Adarshpal Sethi[†]   Errol L. Lloyd[‡]

Abstract

While there have been many register construction algorithms in the recent years, results establishing the intrinsic complexity of such constructions are scarce. In this paper, we establish the minimum shared memory necessary to construct an atomic single-writer, single-reader, N-bit register A from single-writer, single-reader, single bit safe registers. The write operation on A is wait-free, but the read operation is not. We also provide constructions that match our lower bounds.

1 Introduction

In a concurrent system, shared objects provide the means for communication and synchronization between processes. There has been a great deal of research on how to realize complex shared objects from simpler ones. In particular, the problem of constructing atomic registers from simpler classes has received much attention. Traditional solutions to this problem employ *critical sections* for the read and write operations on the buffer[CHP71]. Critical sections, however, are not suitable for fault tolerant, asynchronous systems. For example, a crash failure of a reader process in the critical solution could permanently disable the progress of a writer process. Even if processes exhibit no faults, a slow process in the critical section can arbitrarily delay other processes from entering the critical section.

Lamport was the first to suggest an alternative to the critical section approach[Lam77]. In [Lam77], the write operation on the buffer is *wait free*, i.e., the operation always completes in a bounded number of steps, irrespective of the state or the rate of progress of the reader processes. Since the appearance of Lamport's paper, the focus has been on *wait free register constructions*. That is, constructions that support wait-free read and write

---

[*]Department of Computer Science, Cornell University, Ithaca, NY 14853. E-mail: prasad@cs.cornell.edu
[†]Department of Computer Science, University of Delaware, Newark, DE 19716. E-mail: sethi@udel.edu
[‡]Department of Computer Science, University of Delaware, Newark, DE 19716. E-mail: elloyd@udel.edu

operations on the constructed register. Denoting a K-writer, M-reader, N-bit register by [K,M,N], the previous results on the construction of atomic registers may be summarized as follows. [Lam86] describes a construction of a [1,1,N] atomic register from [1,1,1] safe registers. [NW87], and [SAG87] describe constructions of a [1,M,N] atomic register from [1,1,N] atomic registers. [Pet83] and [Vid88] describe the construction of a [1,M,N] atomic register from [1,M,1] atomic registers. [BP87] and [Vid89] construct a [1,M,N] atomic register from [1,M,N] regular registers. [VA86], [Blo87], [PB87] and [Sch88] describe the construction of a [K,M,N] atomic register from [1,M,N] atomic registers. Together, these results make the giant leap from a [1,1,1] safe register to a [K,M,N] atomic register possible.

Although, as noted above, research on algorithmic aspects of atomic buffer constructions progressed rapidly, the complementary aspect of lower bound analysis of the resources required for these constructions has not received the attention it deserves. An example of a resource is *shared memory space*, the number of bits of shared memory necessary for the construction. Another example is *time*, measured by the number of lower level operations corresponding to a read or write operation on the constructed register. Discovering when a construction is possible and when it is not will surely enhance our understanding of concurrent systems in general.

In this paper, we study the construction of an N-bit (single reader, single writer) atomic register from single bit safe[1] registers. We require the write operation on the constructed register to be wait-free. The read operation is not wait-free. Following [Lam77], in which the problem was first studied, we call this problem *Concurrent Reading And Writing*, hereafter abbreviated as *CRAW*. Algorithms to solve *CRAW* were presented before in [Lam77] and [Pet83]. The algorithm in [Lam77] requires a shared memory of unbounded size. That result was improved upon in [Pet83], where a solution was given that requires only four boolean variables (besides the N bits of the constructed register) to be shared by the reader and the writer. The primary goal of this paper is to establish the lower bounds on the size of the shared memory between the reader and the writer necessary to realize *CRAW*. Algorithms matching these lower bounds are also presented.

To the best of our knowledge, the only previous works analyzing the complexity of register constructions are [Lam86] and [Tro89]. In [Lam86], it was proved that the reader

---

[1]Lamport defined three classes of registers, *safe, regular and atomic*, based on the consistency guarantees they provide [Lam86]. Roughly speaking, a register is *safe* if every read that does not overlap with a write returns the last value written to the register. A read that overlaps with one or more writes, on the other hand, may return any value from the domain of the register. A register is *regular* if every read returns either the last value written to the register (by a write that completed before the start of this read) or a value written by one of the overlapping writes. A register is *atomic* if it is regular with the additional property that no read returns an older value than the one returned by the previous read. A read or a write operation on an atomic register appears to take effect at some instant in the interval of time in which it actually occurs.

must write in any construction of an atomic register from regular registers. In [Tro89], a space optimal construction of an atomic bit from three safe bits was presented. Concurrent with our work (first reported in [JSL90]) was [CW90] where the space and time complexity trade-offs were studied for the problems of constructing a [1,M,N] safe register from [1,M,1] safe registers and [1,M,N] regular register from [1,M,1] regular registers. Recently [Jay91] presented a space and time optimal construction of an atomic register from regular registers.

The rest of the paper is organized as follows. In section 2, we state the model and define the problem. In section 4, we prove that communication *from* the reader *to* the writer is necessary. We use this result to prove in section 5 that a single shared variable of *any* fixed size is not enough to realize *CRAW*. The sufficiency of two shared *boolean* variables is demonstrated in section 6 through an algorithm.

Although the algorithm of section 6 correctly solves CRAW, it requires a shared variable which both the reader and the writer must write. In section 7, we restrict our attention to shared variables that may be written by only one process. Here we first prove that it is impossible to realize *CRAW* if the set of shared variables that the writer may write consists of a single boolean variable (irrespective of the number of shared variables the reader may write). We then present two optimal algorithms: one that uses three single writer shared boolean variables, and one that uses two single writer shared variables, one boolean and one three-valued.

The impossibility results of sections 4, 5 and 7 use a common proof technique that we call *fooling the reader process*. This technique is adversarial in its approach and is the subject of section 3. The impossibility results for the concurrent reading and writing problem discussed in subsequent sections are obtained through applications of this technique.

2 Problem specification

We consider concurrent reading and writing by two processes, a *reader* and a *writer*. These two processes share a data structure known as the *buffer*. The writer may modify the buffer, while the reader may only read the buffer. In addition to the reader and the writer, there are two user processes, *user-R* and *user-W*. Of these, user-W may present the writer with a *value* from a set called the *value set* and request that the value be stored in the buffer. User-R, on the other hand, may request the reader to present it with the value stored in the buffer. The activities of the reader and the writer are shown in Figure 1. Both the reader and the writer are in an infinite loop and the steps within the loop are grouped into phases. For the writer, these phases are labeled *Wait-For-User-W (WAIT)* and *writing*. In the *WAIT phase*, the writer waits for user-W to present a value from the value set. The *WAIT phase* ends as soon as the writer has received a value from user-W. This value is then stored in the buffer in the *writing phase*. The phases of the reader are grouped in a similar fashion. In the *WAIT phase*, the reader waits for a request from user-R for the value stored in the buffer. In the *reading phase*, the reader reads the value stored in the buffer. In the *REPORT phase*, the reader reports to user-R the value read from the buffer during the *reading phase*.

The buffer is an array of n bits of memory, denoted by $b_1, b_2, \ldots, b_n$. The two atomic operations, *read* b_i and *write* b_i, provide the only means for a process to access/modify the buffer. We define the *value set* to be $\{[u_1, u_2, \ldots, u_n] \mid u_i \in \{0,1\}\}$. This means that the writer receives (in the WAIT phase) a value of the form $[u_1, u_2, \ldots, u_n]$ from user-W. That value is then stored in the buffer by the writer in the writing phase. With respect to processes, we assume nothing about their relative speeds. We further assume that the use of shared variables is the only means of communication between the reader and the writer. In this paper, the term *shared variable* refers to any variable (besides the n bits of the buffer) that the reader and the writer may share.

As in [Lam77] and [Pet83], we require the writing-phase to be *wait-free*, *i.e.*, an execution of the writing phase must terminate in a finite number of steps irrespective of the state of the reader process. We do not require the reading phase to be wait-free. This implies that starvation of the reader is possible. However, in the absence of an overlap with the writing phase, any partial execution of the reading phase must complete in a finite number of steps.

We now define the concepts necessary to state the correctness condition for an algorithm for *CRAW*. A *state* is an assignment of values to the variables of the algorithm (variables include the program counter of the reader and the writer). An *event* is an execution of a step of the algorithm. Let u be the state that results when the reader or the writer executes a step at state t. If e denotes the event corresponding to the execution of that step, then we say $t \overset{e}{\to} u$. A *history* of an execution of the algorithm is a sequence $s_0 \overset{e_0}{\to} s_1 \overset{e_1}{\to} s_2 \ldots$. An event e *precedes* an event e' in a history iff e occurs before e' in that history. Note that there is a set of events corresponding to an execution of a phase of the algorithm. Let p denote an execution of some phase and q denote an execution of (a possibly different) phase. We say p *precedes* q iff each event of p precedes all events of q. We denote the ith execution of the writing phase by W_i, and the ith execution of the reading phase by R_i. We now state the correctness criterion along the lines of [Lam86].

Definition 1 *An algorithm for CRAW is correct iff, for every history h of the algorithm, there exists a function f that maps each execution of the reading phase to some execution of the writing phase such that the following conditions hold:*

- *for all i, if R_i is in h, the buffer value read by R_i is the buffer value written by $W_{f(i)}$.*

- *for all i, R_i does not precede $W_{f(i)}$, and $W_{f(i)+1}$ does not precede R_i.*

- *for all i and j, if R_i and R_j are in h and R_i precedes R_j, then $f(i) \leq f(j)$.*

[Lam86] classifies registers as safe, regular, and atomic, based on the consistency guarantees they provide. The first two conditions above ensure that the buffer is *regular*. The third condition makes the buffer *atomic*. In this context, our lower bound results are exceptionally strong in that we prove the non-realizability of a regular buffer even if the shared variables are atomic. In contrast, our algorithms employ only safe variables and construct an atomic buffer.

3   Fooling the reader process- a proof technique

The theorems in sections 4, 5 and 7 state the impossibility of realizing CRAW under certain constraints such as a limit on the size or the number of shared variables, or restrictions on the allowable operations on the shared variables by the reader and the writer. The proofs of these theorems utilize a common technique: In each instance, we exhibit two (globally distinct) scenarios which are indistinguishable to the reader, each requiring the reader to act differently. Although various versions of this general technique have been employed to establish a number of impossibility results in distributed computation [Lyn89], we use the remainder of this section to provide the specific framework of our approach.

Consider the sequence of values of the words of the buffer and the shared variables as read by the reader at different points in the reading phase. This sequence is called the *observation vector*. The observation vector is empty at the start of an execution of the reading phase and grows by a word each time a word of the buffer or a shared variable is read by the reader. The flow of control in an execution of the reading phase depends, for a given local state of the reader at the time of entry into the reading phase, solely on the observation vector. Further, the reader exits the reading phase only when it knows that it has successfully read the value in the buffer. This value read by the reader is completely determined by the observation vector of the reader at the point of exit from the reading phase. This implies that for any algorithm realizing *CRAW*, there must exist a *value extraction function* Ψ such that, if $\vec{ov}$ is the observation vector when the reader exits the reading phase, then $\Psi(\vec{ov})$ is the value of the buffer read by the reader in that execution of the reading phase.

Now, if a situation is shown to exist in which the reader exits the reading phase with the belief that it has successfully read the value of the buffer to be V, and V is not acceptable as the value of the buffer (because acceptance of V implies that the buffer is not regular), then the reader process is *fooled*. An algorithm that permits the reader to be fooled is not a correct algorithm for *CRAW*.

We implement this idea as follows. We begin by proposing two situations. Assuming their existence, we show that the reader is fooled in one of these situations.

Imagine two different sites running the same algorithm with situation 1 manifesting at one site and situation 2 at the other.

SITUATION 1

- The reader and the writer are both in the WAIT phase. Some value $V = [v_1, v_2, \ldots, v_n]$ is in the buffer. The shared memory between the reader and writer is in some state S.

- The reader enters the reading phase, and performs the finite number of steps necessary to exit the reading phase (Recall that the number of steps that need to be executed to exit the reading phase must be finite when there is no interference from the writer). Let $o\vec{v}_1$ be the observation vector when the reader exits the reading phase.

SITUATION 2

- The reader and the writer are both in the WAIT phase. The same value $V = [v_1, v_2, \ldots, v_n]$ is in the buffer as in situation 1. The shared memory between the reader and the writer is in the same state S as in situation 1.

- The writer enters the writing phase, changes the value in the buffer to some $V_1 \neq V$ and exits the writing phase, returning to the WAIT phase. Let the state of shared memory at this point be S_1, which may be different from S.

- The reader enters the reading phase. As this execution of the reading phase progresses, the writer overlaps its executions of the writing phase as many times as necessary such that the following two conditions are met:

  1. V is not the value stored in the buffer in any of these executions of the writing phase.

  2. The observation vector of the reader at any point in the execution of the reading phase is identical to the observation vector at the corresponding point in situation 1.

Situation 2 may not exist; however, if it does, then it corresponds to a situation in which the reader gets fooled, as can be seen from the following argument. From the fact that the observation vectors in situations 1 and 2 are identical at each point in the respective executions of the reading phase, and the fact that the reader exits the reading phase in situation 1 at some point, it follows that the reader exits the reading phase, even in situation 2, at the same point in the execution of the reading phase. Further, if $o\vec{v}_2$ is the observation vector when the reader exits the reading phase in situation 2, then $o\vec{v}_2$ is identical to $o\vec{v}_1$. If Ψ is the value extraction function, then $\Psi(o\vec{v}_1)$ must be V since V is the only acceptable value in situation 1. Since $o\vec{v}_1 = o\vec{v}_2$, it follows that $\Psi(o\vec{v}_2) = V$. But, V is not acceptable in situation 2 because acceptance of V implies buffer is not regular.

From the above arguments, a proof of the impossibility of realizing $CRAW$ under a given set of constraints reduces to establishing the existence, under those constraints, of two situations analogous to the ones above.

We close this section with a few words about the notation we use in the proofs to express the interleaving of the reader and the writer.

Just prior to a step X by the reader

The writer performs the steps of a procedure P

should be understood to mean the following: If the operations of the reader and the writer are marked on a global time line, then whenever the operation X by the reader appears on the line, it is *immediately* preceded by the steps of P by the writer.

4   When the reader does not write into shared variables

In this section, we prove a rather surprising result that communication *from* the reader *to* the writer is necessary to realize $CRAW$. The relevance of this result to establishing a lower bound on the shared memory necessary to realize $CRAW$ will be clear in section 6.

In the following, we first present a few definitions introducing the concepts needed in this section. We then prove a substantial result on the existence of a critical write-state sequence, a concept that will shortly be introduced. Following that, we prove the main result of this section.

Definition 2 *Let V, V', V'' be n bit vectors. We say V is **constructible** from $\{V', V''\}$ if $V' \neq V \neq V''$ and for all $1 \leq i \leq n$, $V[i] = V'[i]$ or $V[i] = V''[i]$.*

Definition 3 *The **write sequence** at time t is $\langle V_0, V_1, \ldots, V_k \rangle$ iff V_0 is the initial value in the buffer and V_i (for $i \geq 1$) is the value written into the buffer by the ith execution of the writing phase and exactly k executions of the writing phase have completed by time t.*

Definition 4 *Let M represent the entire shared memory space [2] (comprised by zero or more shared variables) into which only the writer may write. Then, the **state sequence** at time t is $\langle M_0, M_1, \ldots, M_k \rangle$ iff M_0 is the initial state of M and M_i (for $i \geq 1$) is the state of M just*

---

[2]Recall that shared memory does not include the buffer according to our terminology.

after the completion of the ith execution of the writing phase and exactly k executions of the writing phase have completed by time t.

Definition 5 *The **write-state sequence** at time t is $\langle (V_0, M_0), (V_1, M_1), \ldots, (V_k, M_k) \rangle$ iff $\langle V_0, V_1, \ldots, V_k \rangle$ is the write sequence at time t and $\langle M_0, M_1, \ldots, M_k \rangle$ is the state sequence at time t.*

Definition 6 *Let Seq be a sequence of elements drawn from some set and let E be some element of this set. **count**(E,Seq) is the number of occurrences of the element E in the sequence Seq. Further, E occurs **infinitely many times** in Seq iff $count(E, Seq) = \infty$.*

Definition 7 *A write-state sequence (at time $t \rightarrow \infty$) given by $\langle (V_0, M_0), (V_1, M_1), (V_2, M_2), \ldots \rangle$ is a **critical write-state sequence** iff there exist a k, V', V'' such that the following hold:*

- *$V_k \neq V_{k+i} \ \forall i > 0$*

- *V_k is constructible from $\{V', V''\}$*

- *Both V' and V'' occur infinitely many times in the sequence $\langle V_{k+1}, V_{k+2}, V_{k+3}, \ldots \rangle$*

- *M_k occurs infinitely many times in the sequence $\langle M_{k+1}, M_{k+2}, M_{k+3}, \ldots \rangle$*

Lemma 1 *Let M represent the entire shared memory space between the reader and the writer into which only the writer may write. If the space represented by M is bounded by some constant, then there exists an n such that, if the buffer is at least n bits long, then, for any initial value V_0 in the buffer and an initial state M_0 for M, there exists a critical write-state sequence.*

Proof Since the space represented by M is bounded by a constant, the number of distinct states possible for this space must be bounded by some constant k. Let $S = \{s_1, s_2, \ldots, s_k\}$ represent the set of these distinct states for M. Consider a buffer that is n words long, where n is any integer such that $2^n > k + 2$. Let $V = \{[u_1, u_2, \ldots, u_n] | u_i = 0 \, or \, 1, 1 \leq i \leq n\}$. Note that the cardinality of V is 2^n.

At the outset, it must be noted that we possess the right to specify the write sequence as we please. However, we have no control over the corresponding state sequence. It is the prerogative of the algorithm to decide what the state of M should be at the end of any execution of the writing phase. In the following, we specify the write sequence in such a way that the resulting write-state sequence must be a critical write-state sequence.

Consider the write sequence $\mathbf{Wseq} = \langle V_0, V', V_1, V'', V', V_2, V'', \ldots \rangle$, where V_0 is the initial value in the buffer, $V' = [0, 0, 0, \ldots, 0]$, and $V'' = [1, 1, 1, \ldots, 1]$. The V_i's (for

$i \geq 1$) will be specified shortly. Let $\mathbf{Sseq} = \langle M_0, M_1', M_1, M_1'', M_2', M_2, M_2'', \ldots \rangle$ be the state sequence (determined by the algorithm) that corresponds to $\mathbf{Wseq}$. Here M_0 is the initial state of M and $M_i', M_i, M_i'' \in \mathbf{S}$ for all i.

We now state, through the procedure in Figure 2, how the V_i's (for $i \geq 1$) in $\mathbf{Wseq}$ are chosen. This makes the specification of the write sequence $\mathbf{Wseq}$ complete. The procedure uses two recursively defined sets - $\mathbf{V_j}$ and $\mathbf{S_j}$ (for $j \geq 0$). Note that these $\mathbf{V_j}$'s are different from the V_j's in $\mathbf{Wseq}$.

This procedure ensures that if, after writing a value V_j in the buffer, the algorithm leaves the shared memory M in a state M_j which is different from all of $M_0, M_1, \ldots, M_{j-1}$, then V_j will never again be written in the buffer, i.e., V_j will not appear in the remainder of the write sequence.

It is easily verified that $\mathbf{V_j}$ is non-empty for all j whenever $2^n > k + 2$. Therefore, the step in which V_j is chosen from $\mathbf{V_{j-1}}$ is always possible.

Now, consider $\mathbf{T} = \langle M_1, M_2, M_3, \ldots \rangle$ which is a sub-sequence of $\mathbf{Sseq}$. Recalling that $M_i \in \mathbf{S}$ for all i and $\mathbf{S} = \{s_1, s_2, \ldots, s_k\}$, we have (trivially) that $\sum_{i=1}^{k} count(s_i, \mathbf{T}) = \infty$. Since k is finite, it follows that $count(s_j, \mathbf{T}) = \infty$ for some $s_j \in \mathbf{S}$. Let l be the least integer such that $M_l = s_j$. Now, the following statements are true.

- V_l does not occur in the sequence $\langle V', V_{l+1}, V'', V', V_{l+2}, V'', \ldots \rangle$. This is due to the procedure above and the fact that M_l is different from all of $M_1, \ldots, M_{l-1}$.

- V_l is constructible from $\{V', V''\}$. This is because $V_l \in \mathbf{V} - \{V', V''\}$. As a result, V_l is a vector of n elements, some of which are 0's and the others 1's, and is trivially constructible from $\{V', V''\}$.

- Clearly, both V' and V'' occur infinitely many times in $\langle V', V_{l+1}, V'', V', V_{l+2}, V'', \ldots \rangle$.

- M_l occurs infinitely many times in $\langle M_{l+1}', M_{l+1}, M_{l+1}'', M_{l+2}', M_{l+2}, M_{l+2}'', \ldots \rangle$. This follows from the fact that $count(s_j, \mathbf{T}) = \infty$ and $s_j = M_l$.

Therefore, the write-state sequence formed by the write sequence $\mathbf{Wseq}$ and the corresponding state sequence $\mathbf{Sseq}$ is a critical write-state sequence. Hence the lemma. $\square$

Theorem 2 *Using only a constant number of variables, each bounded in size by some constant, it is impossible to realize CRAW for an arbitrary buffer size if the reader never writes into any shared variable.*

Proof By the application of the technique of section 3 and lemma 1. Details are omitted due to space constraints. $\square$

A theorem similar to theorem 4.1 above can be found in [Lam86], where it was proved that the reader must do some writing in any construction of an atomic register from regular registers. In contrast, our result asserts that in any construction of an n-bit regular register from n atomic bits and a constant number of fixed sized shared atomic variables, the reader must do some writing even if the read operation on the constructed register is not required to be wait-free. It is important to note that our theorem does not contradict the construction in [Lam86]. There, an n-bit regular register is constructed from regular bits without the reader having to write. However, that construction employs 2^n regular bits to construct an n-bit regular register and therefore is not in contradiction with our theorem.

5 With a single shared variable

Theorem 3 *It is impossible to realize CRAW using only a single shared variable of constant size, however large that constant may be.*

Proof By the application of theorem 2 and the proof technique of section 3. Details are omitted due to space constraints.   □

6 Algorithm with minimal shared memory

We now present an algorithm for *CRAW*. The algorithm employs two shared variables: **flag** is a two writer, single reader boolean regular register and **in** is a single writer, single reader boolean safe register.

Theorem 4 *The algorithm in Figure 3 realizes CRAW. It is minimal in the number and the size of shared variables necessary to realize CRAW.*

Proof The optimality of the algorithm follows from theorem 6.1 and the fact that the shared variables are booleans. The proof of correctness of the algorithm is omitted due to space constraints.   □

7 With single-writer shared variables

The variable **flag** in the algorithm of Figure 3 is a two writer register. In this section, we study the shared memory requirements of *CRAW* if only single writer, single reader registers may be used. The main result of this section is that it is impossible to realize *CRAW*, using only single-writer shared variables, if the set of shared variables that the writer may write into consists of a single boolean (Note that the size or the number of shared variables the reader may write into is irrelevant). Following the impossibility result, we present two optimal algorithms for *CRAW* that use only single-writer shared variables.

Let w represent the shared boolean into which only the writer may write. Let $r_1, r_2, \ldots, r_m$ (for some $m \geq 0$) represent the shared variables into which only the reader may write. Let the buffer be n bits long $(n \geq 3)$. Let V_i $(0 \leq i \leq 2^n - 1)$ denote the buffer value $[b_1, b_2, \ldots, b_n]$ where $b_1 b_2 \ldots b_n$ is the binary representation of i. Let $Wseq = \langle V_{init}, V_1, SS_1, SS_2, \ldots \rangle$ where V_{init} is the initial value in the buffer and SS_i (for all i) is an abbreviation for the subsequence $\langle V_2, V_3, \ldots, V_{2^n-1}, V_0 \rangle$ (note the absence of V_1 in SS_i). The only purpose of the subscript i in SS_i is to make it possible to reference a particular "SS" subsequence in $Wseq$.

Definition 8 state-set$(SS_i) = \{w \mid w$ *assumes the value w at some point during (at least) one of the $2^n - 1$ executions of the writing phase that correspond to $SS_i\}$*

Clearly, since w is a boolean, state-set$(SS_i) \in \{\{0\}, \{1\}, \{0,1\}\}$, for all i.

Definition 9 *An interleaving of the reader and the writer is* **special** *iff*

- *$Wseq$, as defined above, is the write sequence and*

- *the writing of any r_j $(1 \leq j \leq m)$ by the reader occurs only when the writer is in the WAIT phase that is reached after the completion of SS_i and before the start of SS_{i+1} (for some i).*

Lemma 5 *If the interleaving of the reader and the writer is special, then state-set$(SS_i) = \{0,1\}$, for all i*

Proof Omitted due to space constraints. □

Theorem 6 *If only single-writer shared variables must be used, then it is impossible to realize CRAW if the set of shared variables that the writer may write into consists of a single boolean variable.*

Proof By the application of lemma 5 and the proof technique of section 3. Details are omitted due to space constraints. □

Corollary 7 *If only single-writer shared variables must be used, then two shared boolean variables are not sufficient to realize CRAW.*

Proof Suppose there are only two single-writer shared booleans. It follows from theorem 5.1 that the reader needs to write into at least one of them. That leaves the writer with at most one shared boolean to write into. According to theorem 8.1, this is not sufficient to realize $CRAW$. □

We now present two algorithms. The algorithm in Figure 4 employs two shared variables: w is a regular three valued register and r is a regular boolean register. The algorithm in Figure 5 employs three shared variables: r and w are regular boolean registers and in is a safe boolean register. Using a simple construction in [Lam86], it is possible to construct a regular boolean register from a single safe boolean register. This fact, together with algorithm in Figure 5, implies that three safe boolean registers are sufficient to realize CRAW.

Theorem 8 *The algorithms in Figures 4 and 5 realize CRAW. When the variables must be single-writer variables, these algorithms are optimal.*

Proof Note that the algorithm in Figure 4 uses two single-writer shared variables (one boolean and one three-valued). On the other hand, the algorithm in Figure 5 uses three single-writer shared boolean variables. Therefore, the optimality of these algorithms follows trivially from the above corollary. The proofs of correctness of these algorithms are similar to the proof of correctness given in section 7, and are omitted. □

8 Conclusion

While there have been many register construction algorithms in the recent years, results establishing the intrinsic complexity of such constructions are scarce. In this paper, we establish the minimum shared memory necessary to construct an atomic single-writer, single-reader, N-bit register A from single-writer, single-reader, single bit safe registers. The write operation on A is wait-free, but the read operation is not. We also provide constructions that match our lower bounds. Our algorithms can be trivially extended to the multi-reader case. Such an extension requires $O(m)$ shared booleans when there are m readers. Whether this is minimal is an open question.

Acknowledgements

Fred Schneider has provided a great deal of encouragement throughout. Especially helpful were the discussions on our algorithms. Leslie Lamport's suggestion motivated the development of section 7. Sarma Jayanthi, Dexter Kozen, Sam Toueg, K. Vidyasankar and Shmuel Zaks provided helpful comments on earlier drafts. We thank all of the above people.

References

[Blo87] Bard Bloom. Constructing two writer atomic registers. In *The 6th Annual Symposium on Principles of Distributed Computing*, pages 249–259, 1987.

[BP87]   J. Burns and G. Peterson. Constructing multi-reader atomic values from non-atomic values. In *The 6th Annual Symposium on Principles of Distributed Computing*, pages 222–231, 1987.

[CHP71] P.J. Courtois, F. Heymans, and D.L. Parnas. Concurrent control with readers and writers. *Communications of the ACM*, 14(10):667–668, 1971.

[CW90]   Soma Chaudhuri and Jennifer Welch. Bounds on the costs of register implementations. Technical report, University of North Carolina at Chapel Hill, Dept. of Computer Science, Univ. of North Carolina at Chapel Hill, Chapel Hill, NC 27599-3175, 1990.

[Jay91] Prasad Jayanti. Optimal atomic register (in preparation). Technical report, Cornell University, Dept. of Computer Science, Cornell University, Ithaca, NY 14853, 1991.

[JSL90] Prasad Jayanti, Adarshpal Sethi, and Errol Lloyd. Minimal shared information for concurrent reading and writing. Technical report, University of Delaware, Dept. of Computer and Information Sciences, University of Del aware, Newark, DE 19716, 1990.

[Lam77] Leslie Lamport. Concurrent reading and writing. *Communications of the ACM*, 20(11):806–811, 1977.

[Lam86] Leslie Lamport. On interprocess communication, parts i and ii. *Distributed Computing*, 1:77–101, 1986.

[Lyn89] Nancy A. Lynch. A hundred impossibility proofs for distributed computing. In *The 8th Annual Symposium on Principles of Distributed Compu ting*, pages 1–27, 1989.

[NW87]   R. Newman-Wolf. A protocol for wait-free, atomic, multi-reader shared variables. In *The 6th Annual Symposium on Principles of Distributed Computing*, pages 232–248, 1987.

[PB87]   G. Peterson and J. Burns. Concurrent reading while writing ii: the multi-writer case. In *The 28th Annual Symposium on Foundations of Computer Science*, 1987.

[Pet83] Gary L. Peterson. Concurrent reading while writing. *ACM TOPLAS*, 5(1):56–65, 1983.

[SAG87] A. Singh, J. Anderson, and M. Gouda. The elusive atomic register, revisited. In *The 6th Annual Symposium on Principles of Distributed Computing*, pages 206–221, 1987.

[Sch88] R. Schaffer. On the correctness of atomic multi-writer registers. Technical report, TR No: MIT/LCS/TM-364, MIT Laboratory for Computer Science, 1988.

[Tro89] J.T. Tromp. How to construct an atomic variable. Technical report, Centrum Voor Wiskunde en Informatica, 1989.

[VA86] P. Vitanyi and B. Awerbuch. Atomic shared register access by asynchronous hardware. In *The 27th Annual Symposium on Foundations of Computer Science*, 1986.

[Vid88] K. Vidyasankar. Converting lamport's regular register to atomic register. *IPL*, 28:287–290, 1988.

[Vid89] K. Vidyasankar. An elegant 1-writer multireader multivalued atomic register. *IPL*, 30:221–223, 1989.

Reader	Writer
Loop	Loop
Wait-For-UserR (WAIT)	Wait-For-UserW (WAIT)
Reading	Writing
Report-Value-To-UserR (REPORT)	Forever
Forever	

Figure 1: The activities of the reader and the writer

$$V_0 := V - \{V', V''\}$$
$$S_0 := S$$
for $j := 1$ to ∞
```
    {
    choose V_j such that V_j ∈ V_{j-1}
    if M_j of Sseq is from S_{j-1}
      then {
            V_j := V_{j-1} - {V_j}
            S_j := S_{j-1} - {M_j}
            }
      else {
            V_j := V_{j-1}
            S_j := S_{j-1}
            }
    }
```

Figure 2: Procedure for determining the write sequence

shared **in, flag** : boolean;

in := false; (* initialization *)

Reading phase	Writing phase
1. repeat	1. **in** := true;
2. **flag** := true;	2. **flag** := false;
3. while (**in**); (* busy wait *)	3. **buffer** := u;
4. v := **buffer**;	4. **in** := false;
5. until (**flag**);	

Figure 3: Algorithm for concurrent reading and writing

shared **r**: $(0, 1)$;

    w: $(0, 1, 2)$;

r := **w** := 0; (* initialization *)

Reading phase

repeat

   repeat

      w := **w**;

   until (w = 0) or (w = 1);

   r := w;

   v := **buffer**; (* Read buffer *)

until (**r** = **w**);

Writing phase

w := 2;

buffer := u; (* write buffer *)

$$\mathbf{w} := \begin{cases} 0 & \text{if } \mathbf{r} = 1 \\ 1 & \text{otherwise} \end{cases}$$

Figure 4: An optimal algorithm for CRAW using single-writer variables

shared **in**, **r**, **w** : boolean;
in := false; (* initialization *)

Reading phase

repeat

    r := **w**;

    while (**in**); (* busy wait *)

    v := **buffer**; (* read buffer *)

until (**r** = **w**);

Writing phase

in := true;

w := ¬ **r**;

buffer := u; (* write buffer *)

in := false;

Figure 5: An optimal algorithm for CRAW using single-writer boolean variables

Reading Many Variables in One Atomic Operation Solutions With Linear or Sublinear Complexity[1]

Lefteris M. Kirousis[2,3] Paul Spirakis[2,3,4] Philippas Tsigas[2,3]

E-addresses: ⟨lastname⟩@grpatvx1.bitnet

Abstract

We address the problem of reading more than one variables (components) $X_1, \ldots, X_c$, all in one atomic operation, by a process called the reader, while each of these variables are being written by a set of writers. All operations (i.e. both reads and writes) are assumed to be totally asynchronous and wait-free. The previous algorithms for this problem require at best quadratic time and space complexity (the time complexity of a construction is the number of sub-operations of a high-level operation and its space complexity is the number of atomic shared variables it needs). We provide a (deterministic) solution which has linear (in the number of processes) space complexity, linear time complexity for a read operation and constant time complexity for a write. Our solution does not make use of time-stamps. Rather, it is the memory location where a write writes that differentiates it from the other writes. Now, introducing randomness in the location where a reader gets the value it returns, we get a conceptually very simple probabilistic algorithm. This is the first probabilistic algorithm for the problem. Its space complexity as well as the time complexity of a read operation are both sublinear. The time complexity of a write is still constant. On the other hand, under the Archimedean assumption, we get a protocol whose both time and space complexity do not depend on the number of writers but are linear in the number of components *only* (the time complexity of a write operation is still constant).

1   Introduction

Afek et al. [1990] and Anderson [1990] give bounded constructions for reading, in a single atomic operation, a shared, array-like variable (called a **composite register**) that comprises a number $X_1, \ldots, X_c$ of variables (the **components**), such that each X_k, $k = 1, \ldots, c$, can be written on by a set of processes (the **writers** of each component). All operations (i.e. either writes on a component or reads) are assumed to be executed in a totally asynchronous and wait-free manner. The building blocks of the constructions are shared, atomic, single-component variables (**subregisters**).

All constructions by Anderson are recursive on the number of components and because at each recursive call the number of steps of an operation is at least doubled, the time complexity of the algorithms (i.e. the number of sub-operations of an operation) is exponential as a function of the number of components. Afek et al. give constructions with polynomial complexity, but in all cases, both time and space complexity are at least

---

[1]This research was partially supported by the ESPRIT II Basic Research Actions Program of the EC under contract no. 3075 (project ALCOM).

[2]Department of Computer Science and Engineering, University of Patras, Patras 26110, Greece.

[3]Computer Technology Institute, P.O. Box 1122, Patras 26110, Greece.

[4]Courant Institute of Mathematical Sciences, NYU, U.S.A.

quadratic in the of processes (space complexity is the number of subregisters used in the construction).

In this paper, we first give a conceptually very simple construction for the single-reader, many-writers per component case assuming that we have an *unbounded* number of memory locations (subregisters). Then, we show how to "recycle" the memory locations in order to obtain a (deterministic) protocol that uses only a bounded number of subregisters. Our construction has linear space complexity and linear time complexity for a read and constant time complexity for a write. We believe that the tool of using uboundedly many memory locations is stronger than the method of unbounded time-stamps which is often used to obtain a first construction with easy semantics (see [Aspnes, Herlihy, 1990] and [Kirousis, Spirakis, Tsigas, 1991]). Moreover, introducing randomness in the choice of memory location where a read gets the value it returns, we obtain a conceptually very simple probabilistic protocol. To our knowledge, this is the first probabilistic protocol for this problem. If m is the number of writers per component and c is the number of components, then the space complexity of our probabilistic algorithm as well as the time complexity of a read operation are $O(cq(mc)^{1/q})$, where q can be chosen by the algorithm designer. A write operation has $O(q)$ time complexity. Finally, under the Archimedean assumption, we give a protocol with space and time complexities that do not depend on the number of processes but are linear in the number of components *only* (the time complexity of a write is still constant).

Thus, we answer the questions posed in [Afek et al., 1990] concerning the improvement of complexity of composite register constructions.

Our notations and definitions closely follow the existing literature (see [Anderson, 1990] or [Afek et al., 1990]; in [Lamport, 1986] one can find a formalism for the notion of atomic registers). We assume that there is a precedence relation on operations which is a strict partial order (denoted by '$\rightarrow$'). Semantically, $a \rightarrow b$ means that operation a ends before operation b starts. If two operations are incomparable under $\rightarrow$, they are said to **overlap**. A construction of a composite register comprises: (i) a descriprion of the set of the subregisters and their initial values and (ii) procedures (protocols) that describe a high-level operation in terms of its sub-operations on the subregisters. A protocol, apart from the shared variables makes use of **local** variables as well (these cannot be shared by concurrent processes). The local variables are assumed to retain their values between invocations of the corresponding procedures, i.e. they are *static*. We adopt the convention to denote shared variables with capital letters and local variables with lower case letters. A **reading function** π_R for a subregister R is a function that to each read operation r on R assigns a write operation w on R, such that the value returned by r is the value written by w. If the value read by r is the initial value of the register, we assume that $\pi_R(r)$ is a write on R that precedes all other operations on R.

A run is an execution of an arbitrary number of operations according to the respective protocols. A run is atomic if the partial order $\rightarrow$ on its operations can be extended to a *total* strict order $\Rightarrow$ and if for each component X_k there is a reading function π_k such that for all high-level reads r: (i) $\pi_k(r) \Rightarrow r$ and (ii) there is no write w on X_k such that $\pi_k(r) \Rightarrow w \Rightarrow r$. A construction is atomic if all its runs are atomic. We assume all subregisters to be atomic, therefore we can assume that the precedence relation $\rightarrow$ is total when restricted to sub-operations on a single subregister.

One obviously necessary condition for a composite register to be atomic is that for any read r and for any component X_k, it is not the case that $r \to \pi_k(r)$ (indeed, otherwise the extension of $\to$ to a total order respecting the reads would be impossible). All our constructions will satisfy this condition for trivial to check reasons. For notational convenience, we call registers satisfying this condition **normal**.

2 The Deterministic Protocols

For the case of a single-reader (where we do not have overlapping high-level reads) we have the following criterion for atomicity of a composite register:

Lemma 1 *A construction of a normal composite register is atomic if and only if for each component X_k, the writes on it can be serialized by a strict total order $\Rightarrow_k$ compatible with the precedence relation $\to$ and such that the following two conditions hold:*

1. *Each $\Rightarrow_k$ is compatible with the respective reading function π_k, i.e. for each read r, it is not the case that there is a write w on X_k so that $\pi_k(r) \Rightarrow_k w \to r$. Moreover, for any two reads r and s and for any component X_k, it is not the case that: $r \to s$ and $\pi_k(s) \Rightarrow_k \pi_k(r)$.*

2. *For any two different components X_k and X_l and for any read r, it is not the case that there are writes v and w on X_k and X_l respectively such that*

$$\pi_k(r) \Rightarrow_k v \to w \stackrel{\equiv}{\Rightarrow}_l \pi_l(r),$$

where $w \stackrel{\equiv}{\Rightarrow}_l \pi_l(r)$ means that either $w \Rightarrow_l \pi_l(r)$ or $w = \pi_l(r)$.

This lemma is essentially the restriction to the single-reader case of the atomicity criteria mentioned in [Anderson, 1990], and so we omit its proof. $\square$

Based on the above, we obtain the following basic lemma which gives *sufficient* conditions for atomicity that refer to *each component separately*. All our algorithms satisfy the conditions of this lemma. Therefore, our algorithms not only implement an atomic register but have stronger, in general, properties described by these conditions.

Basic Lemma *A construction of a normal composite register is atomic if for each component X_k, the writes on it can be serialized by a strict total order $\Rightarrow_k$ that is compatible with the precedence relation $\to$ and such that the following two conditions hold:*

1. *Each $\Rightarrow_k$ is compatible with the respective reading function π_k, i.e. for each read r, it is not the case that there is a write w on X_k so that $\pi_k(r) \Rightarrow_k w \to r$ and moreover, for any two reads r and s and for any component X_k, it is not the case that: $r \to s$ and $\pi_k(s) \Rightarrow_k \pi_k(r)$.*

2. *For any read r and for any component X_l, if a is either the first sub-operation of $\pi_l(r)$ or the first sub-operation of any write w on X_l for which $w \Rightarrow_l \pi_l(r)$, and if b is the first sub-operation of r then a and b take place on the same (atomic) subregister and a precedes b.*

Proof It suffices to prove the second condition of Lemma 1. Indeed, let X_k and X_l be two distinct components. Suppose, towards a contradiction, that there is a v and a w on X_k and X_l, respectively, such that:

$$\pi_k(r) \Rightarrow_k v \to w \stackrel{\cdot}{\Rightarrow}_l \pi_l(r).$$

Then, since by hypothesis the first sub-operation of w precedes the first sub-operation of r, we get that $\pi_k(r) \Rightarrow_k v \to r$, a contradiction. $\square$

2.1   Unbounded Memory-Space

In this subsection we are going to describe a single-reader, multi-writer per component construction that uses unbounded memory-space (i.e. the number of subregisters used may be equal to the number of operations to be performed). We then show in the next subsection how to "recycle" the memory space in order to obtain a construction with bounded space, i.e., independent from the number of operations (actually, the space will be linear in the number of components). Let us point out that in the unbounded memory-space construction, there is a subregister that holds addresses of memory locations as a value. Therefore, this subregister must be assumed to have an unbounded number of bits. This is not the case in the bounded memory-space construction, where there is only a bounded number of addresses.

In the unbounded construction, for each component $k = 1, \ldots, c$, we introduce an unbounded number of subregisters $ML[k][l]$, $l = 0, \ldots, \infty$ which are written on by the writers of the corresponding component and are read by the reader. We call these subregisters **memory locations**. The second index of each memory location $ML[k][l]$ is its address (the first indicates the corresponding component). A memory location holds a value that belongs either to the set of values of the corresponding component or is a special new value denoted by nil. The type of all these values is denoted by *valtype*. Initially, the subregisters $ML[k][l]$ for $k = 1 \ldots, c$ and $l = 1, \ldots, \infty$ hold the value nil, while the subregisters $ML[k][0]$, $k = 1, \ldots, c$ hold a value from the set of values of the corresponding component. Moreover, we introduce a subregister PTR which holds as value an integer (a pointer to a memory location). This subregister can be written on by the reader and can be read by all writers. It is initialized with the value 0. The protocol now is the following: A writer first reads PTR and then writes its value on the memory location of the corresponding component whose address is the value of PTR. The reader, on the other hand, first increments $P\dot{T}R$ by one; stores its new value into a local variable *ptr* and then for each component $k = 1, \ldots, c$ gets the value to be returned by reading $ML[k][ptr-1], \ldots, ML[k][0]$ in this order until it gets a value which is not nil. The protocol is given formally in Figure 1.

Correctness Proof We will show that the above construction satisfies the two conditions of the Basic Lemma. To show that condition 1 of the Basic Lemma is satisfied, define the relation $\Rightarrow_k$ between writes on the same component as follows: $w \Rightarrow_k v$ if and only if either w and v write their value onto the same memory location and the last sub-operation of w precedes the last sub-operation of v or w writes its value onto a memory location with address less than the address of the corresponding memory location of v. It is clear that this relation is compatible with the reads. Condition 2 of the Basic Lemma is obviously satisfied. Indeed, both $\pi_k(r)$ and r have their first sub-operation performed on PTR. If var PTR : $integer$; ML : array$[1..c][0..\infty]$ of $valtype$; /*Shared variables declaration*/

procedure $reader$ /*returns array$[1..c]$ of $valtype$*/
var ptr, b : $integer$; a : array$[1..c]$ of $valtype$;
begin
    write $ptr + 1$ to PTR; $ptr := ptr + 1$;
    for $k := 1$ to c do
        $b := ptr$;
        repeat
            $b := b - 1$; read $a[k]$ from $ML[k][b]$;
        until $a[k] \neq$nil; od;
    return $(a[1], \ldots, a[c])$;
end

procedure $writer$ /*writes u : $valtype$ on component k*/
var $w\text{-}ptr$: $integer$; u : $valtype$;
begin
    read $w\text{-}ptr$ from PTR; write u to $ML[k][w\text{-}ptr]$;
end

Figure 1: The unbounded memory-space protocol.

now the first sub-operation of $\pi_k(r)$ followed that of r, then $\pi_k(r)$ would write its value on a memory location not visited by r. We get a similar contradiction if there is a w such that $w \Rightarrow_k \pi_k(r)$ and the first sub-operation of r precedes that of w. $\Box$

2.2 Bounded Memory Space

As mentioned, the idea for the bounded construction is to recycle the memory space. Below, we first informally describe the protocol and we explain the reasons it works. To make this informal presentation clear, we assume that we have only one component and only one writer on it. Although these restrictions may seem severe, they are not essential, mainly because the interaction of the reader with each writer is on a separate basis and because of the separability of the conditions of the Basic Lemma. The protocol for the single-component, single-writer, single-reader case is given in Figure 2. The description of the multi-component, multi-writer (single-reader) protocol is given in Paragraph 2.2.2.

2.2.1 One Writer and One Component

Intuitively, the reader, during a read operation r, instead of dictating to an upcoming writer to move to an unused memory location, as it does in the unbounded case, it designates for it a possibly used (i.e. a "recycled") location. This will be a location never to be read by r and thus it is guaranteed that a write that starts after the start of r

is not read by r. The address of this used memory location, is written on PTR at the first sub-operation of r. Its value will be calculated by the procedure named 'recycle' during the read operation immediately preceding r (since we assume that there is only one reader, the reads are linearly ordered). The procedure 'recycle' apart from computing the address of a memory location, makes also sure that this location looks unused to the next high-level read. This is done by having the procedure 'recycle' write the value nil on the location to be recycled. But then there are two conditions that the protocol must guarantee:

Condition A If the location to be recycled already has a value, this must be an old value not needed anymore.

Condition B Between the start of the procedure 'recycle' and the start of the next read r, no new value appears on the location to be recycled.

The reader maintains an array $ma[1..dim]$ of memory addresses. The constant dim is the number of memory locations needed. For m writers per component, $dim = 2m + 3$, therefore $dim = 5$ in the case under examination. A read operation gets the value it returns by reading the locations with addresses $ma[dim-1], \ldots, ma[1]$ in this order until it finds a value$\neq$nil (it does not read $ma[dim]$). The procedure recycle that follows these subreads stores into $ma[dim]$ the address of the location that is to be recycled. This address is chosen from the values $ma[dim-1], \ldots, ma[1]$ (a left shift is made on part of this array to compensate for the deletion of the address to be recycled). To guarantee Condition A, a read never recycles the location where it obtained the value it returns. To guarantee Condition B, the reader during the operation recycle must "know" which are the memory locations where a new value may appear until the starting sub-operation of the next read. To implement this requirement, instead of having PTR carry single memory address, as in the unbounded case, we assume that PTR is a record carrying an array $ptr[0..1]$ of two memory addresses as well as a boolean $flag$. The writer first reads the $flag$, writes it into a shared variable $WFLAG$ and then reads $ptr[flag]$ and moves—in order to write its value—to the memory location with address $ptr[flag]$. Observe that this entails two reads of PTR with a write on $WFLAG$ in between. The reader, on the other hand, gets the value of $flag$ from $WFLAG$ and changes it. Thus, the reader knows that the memory addresses where the writer could write until the start of the next read operation are either $ma[dim]$ or the last two values it wrote on $ptr[flag]$. These memory locations should not be among the positions to be recycled. To implement this, the reader maintains two local variables vb_0 and vb_1 (one for each value of $flag$) and the address to be recycled is chosen not to be in vb_{flag} (vb_{flag} contains at most two elements). Moreover, it should be different from $ma[5]$ and from the address where the read gets its value. In other words, each time four are the memory locations that should not be recycled (that necessitates five memory addresses). Thus, we have that both Conditions A and B are satisfied.

The initialization of the shared variables is the following: $ML[4] := a$ ($\neq$nil) value from the set of possible values of the component, and $ML[i] :=$nil, for $i = 1, 2, 3, 5$. $PTR.flag := 0$ and $PTR.ptr := (4, 4)$. $WFLAG := 0$. Reader's local variables are initialized as follows: $vb_0, vb_1 := \{4, 5\}$, $ma[1], \ldots, ma[5] := 1, \ldots, 5$, $flag := 0$ $ptr := (5, 5)$. All other variables are arbitrarily initialized.

type *Rtype* = **record** *flag* : 0..1; *ptr* :**array**[0..1] **of** 1..5; **end**;
var *PTR* : *Rtype*; *WFLAG* : 0..1; *ML* : **array**[1..5] **of** *valtype*;
/*Shared variables declaration*/

procedure *writer* /*writes u : *valtype**/
var d, e : *Rtype*; m : 1..5; u : *valtype*;
begin
    read d **from** *PTR*; **write** *d.flag* **to** *WFLAG*;
    read e **from** *PTR*; $m := e.ptr[d.flag]$; **write** u **to** $ML[m]$;
end

procedure *reader* /*returns a : *valtype**/
var *ptr* : **array**[0..1] **of** 1..5; *flag* : 0..1;
    ma : **array**[1..5] **of** 1..5; i : 1..5;
    vb_0, vb_1 : **set of** 1..5; a : *valtype*;

    procedure recycle
    var j : 1..5;
    begin
        choose j **such that** $ma[j] \notin vb_{flag} \cup \{ma[i], ma[5]\}$;
        rotate one position to the left $ma[j], \ldots, ma[5]$;
        $ML[ma[5]] :=$**nil**; $flag :=$ **not** $flag$;
        $vb_{flag} := \{ptr[flag], ma[5]\}$; $ptr[flag] := ma[5]$;
    end;

begin
    write $(flag, ptr[0..1])$ **to** *PTR*; $i := 5$;
    read *flag* **from** *WFLAG*;
    repeat
      $i := i - 1$; **read** a **from** $ML[ma[i]]$;
    until $a \neq$**nil**;
    recycle; **return** a;
end

Figure 2: The protocol for the single-writer, single-component, bounded-space case.

2.2.2 Many Components with Many Writers on Each

In this case, the same protocol as in the single-component, single-writer case works, only the reader must interact with the writers of each component separately. This necessitates an increase in the number of dimensions of the arrays that appear in the protocol. Moreover, the sets vb should have $2m$ elements (m is the number of writers per component), because during the execution of the procedure recycle, the reader must maintain *two* memory locations for each writer, which together with the $ma[dim]$ of the corresponding component are the locations where a new value written from this writer may appear. We give the formal protocol for this case in Figure 3. The initialization of the variables is, for each component and for each writer, analogous to the single-component, single-writer case (with $2m + 3$ in place of 5).

Theorem 1 *A single-reader, c-component, m-writer per component composite register can be constructed using $O(m \cdot c)$ (i.e., linear in the number of processes) atomic, single-reader, multi-writer subregisters and one atomic, multi-reader, single-writer subregister. The number of steps for a read operation is linear in the number of processes, while a write operation has only four sub-operations.*

Proof The complexity claims follow immediately inspecting the protocols. To prove the correctnesss of the protocols, we show that the two conditions of the Basic Lemma are satisfied. To show that condition 1 of the Basic Lemma is satisfied, we must give a linear order for the writes on each component. Consider an arbitrary component, say numbered by k. We assign to each value written on the local variables $ma[k][1..2m+3]$ an integer (a tag) as follows: the initial values of $ma[k][i]$ have the tag i ($i = 1, \ldots, 2m+3$). Each time a value is written on $ma[k][2m+3]$, it gets the tag of the previous value of $ma[k][2m+3]$ incremented by one. This tag is retained each time when, because of a shift to the left, this value is moved to another $ma[k][j]$, $j = 2m+2, \ldots, 1$. Now, observe that a high-level write w on component k decides the memory location where it writes its value by reading from PTR an address that at some point appeared on $ma[k][2m+3]$ (w gets this address at its second reading of the subregister PTR). We associate the tag of this address with the high-level write w as well. We say now that a write w precedes a write v if either the tag of w is less than the tag of v or they have the same tag (and consequently, they write on the same memory location as well) and the subwrite of w that writes its value precedes the corresponding subwrite of v. It is now easy to see that this linear order on the high-level writes of a component satisfy the requirements of condition 1 of the Basic Lemma (see also the informal remarks in the Paragraph 2.2.1). The fact that condition 2 is satisfied is immediate because the first sub-operation of both reads and writes take place on PTR, and moreover because a write that either starts after the start of a read r or, alternatively, follows (in the sense of $\Rightarrow_k$) a write that starts ater the start of r, writes in a memory location never visited by r. □

3 A Probabilistic Protocol

In this section, we describe a randomized protocol which will satisfy the atomicity requirements, except that for each high-level write w, the probability that w is not read, while it should, is overwhelmingly small and controllable.

```
type Rtype = record flag : array[1..c][1..m] of 0..1; ptr : array[0..1][1..c][1..m] of 1..2m + 3; end;
var   PTR : Rtype; WFLAG : array[1..c][1..m] of 0...1; ML : array[1..c][1..2m + 3] of valtype;
      /*Shared variables declaration*/

procedure reader /*returns array[1..c] of valtype*/
var   vb₀, vb₁ : array[1..c][1..m] of sets of 1..2m+3; i :  1..2m + 3; k :  1..c;
      ptr : array[0..1][1..c][1..m] of 1..2m+3;          flag : array[1..c][1..m] of 0..1;
      ma : array[1..c][1..2m + 3] of 1..2m+3;            a : array[1..c] of valtype;

      procedure recycle(k :  1..c)
      var j :  1..2m + 3;
      begin
          choose j such that ma[k][j] ∉ ⋃ᵐₗ₌₁ vb_{flag[k][l]}[k][l] ⋃{ma[k][i], ma[k][2m + 3]};
          rotate one position to the left ma[k][j], . . . , ma[k][2m + 3];
          ML[k][ma[k][2m + 3]] :=nil;
          for l := 1 to m do
              flag[k][l] := not flag[k][l];
              vb_{flag[k][l]}[k][l] := {ptr[flag[k][l]][k][l], ma[k][2m + 3]};
              ptr[flag[k][l]][k][l] := ma[k][2m + 3]; od;
      end

begin
      write (flag[1..c][1..m], ptr[0..1][1..c][1..m]) to PTR;
      for k = 1 to c do
          for l := 1 to m do read flag[k][l] from WFLAG[k][l] od;
          i := 2m + 3;
          repeat
              i := i − 1; read a[k] from ML[k][ma[k][i]];
          until a[k] ≠nil;
          recycle(k); od;
      return (a[1], . . . , a[c]);
end

procedure writer i /*writes u :  valtype on component k*/
var d, e :  Rtype; m : 1..2m + 3; k : 1..c; u :  valtype
begin
    read d from PTR; write d.flag[k][i] to WFLAG[k][i];
    read e from PTR; m := e.ptr[d.flag[k][i]][k][i]; write u to ML[k][m];
end
```

Figure 3: The bounded memory-space protocol.

var PTR : array$[1..c]$ of $1..l$; ML : array$[1..c][1..l]$ of *valtype*;

procedure *reader* /*returns array$[1..c]$ of *valtype*/
var *ptr* : array$[1..c]$ of $1..l$; i : $1..l$; *ma* : array$[1..c][1..l]$ of $1..l$; a : array$[1..c]$ of *valtype*;
begin
    write $ptr[1..c]$ to PTR;
    for $k := 1$ to c do
      $i := l$;
      repeat
        $i := i - 1$; read $a[k]$ from $ML[k][ma[k][i]]$;
      until $a[k] \neq$nil;
      randomly choose j such that $ma[k][j] \notin \{ma[k][i], ma[k][l]\}$;
      rotate one position to the left $ma[k][j], \ldots, ma[k][l]$;
      $ML[k][ma[k][l]]$:=nil; $ptr[k] := ma[k][l]$; od;
    return $(a[1], \ldots, a[c])$;
end

procedure *writer* /*writes u : *valtype* on component k*/
var *w-ptr*: array$[1..c]$ of $1..l$; u : *valtype*;
begin
    read *w-ptr* from PTR; write u to $ML[k][w\text{-}ptr[k]]$;
end

Figure 4: The probabilistic protocol.

The idea is (again) to recycle the memory space (which is assumed bounded). The protocol works essentially as in the deterministic case, except that the value to be written on PTR is chosen randomly rather than through the subroutine 'recycle'. The protocol is formally given in Figure 4.

Analysis of the protocol's behaviour Note that a high-level read never recycles the location where it got the value it returns. Thus, Condition A of Subsection 2.2.1 is satisfied. Now observe that Condition B would also hold (as in the deterministic protocol) if a read r could "know" which are the memory locations where a new value might appear until the start of the next read. Clearly such positions are of two kinds: (i) $ML[ptr]$ (where a writer would write after reading PTR) and (ii) $ML[x]$ where x is an "old" value of PTR that a "slow" write w (i.e. one that overlaps r and which started before the start of r) read in the past. In our randomized protocol, it is possible for r to choose an x as in (ii) above to be the next value of *ptr*. In that case, the value of w may never be read, since r prints nil on this memory location. We say then that r erased the write w. A run following our randomized protocol is atomic if we ignore the erased writes. Notice, however, that each read can erase at most one write and in the case of a single component and one writer on it, that could happen only with probability $\leq \frac{1}{l-2}$, where l is the number of memory locations (the value of l is decided by the algorithm designer).

For the c-component, m-writer per component case, the reader chooses independent random numbers in each component's window of memory. Taking into account that for each component there are m locations where a "slow" write may appear, we get that the probability for a read r not to erase a write on any one of the c components is at least $(1 - \frac{m}{l-2})^c$. Choosing $l = m \cdot c \cdot \omega + 2$, we get (by the Bernoulli Inequality) that:

Theorem 2 *Our randomized protocol (for the case of c-components, m-writers per component) has the property that for each read r the probability that r does not erase a write is at least $1 - \frac{1}{\omega}$, where $\omega = \frac{l-2}{m \cdot c}$ and l is chosen by the algorithm designer. Moreover, a run is atomic if we ignore the erased writes.*

We can further improve our probabilistic algorithm as follows: the reader instead of keeping one memory address $(ptr[k])$ for each component, keeps a sequence $ptr_1[k], \ldots, ptr_q[k]$ of them, where q is a constant to be chosen by the algorithm designer. These addresses are chosen randomly and uniformly so that they are all different from the location where the read gets the value it returns and from the previous values of the $ptr[k]$'s. On the other hand, a writer of the component k writes its value on the q memory locations it reads from PTR. Now observe that a write can be erased by the one or more reads that overlap its subwrites. By an easy counting argument, it can be proved that for any particular component, the probability for such an error to take place is $O(m(q/l)^q)$ (l is the number of memory locations). Therefore by the Bernoulli Inequality, the probability not to erase a write on any component is $\Omega((1 - mc(q/l)^q))$. From that we get that:

Theorem 3 *The space complexity of our improved probabilistic protocol as well as the time complexity for a read operation are $O(lc)$. The time complexity for a write is $O(q)$ and the probability of error is $\Omega(1 - \frac{1}{\omega})$, where $\omega = (l/q)^q/(mc)$ (q and l are chosen by the algorithm designer).*

4 A Protocol for Archimedean Systems

It has been pointed out (see, e.g., [Reif and Spirakis, 1984] or [Vitányi, 1984]) that in real distributed systems, it is reasonable to assume that the ratio of the rates of execution of elementary instructions for arbitrary pairs of processes is bounded by a fixed constant. In other words, it is assumed that the clocks of any two processes have a bound on their running rates. Systems complying with such a restriction are called **Archimedean** (this assumption does not imply any restriction on the idle time intervals between two high-level operations by the same process). In this section, we give a protocol for a composite register under the Archimedean assumption. Our construction has the interesting property that both its space and time complexity are independent of the number of processes and are both linear only on the number of components. Moreover, the time complexity of a write operation is an absolute constant.

To formalize the above notions, we assume that there is a global time-reference system, which however is not known to the processes (this is not an essential restriction; it is proved in [Lamport, 1986] that under some quite general assumptions any system has such a global-time model). Therefore, with every operation (low- or high-level) there is associated a finite time interval, its duration. Now, our assumption of Archimedean time states that there is a fixed integer A_0 such that for any two high-level operations a and

var PTR : **array**[1..c] **of** 1..3; ML : **array**[1..c][1..3] **of** *valtype*;

procedure *reader* /*returns **array**[1..c] **of** *valtype**/
var *ptr* : **array**[1..c] **of** 1..3; i : 1..3; *ma* : **array**[1..c][1..3] **of** 1..3; a : **array**[1..c] **of** *valtype*;
begin
    write *ptr*[1..c] to PTR;
    busy wait for A steps;
    for $k := 1$ **to** c **do**
      $i := 3$;
      repeat
        $i := i - 1$; read $a[k]$ from $ML[k][ma[k][i]]$;
      until $a[k] \neq$ nil;
      choose j such that $ma[k][j] \notin \{ma[k][i], ma[k][3]\}$;
      rotate one position to the left $ma[k][j], \ldots, ma[k][3]$;
      $ML[k][ma[k][3]] :=$ nil; $ptr[k] := ma[k][3]$; **od**;
    return $(a[1], \ldots, a[c])$;
end

Figure 5: The protocol for the Archimedean case.

b and for any time interval I within which a completes the execution of A_0 elementary instructions, if b starts before the start of I and if the duration of b intersects I, then b completes the execution of at least one elementary instruction within I. It must be pointed out that by elementary instructions we mean instructions on the lowest level like assignments of variables, tests, calculations of logical or arithmetical expressions, etc. Observe, however, that for any particular implementation of subregisters, a constant A can be found, (depending on A_0 and this implementation) such that if within I a executes A elementary instructions, then b will complete within I at least one sub-operation (sub-read or sub-write).

The idea of our construction is the following: As explained in the previous section, a basic difficulty for a read r in selecting a memory location to be recycled is to avoid an $ML[x]$ where x is an old value of PTR that a "slow" write w (i.e. one that overlaps r and which started before the start of r) read in the past. If such an x is chosen, then the value of $ML[x]$ can be erased *after* w writes on it, thus the next high-level read may miss values. Notice, however that such a w must have started before the start of r. So, by the Archimedean-time assumption, if we require from r to do busy-waiting for a sufficient number of its clock ticks, *before* it starts reading the $ML[x]$'s and *after* it has written on PTR, we can guarantee that r will see all writes that are to write on an $ML[x]$ with $x \neq ptr$. We give the formal protocol for the reader in Figure 5. The protocol for the writer is exactly the same as in the probabilistic case. So, we have:

Theorem 4 *Under the Archimedean assumption, a single-reader, c-component, m-writers per component composite register can be constructed with time and space complexities in-*

dependent of the number of processes. Specifically, the number of subregisters is $3c + 1$, the number of sub-operations of a read operation is at worst $2c + 1$, while a write has only two sub-operations.

References

Y. Afek, H. Attiya, D. Dolev, E. Gafni, M. Merritt and N. Shavit (1990): Atomic snapshots of shared memory, *Proceedings of the 9th ACM Symposium on Principles of Distributed Computing, Quebec City, Quebec, Canada.*

J.H. Anderson (1990): Composite registers, *Proceedings of the 9th ACM Symposium on Principles of Distributed Computing, Quebec City, Quebec, Canada.*

J. Aspnes and M.P. Herlihy (1990): Wait-free data structures in the asynchronous PRAM model, *Proceedings of the 7th ACM Symposium on Parallel Algorithms and Architectúres*, Greece, 1990.

L.M. Kirousis, P. Spirakis, Ph. Tsigas (1991): Simple atomic snapshots: a linear complexity solution with unbounded time-stamps, *Proceedings of the International Conference on Computing and Information*, Ottawa, Canada, 1991.

L. Lamport (1986): On interprocess communication, part i: basic formalism, part ii: basic algorithms, *Distributed Computing* 1, 77-101.

J.H. Reif and P. Spirakis (1984): "Real-time synchronization of interprocess communication", *ACM Transactions on Programming Languages and Systems* 6, 215-238.

P. Vitányi (1984): Distributed elections in an Archimedean ring of processors, *Proc. 16th Ann. ACM Symp. on Theory of Computing*, 542-547.

Analysis of Distributed Algorithms based on Recurrence Relations
(Preliminary Version)

Yossi Malka*
IBM Science & Technology
Technion City
Haifa 32000, Israel

Sergio Rajsbaum†
Instituto de Matemáticas
U.N.A.M.
D.F. 04510, México

Abstract

Recurrence relations of a certain type and their connection to Marked Graphs are studied. We show that these recurrence relations provide a paradigm which unifies distributed algorithms like synchronizers and distributed schedulers under a common formalism. This paradigm provides a technique for studying the properties of these algorithms when they are used in networks where link delays are not necessarily equal. We use the paradigm to analyze the performance of these algorithms. In particular it is shown that the behavior of algorithms which can be described by the recurrence relations is periodic after a short transitory phase and that the rate of computation can be computed efficiently.

1 Introduction

A *network* is represented by a directed and strongly connected graph $G = (V, E)$, where $V = 1, 2, \ldots n$ is the set of *processors* and E is the set of one-way communication *links*. All processors have local memory and communicate only by sending messages along the links. The links are assumed to be error-free.

Usually, the analysis of the time complexity of a distributed algorithm is based on the assumption that it is run in a network where all link delay bounds are equal, and processing times are negligible. The design of distributed algorithms most often disregards the possibility

---

*E-mail: yossi@haifasc3.vnet.ibm.com

†E-mail: rajsbaum at unamvm1.bitnet. Part of this work was done while the author was at the Department of Computer Science, Technion.

that link delays may be different. Actually, more than one type of information may be available regarding link delays. A number $d(e)$ associated with each link e may represent the link's constant delay, its average delay, or a bound on the delay. Such information may be useful in the design and analysis of distributed algorithms.

Recently some work has been done to take different link delays into account in the design and analysis of algorithms. The approach taken by Awerbucha, Bartaz and Peleg in [2] is to design an algorithm for a specific network using the topology and the bounds on link delays. Thus, they assume that some measure of the delay of each link is known in advance. Different delays indeed yield different algorithms. In Even and Rajsbaum ([9]) the effect of different delays on the performance of a distributed synchronizer is analyzed. In this paper an approach similar to theirs is taken and the results can be interpreted when the delays are fixed, bounded or random. For example, we compare two implementations of a distributed algorithm (a synchronizer). In one implementation time is proportional to the largest link delay, while in the other time is proportional to some measure of the average link delay (see Section 7.2).

Designing correct distributed algorithms is a difficult task, and very few general paradigms exist to assist the designer. However, for sequential algorithms several paradigms exist: Branch and Bound, Dynamic Programming, Divide and Conquer, etc. This paper intends to be a modest contribution in this direction. We present a paradigm which is useful for the performance analysis of a class of algorithms defined by a type of recurrence relations. Using these recurrence relations it is possible to analyze distributed algorithms such as marked graphs (e.g. [6]), synchronizers (Even and Rajsbaum [9] and α, β, γ in [1]), distributed schedulers ([3], [13]), clock synchronization ([14]), snapshots [5], resource allocation (e.g. [3]), broadcast, channel access in radio networks (e.g. [7]) and simulated annealing [11]. The paradigm provides a general framework for analyzing the performance of these algorithms in networks where link delays are not necessarily equal. The main performance measure is the *rate of computation*, i.e., the number of computation steps performed per time unit. Furthermore, the proposed paradigm helps in clarifying differences between various algorithms in the class. Using the paradigm we are able to generalize properties which have been proven for specific algorithms to all algorithms in this class. Several examples are provided.

Unifying the analysis of the algorithms is based on a generalization of a theorem by Even and Rajsbaum [9] regarding the behavior of a synchronizer, to any algorithm in the class. This is done by showing that any algorithm that can be described by the recurrence relations can be reduced to a synchronizer. A few applications of this result are presented in this paper as follows. In a network with fixed delays, algorithms in this class are shown to enter a periodic state (self-synchronize) operating at the maximum rate of computation, regardless of the order in which the processors initiate the computation. The time required for the algorithm to enter the periodic state is shown to be polynomialy bounded. The rate of computation of such algorithms can be computed by an $O(|V||E|\log|V|)$ algorithm. In the case where the quantities $d(e)$ represent upper bounds on the delays, the algorithm computes a lower bound on the rate of computation. Previous work has shown that if $d(e)$ is the average delay then the algorithm computes an upper bound on the rate ([16] and [17] study the case of random

delays). Lastly, it has been shown in [3] that computing an initial state that maximizes the rate is NP-complete. This result applies to the algorithms discussed here since we show that their scheduler can also be transformed into a synchronizer.

The results obtained in this paper are shown to improve previous results for certain distributed schedulers. The performance of such schedulers in networks with unit link delays has been previously studied by Barbosa and Gafni [3] and Malka et.al. [13]. No bound on the length of the transitory stage was given, and the algorithm proposed in [3] for computing the rate is of time complexity $O(|V|^5)$. Our work implies a tight (polynomial) upper bound on the length of the transitory stage of the scheduler and present an $O(\min\{|V| \cdot |E| \log |V|, |V|^3\})$ algorithm to compute the rate. These results hold for arbitrary link delays as well.

This paper is organized as follows. Sections 2, 3 and 4 describe recurrence relations, timed marked graphs and the close relation between them which is the basis of our paradigm. In Section 5 we define a canonical form of recurrence relations (which corresponds to the synchronizer in [9]) and show that any given set of relations can be transformed to this form. The behavior of the recurrence relations is analyzed in Section 6. Finally, Section 7 presents some applications. It describes a way to model specific distributed schedulers and synchronizers using recurrence relations. In this reduced version of the paper the proofs are omitted.

2 Recurrence Relations

In many algorithms a processor v waits until it knows that a computation has occurred in another processor u before it performs its own computation. This may be needed for synchronization purposes, or because v uses the results of a computation performed by u. If we denote by $t(v)$ $(t(u))$ the time when v (u) performs such a computation, then $t(v) \geq t(u) + d(u, v)$, where $d(u, v)$ denotes the time it takes to transmit the information from u to v in the network G. It may be the case that v waits for computations to be completed by several processors. For example, if v waits for a processor w as well, then $t(v) \geq \max\{t(u) + d(u, v), t(w) + d(w, v)\}$.

For nodes i, j of G, let $t_k(i)$, $k \geq 0$ denote the time at which i performs a computational step for the k-th time, and $d_k(i, j)$ the time it takes the information to reach j. In this work we assume that $d_k(i, j) = d(i, j)$, for some constant $d(i, j)$. Thus, if $e = i \to j$ is a link in E, then $d(i, j) = d(e)$ is interpreted as the delay of the link. Alternatively, $d(e)$ can be interpreted as a bound on the link delay, or as the average link delay. In the former case, $t_k(i)$ is an upper bound on the time at which i performs the k-th step. In the latter case, $t_k(i)$ is a lower bound on the expected time at which i performs the k-th step (see [16] and [17]). We are interested in distributed algorithms whose behavior can be represented by the following kind of recurrence relations. Let $G = (V, E)$, $V = 1, \ldots, n$ be a digraph, and let d and s be functions from E to the integers. The system of n *recurrence relations* defined by G, d and s, denoted by $R(G, d, s)$, is

$$t_k(v) = \max_{e=u \to v \in E}\{t_{k-s(e)}(u) + d(e)\}, \quad v = 1, \ldots, n, \ k \geq 0. \tag{1}$$

The terms $t_{k-s(e)}(u)$ where $k - s(e) < 0$ are assumed to be specified by the *initial conditions* of the system. Hence, for each node v there exists a relation, which will be denoted by $r(v)$.

The relations above can be used to model the performance of a distributed algorithm, with the function d specifying the delays, and $t_k(v)$ representing the time of the k-th step of v. In this case $d > 0$. Furthermore, a term $t_{k-1}(v) + d(v, v)$, $d(v, v) > 0$, can be included in the relation of v, to represent a lower bound on the time between two consecutive operations of v, since it implies $t_k(v) \geq t_{k-1}(v) + d(v, v)$.

In equation 1 two operations are used: $+$ and max, which satisfy the axioms of a ring. These operations are used for the sake of concreteness, but they could be replaced by other operations satisfying the same axioms. A natural alternative is to use $+$ and min, instead. Some protocols can be represented by recurrence relations in which both min and max are used, but they are outside of the scope of this paper. An algebraic study of similar structures was pursued by Cuninghame-Green [4]; our approach is graph theoretic.

3 Timed Marked Graphs

Marked graphs (e.g. Commoner et.al. [6]), a special case of Petri Nets, provide a useful model for studying a number of distributed algorithms. The model of a timed marked graph, or TMG (e.g. Ramamoorthy and Ho [15]) is obtained from a marked graph by adding durations to the events in the system. As shown in the following sections, a TMG is an instance of recurrence relations, where the operations used are $+$ and max, and the functions d and s are non-negative. Thus, recurrence relations can be used to study marked graphs, and TMG's constitute a graph theoretic setting for studying recurrence relations even when d and s may be negative.

A *timed marked graph* $TMG = (V, E, d, s_0)$, consists of a finite, directed graph $G = (V, E)$, a function d which assigns positive, integral *delays* to the edges of G, and an *initial state* s_0. The initial state defines the number of messages or *tokens* in each of the edges of G at time 0.

Intuitively, the dynamic operation of the TMG is as follows. When a node receives a token on all its incoming links, it performs some internal computation and then sends one token along each of its outgoing links. The time it takes a token to travel along a link e is $d(e)$.

To describe the operation of a timed marked graph formally, assume that the length of an edge $e = u \to v$ is equal to its delay $d(e)$, or simply d, and mark d *points* on e, $b_1(e), b_2(e), \ldots, b_d(e)$, such that $b_k(e)$ is k units away from the start-point of e. Using the points on e we define the meaning of delays as follows. If u sends a token M along e at time t, then M will be in point $b_k(e)$ at time $t + k$. Thus, at time $t + d$, M arrives at point $b_d(e)$, and remains there until v reads it; only when M is the last point of e, v may remove it from e. Hence, each end-point $b_d(e)$ has a *buffer* to store messages.

Let $B = \{b_k(e) | e \in E, 1 \leq k \leq d(e)\}$ be the set of all points of G. A *global state* of TMG is a function s from B to the non-negative integers, such that $s(b_k(e)) \leq 1$ if $k < d(e)$. The integer, $s(b)$ denotes the number of tokens at point b in the global state. The *initial state, s_0*, is the global state at time 0.

Let s_t denote the state at time t. A node v is *enabled* in s_t, if $s_t(b_d(e)) > 0$ for every edge e ending at v. An enabled node v *fires* by removing one token from the buffer of each of its incoming edges, and adding one token to the first point of each of its outgoing edges; namely, if e is an edge ending at v, then $b_d(e)$ is decremented by 1, and if e' is an edge starting at v, then $b_1(e')$ is incremented by 1. The state s_{t+1} reached from s_t is obtained by firing every enabled node in s_t and advancing tokens not in a buffer by one point.

It follows that there exists a lower bound of one unit on the time between two consecutive firings (message transmissions) of a node v. Therefore, we may assume (for notational convenience), without affecting the behavior of the TMG, that for every node v there is an edge $e = v \rightarrow v$, of delay 1, with 1 message in the initial state, e.g. $s_0(b_1(e)) = 1$.

4 Recurrence relations and TMG's

In this section we show that the behavior of a TMG is described by a set of recurrence relations and that for any set of recurrence relations (with operations $+$ and max) there exists a TMG whose behavior satisfies the relations.

For a global state s, define the following related quantities:

$$s(e) = \sum_{k=1}^{d(e)} s(b_k(e)), \quad \hat{s}(v) = \max_{u \rightarrow v \in E} s(u \rightarrow v), \quad \hat{s} = \max_{v \in V} \hat{s}(v).$$

For a path P, $s(P)$ denotes the number of tokens on the path, $l(P)$ the number of edges, and $d(P)$ the delay of the path.

The most interesting quantities are the firing times $t_k(u)$, $k \geq 0$, which denote the time at which u fires for the $k + 1$-st time. A token sent on $t_k(u)$ is denoted M_k. Thus, the first token sent by a node u is M_0. If in the initial state tokens are present on an edge $e = u \rightarrow v$, these tokens will be consumed by v one after the other. Let us extend the notation to include tokens of the initial state, denoting them by M_{-i}, $i > 0$, as follows. The tokens on e, starting with the one nearest u, are $M_{-1}, M_{-2}, \ldots, M_{-s}$, where $s = s_0(e)$. For example, if there are two tokens on e they are denoted M_{-1}, M_{-2}; and M_{-2} is consumed first while M_{-1} is consumed second. In the following theorem the initial conditions are given by $t_{-i}(u) + d(e)$, $i > 0$, which is equal to the time it takes M_{-i} to reach v. Thus, a misuse of notation is done to simplify the presentation. Hence the quantities $t_{-i}(u) + d(u \rightarrow v) < d(u \rightarrow v)$ are defined by s_0.

Theorem 4.1 *The firing times of a TMG are described by the relations*

$$t_k(v) = \max_{u \rightarrow v \in E}\{t_{k-s_0(u \rightarrow v)}(u) + d(u \rightarrow v)\}, \quad \text{for every } v \in V, \ k \geq 0. \tag{2}$$

The proof of this theorem is by induction on k.

Let Δ be equal to the maximum edge delay.

It is well known and easy to see, that the number of tokens in a cycle does not change by node firing. Also, the times $t_k(v)$ are finite for any k (there is no deadlock), if and only if, in a global state, the number of tokens in every cycle of G is positive (e.g. Commoner et.al. [6]). We shall assume that a TMG is deadlock-free and strongly connected. Thus, all the firing times $t_k(v)$, $k \geq 0$, are finite, non-negative, and uniquely determined by the initial state s_0.

Theorem 4.1 shows that the behavior of a $TMG = (G, d, s_0)$ is characterized by an associated set of n relations of the form 2, with $d(e) > 0$ and $s(e) \geq 0$, $e \in E$, and with corresponding initial conditions determined by an initial state s_0. Namely, the set of n recursive relations defined by G, d and s_0. Such a set of relations is said to be of *dimension* n.

Similarly, theorem 4.2 shows that given a set of relations $R(G, d, s)$, $d > 0$ and $s \geq 0$, interpreting d as delays, and s as the initial marking, it is straightforward to obtain the *associated timed marked graph*, (G, d, s) (for a term $t_{k-s_0(u \to v)}(u) + d(u \to v)$ add an edge $u \to v$ with delay $d(u \to v)$ and $s_0(u \to v)$ tokens. For the case where d and s are not non-negative, we can still show that there exists a TMG with the required behavior provided that R is *deadlock-free*, i.e., the sum of the $s(\cdot)$ along any cycle is positive. Namely, we can transform R into R' s.t. in R' $d > 0$ and $s \geq 0$. Observe that the relations specify the graph G, the delays and the number of tokens on an edge in the initial state s_0, but not the position of the tokens on the edge. To specify the position of messages on an edge $e = u \to v$, the initial conditions $t_i(u) + d(e)$, $i = -1, \ldots, -s_0(e)$ are needed. We say that the a set of initial conditions is *valid* if they are non-decreasing and $0 \leq t_i(u) + d(e) \leq d(e) - 1$. In the absence of this information, we may assume that all the messages on an edge are in its buffer. If G is deadlock-free, we say that the set of relations is deadlock-free. In the sequel we assume that the relations are deadlock-free. The proof if this theorem follows from theorems

Theorem 4.2 *For every set of n recurrence relations and any set of valid initial conditions, there exist a timed marked graph and a constant A, such that for $k \geq 0$ every node v fires for the $k + 1$-st time at time $t_k(v) + kA$ specified by the set of relations.*

It follows from the two previous theorems that there exists a close relationship between TMG's and recurrence relations. In our study of these structures we shall use the more convenient representation of the two, according to the particular problem studied.

Lemma 4.3 *For a deadlock-free TMG,* $|t_0(u) - t_0(v)| \leq \Delta|V|$.

5   Canonical Representation of Recurrence Relations

Our aim is to represent a set of recurrence relations in a canonical form in which $s(e) = 1$, for $e \in E$. This is done in two steps. First we show that any set of relations can be transformed into normal form in which $s(e) > 0$. Then we show that a normal set of relations can be transformed into canonical form. By theorem 4.1 and the discussion in the previous section, we will show that any TMG can be transformed into a TMG of a canonical form with the same behavior. This canonical form corresponds to the synchronizer studied by Even and Rajsbaum in [8], [9]. Thus, this transformation will enable us to extend some of the results regarding synchronizers to any marked graph.

5.1 Normal Form

A set of recurrence relations is *normal* if $s(e) > 0$, for every edge e. Our purpose is to derive an algorithm to transform a set of deadlock-free (every cycle has a positive sum of the $s(\cdot)$ values) relations into normal form. This implies that for relations with $s(e) < 0$ there exists an equivalent TMG also.

The operation *expand(v)* used in the algorithm is described as follows. Consider the recursion $r(v)$ for v's firing times:

$$t_k(v) = \max_{u \to v} \{ t_{k-s(u \to v)}(u) + d(u \to v) \}, \tag{3}$$

where for some edge $e_1 = v_1 \to v$, $s(e_1) \leq 0$ (if no such edge exists, then expand$(v) = r(v)$). Then a term $t_{k-s(e_1)}(v_1) + d(e_1)$, in the right-hand side of equation 3, can be replaced by

$$\max_{v_2 \to v_1} \{ t_{k-s(v_2 \to v_1)-s(e_1)}(v_2) + d(v_2 \to v_1) \} + d(e_1),$$

using the equation $r(v_1)$ corresponding to v_1. The result of expand(v) is obtained by performing this replacement for every edge $e = u \to v$ for which $s(e) \leq 0$, leaving only one *max* operator, and eliminating redundant terms. A term $t_{k-c}(u) + d$ is *redundant* if there is another term $t_{k-c}(u) + d'$ with $d' \geq d$. Clearly, the expand operation does not modify the solutions of the set of relations.

It follows that in terms of the timed marked graph TMG, the operation expand(v) transforms TMG into TMG_1 by removing every edge $v_1 \to v$, such that $s(v_1 \to v) \leq 0$, and adding an edge $v_2 \to v$ for every edge $v_2 \to v_1$ with delay $d(v_2 \to v) = d(v_2 \to v_1) + d(v_1 \to v)$, and $s(v_2 \to v_1) + s(v_1 \to v)$ tokens, for every edge $v_2 \to v_1$. The equivalent of removing redundant terms is to eliminate parallel edges with the same number of tokens, by leaving only the one with largest delay. The positions of the tokens on new edges are again determined by the initial firing times, and since $\hat{s}$ (the maximum number of tokens on an edge) does not increment after an expand operation, then the same set of initial firing times suffices. Note that TMG_1 is *equivalent* to TMG in the sense that they both have an associated set of relations with the same solutions, namely, the vertices in both graphs fire at exactly the same times. As we shall later see, as far as the ultimate steady state rate of computation is concerned, the position of the tokens on the new edges is immaterial.

For a set of relations R, expand(R) is the set of relations obtained by simultaneously expanding every relation in R. The idea behind the algorithm to normalize the deadlock-free

R is that by performing a sequence of expand operations on R, a normal set of relations is eventually obtained.

In the full version we describe a normalizing algorithm in terms of the TMG associated with the relations, for the case that $s(e) \geq 0$. If there are edges with negative number of tokens, a similar algorithm should be repeated n times. Let $free(TMG)$ be the (acyclic) subgraph of TMG induced by all the token-free edges. The transitive closure of free(TMG) has the same vertices as free(TMG), and there exists an edge e from u to v if there is a path from u to v in free(TMG). The value $d(e)$ is equal to that of the largest $d(P)$ among all paths P from u to v in free(TMG), where $d(P)$ is equal to the sum of the delays of the edges of P.

Let Δ_1 be the maximum delay of an edge in E_1.

Theorem 5.1 *There exists an algorithm that computes a normal marked graph $TMG_1 = (V, E_1, d_1, s_1)$ equivalent to TMG in $O(|V| \cdot |E|)$ time, with the following properties.*

(i) $|E_1| = O(|V| \cdot |E|)$.

(ii) *If $\hat{s} = 1$, then $|E_1| \leq |V| + |V|^2$.*

(iii) $E \subseteq E_1$.

(iv) $\Delta_1 \leq \Delta |V|$

(v) *For every edge $e = u \rightarrow v \in E_1 - E$ there exists a path P from u to v in TMG such that the first edge e_1 in P has $s(e_1) > 0$, all other edges in P have zero tokens, and $d_1(e) = d(P)$, $s_1(e) = s(P) = s(e_1)$.*

Theorem 5.1 is used in the following theorem, which implies that a set of recurrence relations with integer delays can be studied by a TMG (with positive delays).

We *shift* a set of relations $R(G, d, s)$ by A to obtain $\bar{R}(G, \bar{d}, s)$ as follows: $\bar{d}(e) = d(e) - s(e)A$. Denote by $\bar{t}_k(\cdot)$ the shifted firing times of $\bar{R}$. The initial conditions are $\bar{t}_{-i}(u) = t_{-i}(u) + iA$. The proof of the following is by induction on k using theorem 4.1.

Theorem 5.2 $\bar{t}_k(v) = t_k(v) - kA$.

5.2 Canonical Form

A set of relations $R(G, d, s)$ is *canonical* if $s(e) = 1$, for every edge e. Similarly, a TMG is canonical if every one of its edges contains exactly one token. We shall now show that for every TMG there is an equivalent canonical TMG .

Let $e = u \rightarrow v$ be an edge of TMG with $s(e) > 1$. The *extension* of e, ext(e), is a directed path $(u =)uv_0 \rightarrow uv_1 \rightarrow uv_2 \rightarrow \cdots \rightarrow uv_c(= v)$, $c = s(e)$. For $i = 1, 2, \ldots, c - 1$, the vertices

uv_i are new vertices not previously in V. There is one token in the buffer of each edge in $\text{ext}(e)$, and therefore $t_0(uv_i) = 0$. The delays along the path are as follows. Suppose that the k-th token of e, going from u to v, is in point $b_i(e)$, $i = i_k$. Note that this token corresponds to the one on the edge entering uv_k. The delay of this edge is equal to $i_k - i_{k-1}$, where $i_0 = 0$. It follows that $d(\text{ext}(e)) = d(e)$, $s(\text{ext}(e)) = s(e)$. Again, we may assume that each of the vertices in $\text{ext}(e)$ has a loop of delay 1 and a single token.

Let $TMG_1 = (V, E_1, d_1, s_1)$ be a normal timed marked graph corresponding to TMG. Let $t_k^{(i)}(v)$, $i = 1, 2$, be the k-th firing time of v in TMG_i.

Theorem 5.3 *For every timed marked graph $TMG = (V, E, d, s)$ there exists a timed marked graph $TMG_2 = (V_2, E_2, d_2, s_2)$ such that*

(i) *For every node $v \in V$, and for every $k \geq 0$, $t_k(v) = t_k^{(2)}(v)$.*

(ii) *For every edge $e \in E_2$, $s(e) = 1$.*

(iii) *For every path P from u to v, $u, v \in V$ in TMG_2 there exists a path P' in TMG from u to v such that $d_2(P) = d(P')$ and $s(P') = l(P)$.*

(iv) *$|V_2| = |V| + \sum_{e \in E_1} s(e) - 1 = O(|V| + (\hat{s} - 1)|V| \cdot |E|)$, $|E_2| = O(\hat{s}|V| \cdot |E|)$.*

Corollary 5.4 *For every set of relations R, there exists an equivalent canonical set of relations. That is, there is an equivalent canonical set of relations of the form:*

$$t_k(v) = \max_{u \to v \in E}\{t_{k-1}(u) + d(u \to v)\}, \quad k \geq 1,$$

and corresponding initial times $t_0(v)$ implied by the initial conditions of R.

6 Rate of Computation and Periodicity

In this section we analyze the behavior of the recurrence relations and their performance. We show that a system described by recurrence relations, always enters a periodic state, with optimum rate of computation, after a transitory stage. We present a tight bound on the time to reach periodicity and an algorithm to compute the rate of computation. The result presented here is a direct consequence of Theorem 5.3 and the results of [9]; we defer a direct proof of a slightly stronger result to an extended version of the paper.

For a set of recurrence relations, described by a strongly connected graph, the *rate R* is defined as

$$R = \lim_{k \to \infty} \frac{k}{t_k(v)},$$

for every node v. As will be shown, the limit exists and is independent of v. Moreover, every node enters a periodic state. We say that v is k/d-periodic at t, if $t_i(v) \geq t$ implies that $t_{i+k}(v) = t_i(v) + d$. Note that if v is k/d-periodic then $R = k/d$.

For a directed cycle C of G, let $A(C) = d(C)/s(C)$. Let $\hat{A}$ be the maximum $A(C)$ over all directed cycles C of G. A cycle C is called *critical* if $A(C) = \hat{A}$. The following theorem is easy to prove, considering any cycle which passes through v; in the context of marked graphs it is well known (e.g. Ramamoorthy and Ho, [15]). This theorem shows that even if vertices do not have to wait for tokens on edges outside the cycle, the rate is at most $1/\hat{A}$.

Theorem 6.1 $R \leq 1/\hat{A}$.

For canonical timed marked graphs, the following theorem by Even and Rajsbaum, [9], shows that the optimal rate of Theorem 6.1 is always achieved, and once achieved, the network enters a periodic state. Moreover, the theorem provides a tight upper bound on the length of the transitory stage.

Theorem 6.2 (ER) *If $s(e) = 1$ for all edges e, then there exists a constant λ, such that any v is $\lambda/\lambda\hat{A}$-periodic by time $O(\Delta^2|V|^3)$, after firing at most $O(\Delta|V|^3)$ times.*

For canonical TMG's, $A(C) = d(C)/l(C)$, and $\hat{A}$ is the maximum average cycle delay. Theorem ER implies that $R = 1/\hat{A}$. For these TMG's, an algorithm by Karp [K78] computes $\hat{A}$ in time $O(|V|\cdot|E|)$. The period of a TMG, λ, depends on the lengths of the critical cycles in G, and as shown in [9] may be exponential. Now, Theorem 5.3 states that any timed marked graph can be reduced to a canonical form, and thus it is possible to see that Theorem 6.2 can be generalized to apply to recurrence relations in general; the proof is postponed to the final version of the paper. Let Δ_2 denote the maximum delay of the normal representation of a given TMG and let $|V_2|$ denote the number of vertices in a canonical representation of the TMG. Thus, $\Delta_2 = O(\Delta|V|)$ and $|V_2| = O(\hat{s}|V| \cdot |E|)$.

Theorem 6.3 *For any TMG there exists a constant λ, such that any v is $\lambda/\lambda\hat{A}$-periodic by time $O(\Delta_2^2|V_2|^3)$, after firing at most $O(\Delta_2|V|^3)$ times.*

The case of graphs with at most one token on each edge and a unit delay was studied by Barbosa and Gafni [3] and Malka et.al. [13]. For this case the previous theorem is restated as follows, and applies also when arbitrary positive delays are allowed.

Theorem 6.4 *If $s(e) \leq 1$ for all edges e, then there exists a constant λ, such that any v is $\lambda/\lambda\hat{A}$-periodic by time $O(\Delta^2|V|^4)$, after firing at most $O(\Delta|V|^4)$ times.*

Theorem 6.3 implies that the bound of Theorem 6.1 is tight. Moreover, an algorithm of Lawler [12] can be used to compute the rate:

Corollary 6.5 *The rate R of any TMG is equal to $1/\hat{A}$. The rate can be computed in time $O(|V||E|\log|V|)$.*

7 Applications

As stated in the Introduction, the recurrence relations can be used to analyze a variety of distributed algorithms. We briefly describe how they apply to distributed schedulers and synchronizers.

7.1 Distributed Schedulers

Distributed schedulers for an undirected graph G were studied by Barbosa and Gafni [3] and Malka et.al. [13]. These schedulers are based on scheduling by *edge reversal*. Initially, orient G by an acyclic orientation. Nodes which are sinks in this initial orientation are the ones which operate initially. After operating, all sinks reverse their incident edges. At the next synchronous step another set of sinks exists, and so on (delays are equal to 1).

Schedulers can be studied by recurrence relations as follows. Replace every edge of G with two antiparallel edges. If in the initial orientation an edge was directed from u to v, then put one token on the edge $u \to v$. One can verify that the marked graph obtained behaves equivalently to the scheduler. Moreover, the scheduler can be analyzed with arbitrary delays, and Theorem 6.4 applies.

7.2 Synchronizers

We refer to the synchronizer in Even and Rajsbaum [9] and to α in Awerbuch [1]; synchronizers β and γ can also be analayzed.

The synchronizer of [9] applied to a network G with delays d is equivalent to a $TMG = (G, d, s)$ with exactly one token on every edge. Thus, the rate of this synchronizer is equal to the inverse of the maximum average cycle delay of G.

Synchronizer α can be analyzed as follows. Each node v waits for a safe–message from each of its neighbors, and then waits c time units. The c units represent the time it takes v to send messages to every neighbor and receive back acknowledgments. After the c units v sends safe–messages to all its neighbors. Thus, the worst-case time operation of α can be represented by relations of the form:

$$t_k(v) = \max_{e=u \to v \in E} \{t_{k-1}(u) + d(e)\} + c, \tag{4}$$

or

$$t_k(v) = \max_{e=u \to v \in E} \{t_{k-1}(u) + d(e) + c\}. \tag{5}$$

Since c is equal to twice the largest round trip delay of an edge incident on v, it follows that in the case of α it holds that $R \leq 1/\hat{c}$, where $\hat{c}$ is the largest round trip delay between two neighbors in G. Thus $R \leq 1/\Delta$.

References

[1] B. Awerbuch, "Complexity of Network Synchronization," J. of the ACM, Vol. 32, No. 4, Oct. 1985, pp. 804-823.

[2] B. Awerbuch, A. Baratz, D. Peleg, "Cost-Sensitive Analysis of Communication Protocols," IEEE FOCS, 1990.

[3] V. C. Barbosa and E. Gafni, "Concurrency in Heavily Loaded Neighborhood–Constrained Systems," ACM Trans. on Programming Languages and Systems, Vol. 11, No 4, Oct. 1989, pp. 562-584.

[4] R. Cuninghame-Green, *Minimax Algebra*, Lecture Notes in Economics and Mathematical Systems, No. 166, Springer-Verlag, 1979.

[5] K. M. Chandy and L. Lamport, "Distributed Snapshots: Determining Global States of Distributed Systems," ACM Trans. on Computer Systems, Vol. 3, No 1, Feb. 1985.

[6] F. Commoner, A.W. Holt, S. Even, A. Pnueli, "Marked Directed Graphs," J. of Computer and System Sciences, Vol, 5, No 5, Oct. 1971.

[7] I. Chlamtac and S. Pinter, "Distributed nodes organization algorithm for channel access in a multi-hop dynamic radio network," IEEE Trans. on Computers, vol. C-36, No. 6, June 1987, pp. 728-737.

[8] S. Even, S. Rajsbaum, "Unison in Distributed Networks," *Sequences, Combinatorica, Compression, Security, and Transmission*, R.M. Capocelli (ed.), Springer-Verlag, 1989. It is included in "Lack of a Global Clock Does not Slow Down the Computation in Distributed Networks," TR-522, Department of Comp. Science, Technion, Haifa, Israel, October 1988.

[9] S. Even, S. Rajsbaum, "The Use of a Synchronizer Yields Maximum Computation Rate in Distributed Networks," Proc. 22th ACM STOC, 1990.

[10] R. M. Karp, "A Characterization of the Minimum Cycle Mean in a Digraph," Discrete Mathematics, Vol. 23, 1978, pp. 309-311.

[11] S. Kirkpatrick, C.D. Gelatt, M.P, Vecchi, "Optimization by Simulated Annealing," Science 220 (4598), pp. 671-680 (May 13, 1983).

[12] E.L. Lawler, *Combinatorial Algorithms: Networks and Matroids*, Holt, Rinehart, and Winston, 1976.

[13] J. Malka, S. Moran, S. Zaks, "Analysis of a Distributed Scheduler for Communication Networks," Technical Report # 495, Department of Computer Science, Technion, Haifa, Israel, Feb. 1988. Also in proceedings of the 3rd Agean Workshop on Computing, AWOC 88, Corfu, June 1988), LNCS, Vol 319, pp. 351-360, Springer Verlag, 1988.

[14] Y. Ofek, I. Gopal, "Generating a Global Clock in a Distributed System," IBM Research Report, 1987.

[15] C.V. Ramamoorthy and G.S. Ho, "Performance Evaluation of asynchronous concurrent systems using Petri Nets," *IEEE Trans. Software Engineering*, vol. SE-6, no. 5, pp. 440–449, (1980).

[16] S. Rajsbaum, M. Sidi, "On the Average Performance of Synchronized Programs in Distributed Networks," 4th Int. Workshop on Distributed Algorithms, Italy, October 1990, LNCS 486.

[17] S. Rajsbaum, "Stochastic Marked Graphs," to appear in 4th Int. Workshop on Petri Nets and Performance Models (PNPM91), Melbourn, Australia, Dec. 1991.

Detection of Global State Predicates

Keith Marzullo[*]   Gil Neiger[†]

Abstract

This paper examines algorithms for detecting when a property Φ holds during the execution of a distributed system. The properties we consider are expressed over the state of the system and are not assumed to have properties that facilitate detection, such as stability.

Detection is done by a monitoring process within the system, which cannot perceive an execution of a distributed system as a total order; because of this, we consider two interpretations for "detecting Φ":

1. There is an execution consistent with the observed behavior such that Φ was true at a point in that execution. We refer to this property as *possibly* Φ.

2. For all executions consistent with the observed behavior, there was some point in real time at which the global state of the system satisfied Φ. We refer to this property as *definitely* Φ.

In this paper, we give formal definitions for these two interpretations and present algorithms for them. We give protocols for both asynchronous and synchronous systems and, for synchronous systems, give upper bounds on the time between the occurrence of the property of interest and the time a monitor detects the property.

1   Introduction

A *reactive system* [6] is characterized by a *control program* that interacts with an *environment*. The control program is input-driven: it monitors the environment and reacts to significant events by sending commands to the environment. There are many examples of reactive systems; for example, most embedded real-time systems are reactive systems, in which case the environment is an instrumented physical process. Non-real-time examples of reactive systems includes monitoring and debugging systems [4,13] and tool integration services [5,14].

---

[*]This author was supported in part by the Defense Advanced Research Projects Agency (DoD) under NASA Ames grant number NAG 2-593 and by grants from IBM, Siemens and Xerox. The views, opinions, and findings contained in this report are those of the authors and should not be construed as an official Department of Defense position, policy, or decision. Author's address: Cornell University Department of Computer Science, Upson Hall, Ithaca, New York   14853-7501, USA.

[†]This author was supported in part by the National Science Foundation under grants CCR-8909663 and CCR-9106627. Author's address: College of Computing, Georgia Institute of Technology, Atlanta, Georgia 30332-0280, USA.

In the *Meta* project [9,11], we have been developing tools that support the management of distributed applications through the use of a reactive system structure. Using Meta, the distributed application and its supporting services (for example, operating system, network servers, and hardware) can be instrumented with *sensors* that access its state and *actuators* that allow its state to be changed. Meta also provides a distributed interpreter of finite state automata that reference these sensors and actuators. Under Meta, control programs are translated into finite state automata that are executed by this distributed interpreter. Each interpreter executes *guarded atomic commands* of the form $\langle \Phi \rightarrow S \rangle$, meaning execute the action S in a state satisfying the global state predicate Φ.

The problem addressed in this paper arises in the context of Meta: how can a set of processes monitor the state of a distributed application in a consistent manner? For example, consider the simple distributed application shown in Figure 1. Each of the three

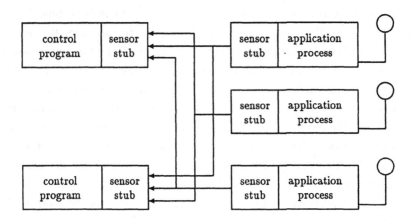

Figure 1: A Monitored Distributed Application

processes in the application has a light, and the control processes would each like to take an action when some specified subset of the lights are on. The application processes are instrumented with *stubs* that determine when the process turns its light on or off. This information is disseminated to the control processes, each of which then determines when its condition of interest is met.

Meta is built on top of the ISIS toolkit [1], and so we first built the sensor dissemination mechanism using atomic broadcast. Atomic broadcast guarantees that all recipients receive the messages in the same order and that this order is consistent with causality [7]. Unfortunately, the control processes are somewhat limited in what they can deduce when they find that their condition of interest holds.

For example, Figure 2 shows a space-time diagram of an execution of the application

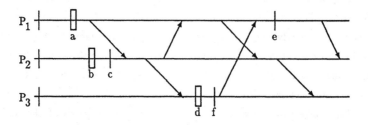

Figure 2: Space-Time Diagram of Application Execution

shown in Figure 1. In this figure, a process turning its light on is represented by a rectangle and the process turning its light off is represented by a vertical line. Assume, for the moment, that this system is *asynchronous*, meaning that there is no bound on message passing delays or on the relative speeds of processes. In this case, the only ordering relations between events that can be determined from within the system are those of potential causality. Two events that are not so related are *concurrent*. In Figure 2, the events a and b are concurrent as are a and c, so the control processes could receive these event notifications (as sent by atomic broadcast) in one of these orders: $(a; b; c)$, $(b; a; c)$ or $(b; c; a)$. Thus, the control processes may or may not determine that p_1's and p_2's lights were on simultaneously, but they will reach the same decision. On the other hand, the events a, d and e *are* causally ordered, so the control processes will determine that p_1's and p_3's lights were on simultaneously.

Given a global property Φ, there are at least two ways that "detecting Φ" can be interpreted:

1. There is an execution (*i.e.* a linear sequence of events) consistent with the observed behavior such that Φ was true at a point in that execution. We will refer to this property as *possibly* Φ. In the space-time diagram shown in Figure 2, the predicate *possibly* (p_1's light on and p_2's light on) holds.

2. For all executions consistent with the observed behavior, there was some point in real time at which the global state of the system satisfied Φ. We will refer to this property as *definitely* Φ. In the execution shown in Figure 2, the predicate *definitely* (p_1's light on and p_3's light on) holds, since the event of p_3 turning its light on happened between p_1 turning its light on and p_1 turning its light off.

Note that *definitely* Φ is stronger than *possibly* Φ. Hence, we will want to guarantee that if a control program determines *possibly* Φ for a set of local states, then no control program will ever determine *definitely* $\neg \Phi$ for the same states. Note that both of these conditions refer to some past state or states.

In this paper, we give formal definitions for these two interpretations and present protocols for them. We first give the protocols for an asynchronous system. These protocols can take an unbounded amount of time to detect their condition of interest; furthermore, they can have substantial running times because they may have to enumerate may possible global states. However, no better is possible, in general, due to the nature of asynchronous systems. We then modify these protocols for a system with approximately synchronized clocks and bounded message delay. These protocols are more practical, and we give upper bounds on the time between the occurrence of the property of interest and the time a control program detects the property. The existence of such a bound makes these protocols more useful in real systems.

Snapshot protocols for computing global states of a distributed system [2] are related to the protocols described in this paper, but they suffer from a limitation similar to that of the atomic broadcast implementation described above. In particular, if S is the global state computed by the snapshot, then there exists a legal execution of the system containing S, and so $\Phi(S)$ implies *possibly* Φ. Snapshot protocols are well-suited for detecting *stable* properties, which are those that, once they become true, the remain so. It may be the case that *possibly* Φ holds of an execution, but the snapshot protocol never detects it (this can happen if Φ is not stable).

A recent dissertation by Spezialetti [16] looks at a broader set of issues than those covered in this paper, such as using semantic information (like relative stability) to determine which local events could make a global property true. This dissertation also presents protocols whose specification is similar to ours. However, her protocols that detects event occurrence suffer from the same limitation as snapshots and the atomic-broadcast-based protocol described above. Additionally, Spezialetti's protocols do not take into account the ordering of events established by the underlying system's communication. We have also looked at the problem addressed in this paper when environments are continuous state transition systems [8]. Such systems have the useful property that physical variables can, in many cases, be interpolated forward. By doing so, the monitor can reason about the current state of the physical process rather than a past state, and so *possibly* Φ and *definitely* Φ can be determined for the current state.

2 Definitions

We first define the notion of an execution of a system. A system is composed of *processes*, some of which are part of the application being run and some of which are part of the monitoring control program. Let $\{p_1, \ldots, p_n\}$ be the set of application processes; for the sake of simplicity, we assume that there is only one monitoring process, denoted p_0. Each pair of processes is connected by a point-to-point, reliable, FIFO communication link, and we assume that processes do not exhibit faulty behavior.

Each process p_i has a *local state* s_i, which changes when an *event* occurs at the process. An event may be completely internal to the process, or it may be the sending or receipt of a message (e.g., "send m_1 to p_j" or "receive m_2 from p_k"). For the sake of simplicity, we assume that all message sent in the system are unique. Process p_i's *local history*, denoted h_i is a (possibly infinite) sequence of states and events

$$s_i^0 e_i^1 s_i^1 e_i^2 s_i^2 \cdots.$$

In this case, s_i^0 is p_i's initial state, and the first event it executes is e_i^1, after which the process's state is s_i^1, etc. A *global state* is a tuple $S = \langle s_0, s_1, \ldots, s_n \rangle$, one for each process. Although the monitor, p_0, is a process in the system, when we refer to a global state, we will usually mean only the global state of the application, $\langle s_1, \ldots, s_n \rangle$. A *global history* (or *history*) is a tuple $H = \langle h_0, h_1, \ldots, h_n \rangle$ of local histories, one per process.

Although a global history does not specify the relative timings of events and states at different processes, it does allow us to draw certain conclusions about these timings. An event e_i^l *happens before* e_j^m (written $e_i^l \rightarrow e_j^m$) if one of the following is true [7]:

- the events are at the same process and occur in the order indicated, that is, if $i = j$ and $l < m$;

- e_i^l is the sending of a message by p_i to p_j and e_j^m is the receipt of that message; or

- there is another event e_k^n such that $e_i^l \rightarrow e_k^n$ and $e_k^n \rightarrow e_j^m$.

The "happens before" relation can be used to reason about the possible executions associated with a global history. We associate with each global history a set of linearizations.

A *linearization* L of a history H is a sequence of global states and local events

$$S^0 e^1 S^1 e^2 S^2 \cdots$$

that contains exactly the events in H such that, if $e^m \rightarrow e^n$ in H, then $m < n$. Notice that no prefix of L contains the receipt of a message whose sending does not appear in that prefix. In synchronous systems (see Section 3.3), there are further constraints on the linearizations of a global history.

(The above definition of linearization assumes that, in the actual execution of a distributed system, no two events can occur simultaneously. This need not be the case; it is possible that events at different processors may occur at exactly the same time. We can easily extend our definition of linearization to include such definitions.)

We use the notion of a cut to represent the global states that could have occurred in the execution. A *cut* [2] of a global history H is a tuple of natural numbers $\langle t_1, \ldots, t_n \rangle$ that represents the state of the system after t_i events have executed at process p_i; that is, the cut represents the global state $\langle s_1^{t_1}, \ldots, s_n^{t_n} \rangle$. Only certain cuts of a global history can truly correspond to global states that took place at some real time. A cut $\langle t_1, \ldots, t_n \rangle$ of H is *consistent* if there is some point in some linearization L of H by which each process p_i executed exactly t_i events. L is said to *pass through* this cut. We will also refer to the associated tuple of events, $\left\langle e_1^{t_1}, \ldots, e_n^{t_n} \right\rangle$ as a consistent cut.

We want to be able to reason about certain facts (such as "possibly Φ") being true in different global histories. To this end, we introduce the following notation. Let H be some global history of the system. To formally define "*possibly* Φ," we introduce the formulas $?\Phi|C$, where C is a consistent cut. Formally, $?\Phi|C$ holds for history H if $C = \langle t_1, \ldots, t_n \rangle$ is a consistent cut of H and Φ holds for the global state $\langle s_1^{t_1}, \ldots, s_n^{t_n} \rangle$. If $?\Phi|C$ holds for H, then it is possible that Φ held during the execution that generated H since it held at some point in some linearization of H.

To formally define "*definitely* Φ," we introduce the formulas $!\Phi|A$, where A is a finite set of cuts. Formally, $!\Phi|A$ holds for H if A is a finite set of consistent cuts of H, every linearization of H passes through some cut in A, and for all cuts $\langle t_1, \ldots, t_n \rangle \in A$, Φ holds for the global state $\langle s_1^{t_1}, \ldots, s_n^{t_n} \rangle$. If $!\Phi|A$ holds for H, then Φ definitely held at some point in the execution that generated H because it held at some point in all linearizations of H.

Note that the definitions of these formulas satisfy two properties discussed earlier. The definitely operator ! is clearly stronger than the possibly operator ? in the following sense: if $!\Phi|A$ holds for H then for any $C \in A$, $?\Phi|C$ holds for H. Furthermore the two operators are, in a certain sense, dual. If $!\neg\Phi|A$ holds for H, then $?\Phi|C$ cannot hold for H if $C \in A$.

Informally, the control process p_0 detects *possibly* Φ when it can determine that there is a consistent cut of H that satisfies Φ, and p_0 detects *definitely* Φ when it can find a finite set of consistent cuts A such that every linearization of H passes through a member of A and such that Φ holds for every member of A. We are investigating the more formal definition of detection, but we do not present such defintions in this version of the paper.

3 Protocols

As noted above, system consists of $n + 1$ processes $\{p_0, p_1, \ldots, p_n\}$ whose only method of interaction is by exchanging messages. The process p_0 monitors the other processes to determine when some state predicate becomes true. This state predicate of interest will be of the form *possibly* Φ or *definitely* Φ, where Φ is a predicate over the states of the processes $p_1, \ldots, p_n$.

Each process p_i will know how to compute Φ and will send a message to p_0 when its local state changes in a way significant with respect to Φ. In particular, a process can determine whether a local event potentially changes Φ. More formally, let Φ be a predicate expressed over a global state; that is, $\Phi(s_1, \ldots, s_n)$ is true or false. Consider some event e_i^t of process p_i; recall that s_i^{t-1} is the value of s_i before e_i^t executes and s_i^t is the value of s_i after e_i^t executes. Event e_i^t *potentially affirms* Φ if the execution of e_i^t could have made Φ true:

$$\exists s_1, \ldots, s_{i-1}, s_{i+1}, \ldots, s_n : \neg\Phi(s_1, \ldots, s_i^{t-1}, \ldots, s_n) \wedge \Phi(s_1, \ldots, s_i^t, \ldots, s_n).$$

Similarly, event e^i *potentially rejects* Φ if the execution of e_i^t could have made Φ false:

$$\exists s_1, \ldots, s_{i-1}, s_{i+1}, \ldots, s_n : \Phi(s_1, \ldots, s_i^{t-1}, \ldots, s_n) \wedge \neg\Phi(s_1, \ldots, s_i^t, \ldots, s_n).$$

An event *potentially changes* Φ if it potentially affirms or rejects Φ; such an event is also called a *relevant event*.

Note that an event can both potentially affirm and reject Φ. For example, if $n \geq 4$ and Φ is "either two or three processes have their lights on," then when a process turns its light on, this action both potentially affirms and rejects Φ even though it is possible that the value of Φ did not change.

Our detection protocols will have the monitored processes periodically send to the monitor its state relevant to Φ; that is, the message will contain the values of the variables of p_i referred to in Φ. For each process p_i $(1 \leq i \leq n)$, process p_0 maintains a sequence Q_i of such messages received from p_i. These messages will also carry information for ordering these states, which is described next.

3.1 Weak Vector Clocks and Enumeration of Global States

Our protocols will have the monitor enumerate possible global states of the system by choosing states from each of the message sequences Q_i. In this section, we describe how this enumeration of global states is performed. We use a slight modification of *vector clocks* [12].

A *logical clock* [7] is a value T that satisfies the *clock condition*: given two events e_1 and e_2 and their associated clock values $T(e_1)$ and $T(e_2)$, if $e_1 \rightarrow e_2$, then $T(e_1) < T(e_2)$. We will find it advantageous to use clocks that also satisfy the converse of the clock condition; that is, clocks that satisfy

$$(e_1 \rightarrow e_2) \Leftrightarrow T(e_1) < T(e_2). \tag{1}$$

In particular, such clocks enable one to determine whether or not two events are concurrent; e_1 and e_2 are *concurrent* if neither $e_1 \rightarrow e_2$ nor $e_2 \rightarrow e_1$. Unfortunately, Lamport's logical clocks of [7] (which are implemented using a single counter) do not satisfy Equation 1.

A logical clock that satisfies Equation 1 can be implemented with a vector V of n counters. If V_i is the logical clock associated with process p_i, then $V_i[i]$ is the number of events that have been executed by p_i and $V_i[j]$, $j \neq i$, is the number of events that p_i "knows" p_j has executed. If e_i is an event at process i, then we use $V(e_i)$ to denote the value of V_i *after* e_i executed. Given this definition, one can easily show that $e_i \rightarrow e_j$ if and only if the vector clock of e_j records the fact that e_i has occurred:

$$e_i \rightarrow e_j \Leftrightarrow V(e_i)[i] \leq V(e_j)[i]. \tag{2}$$

Similarly, if e_i and e_j are concurrent, then

$$V(e_i)[i] > V(e_j)[i] \wedge V(e_j)[j] > V(e_i)[j].$$

If the set of processes is static, then vector clocks are not hard to implement. Initially, $V_i[j]$ is set to zero for all i and j. $V_i[i]$ is incremented whenever p_i executes an event. Every message sent by p_i is timestamped with V_i (let $V(m)$ refer to the timestamp on message m). If e_i is the receipt of message m, then each $V_i[k]$ ($k \neq i$) is set to the maximum of $V_i[k]$ and $V(m)[k]$. As an example, Figure 3 shows the values of vector clocks for the events of

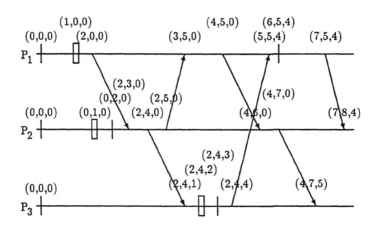

Figure 3: Vector Clocks for Events of Figure 2

the execution shown in Figure 2.

We can use vector clocks to determine whether or not a set of local states represents a consistent cut. The set of local states $S = \langle s_1, \ldots, s_n \rangle$ is a (consistent) global state if every pair of local states s_i and s_j is *potentially concurrent*. In terms of vector clocks, s_i and s_j are potentially concurrent if $V(s_i)[i] \geq V(s_j)[i]$ and $V(s_j)[j] \geq V(s_i)[j]$. Thus, the global state S is a consistent cut if and only if

$$\forall i, j : 1 \leq i, j \leq n : V(s_i)[i] \geq V(s_j)[i]. \tag{3}$$

Because we are interested only in the causal relationship of events that potentially change Φ, we can use a slight weakening of vector clocks [10]. With our clocks, process p_i will increment its local counter $V_i[i]$ only when it executes an event that potentially changes Φ. It will send a message to p_0 whenever its vector clock changes—that is, either when it executes a relevant event or when it executes a receive event through which it learns that another process has potentially changed Φ. The message sent from p_i to p_0 will contain p_i's state s_i after such an event is executed and the vector time $V(s_i)$.

Figure 4 illustrates such vector clocks. These clocks are a weakened version of normal

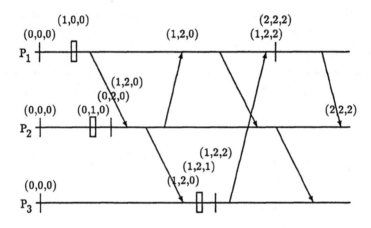

Figure 4: Weak Vector Clocks for Events of Figure 2

vector clocks—for example, if $i = j$, they need not satisfy Equation 2. They do, however, satisfy Equation 3, and this is all that our protocols need. For the sake of simplicity, the remainder of this paper assumes that all events—including send and receive events— are relevant events and thus that our weak vector clocks are true vector clocks.

3.2 Asynchronous Systems

In this section, we assume that processes do not possess local real-time clocks, that there is no global clock, and that there is no upper bound on message delays. We note in advance that there is no way to bound the amount of time between the time a condition becomes true and the time the monitor detects the condition. This is because messages sent to the monitor may be arbitrarily delayed.

The protocols for detecting *possibly* Φ and *definitely* Φ are based on the same data structure: the lattice of consistent global states that correspond to an observed execution. Such a lattice consists of n orthogonal axes, with one axis for each monitored process. A point $\bar{t} = \langle t_1, t_2, \ldots, t_n \rangle$ in this lattice corresponds to a consistent global state in which process p_i has executed t_i events. Of course, not all tuples $\langle t_1, t_2, \ldots, t_n \rangle$ appear in the lattice; this depends on the causal dependencies among the local states of P. Define the *level* of a point $\bar{t}$ to be the sum of its indices $t_1 + t_2 + \ldots + t_n$.

Consider some global history. A linearization of this history is a total order of (consistent) global states in which exactly one process executes one event between adjacent global states In terms of the lattice corresponding to the history, a linearization corresponds to a path in the lattice, where the level of each subsequent point in the path increases by one. A space-time diagram of a two-process system and the corresponding lattice of global states is illustrated in Figure 5. A point S_{ij} represents a state in which process p_1 is in its ith state and process p_2 is in its jth state. From the lattice, it is easy to see that one possible execution corresponds to the sequence of global states

$$S_{00}; S_{01}; S_{11}; S_{21}; S_{22}; S_{23}; S_{33}; S_{43} \cdots.$$

For every point $\bar{t}$ in a lattice, there exists at least one linearization that passes through $\bar{t}$. Hence, if any point in the lattice satisfies Φ, then *possibly* Φ holds. The property *definitely* Φ requires all linearizations to pass through a point that satisfies Φ. For example, suppose in Figure 5 that the points S_{43} and S_{34} both represent states that satisfy Φ; then *definitely* Φ holds. This is because S_{43} and S_{34} are the only points in level 7 and all linearizations must pass through some point in that level. *Definitely* Φ also holds if instead the states represented by points in the set $\{S_{53}, S_{35}, S_{54}, S_{45}\}$ all satisfy Φ. This is because if a linearization does not pass through S_{53} or S_{35}, then it must pass through S_{44} and hence through either S_{54} or S_{45}.

Figures 6 and 7 give the high-level algorithms that a monitoring process uses to detect *possibly* Φ and *definitely* Φ, respectively, in an n-processor system. Each algorithm begins by having the monitor distribute the predicate Φ to all processes and then construct the initial global state of level 0. (It is assumed that the monitor knows *a priori* each process's initial state relevant to Φ; if this is not the case, the processes begin by performing a two-phase synchronization protocol.)

The *possibly* Φ algorithm is straightforward: using the messages it receives, the monitor iteratively constructs levels of the lattice, using the vector timestamps accompanying each

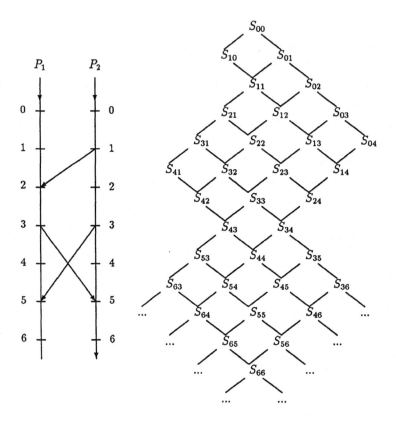

Figure 5: An execution and the corresponding lattice of global states.

message (see below). If it ever finds a global state in the current level satisfying Φ, then it reports *possibly* Φ and halts. Note that this protocol is not optimal in its reporting time because it always waits for a level to be completely enumerated. This restriction is not necessary and is only done to simplify the presentation of the algorithm and the one that follows.

The *definitely* Φ algorithm also iteratively constructs one level at a time. It attempts to prove that all paths in the lattice pass through a state supporting Φ. To this end, when constructing a new level, it adds only states that do not support Φ; call the resulting level a *reduced level*. If the monitor ever finds an empty reduced level, then the monitor halts and reports *definitely* Φ (in fact, it can report that Φ definitely holds by the time processes execute a total of *lvl* relevant events, where *lvl* is the last level enumerated).

As stated earlier, the implementations of the algorithms in Figures 6 and 7 require a monitored process to send the relevant part of its local state to the monitor whenever its

Possibly(Φ): **begin**
    current := global state $\langle s_1^0, s_2^0, \ldots, s_n^0 \rangle$;
    lvl := 0;
    do no state in *current* satisfies Φ →
        last := *current*
        lvl := *lvl* + 1;
        current := states of level *lvl* reachable from a state in *last*;
    od
    end;
    report *Possibly* Φ

Figure 6: Algorithm for detecting *Possibly* Φ.

vector clock changes. The monitor maintains sequences of these states, one per process, and assembles them into the necessary global states. Thus, the monitor must be able to determine when it can assemble all the reachable global states of a given level and when it can drop a local state from its sequence because the local state cannot appear in any further global states of interest. To achieve this, we use weak vector timestamps developed in the previous section.

Let Q_i be the sequence of messages that p_0 has received from p_i stored in FIFO order. Each state s_i in a message stored in Q_i is labeled with the weak vector timestamp $V(s_i)$ of the event that generated that state. Equation 3 defines when a set of states $\langle s_1, s_2, \ldots, s_n \rangle$, with s_i from process p_i, comprise a consistent cut. Note that the level of this global state is $\sum_{u=1}^{n} V(s_u)[u]$.

Consider some point $\bar{t} = \langle t_1, \ldots, t_n \rangle$ in the lattice that corresponds to the state $\langle s_1, \ldots, s_n \rangle$. The monitor can enumerate points of the next level in the lattice as follows. For each process p_i, the monitor checks to see if s_i', the state in the $(t_i+1)^{st}$ message of Q_i, is potentially concurrent with the other s_j's (if there is no such state in Q_i, the monitor cannot complete the next level until it receives that state). Thus, if

$$\forall j : j \neq i : V(s_i')[i] \geq V(s_j)[i] \wedge V(s_j)[j] \geq V(s_i')[j] \tag{4}$$

then point $\langle t_1, \ldots, t_i + 1, \ldots, t_n \rangle$ is in the lattice at the next level. (Although many such states will have be checked, it should be clear that a state at some level in the lattice may follow from several in the previous level; it only has to be checked once and not for each possible predecessor.)

We can also use vector timestamps to determine when a message containing state s_i can be eliminated, in the interest of saving space, from a queue Q_i. If the last state

Definitely(Φ): **begin**
  last := global state $\langle s_1^0, s_2^0, \ldots, s_n^0 \rangle$;
  remove all states in *last* that satisfy Φ;
  $lvl := 1$;
  % Invariant: *last* contains all states of level $lvl - 1$ that are accessible
  % from $\langle s_1^0, s_2^0, \ldots, s_n^0 \rangle$ without passing through a state satisfying Φ.
  do $last \neq \{\ \} \rightarrow$
      current := states of level lvl reachable from a state in *last*
      remove all states in *current* that satisfy Φ;
      $lvl := lvl + 1$; $last := current$
      od
  end;
  report *Definitely* Φ

Figure 7: Algorithm for detecting *Definitely* Φ.

in each other queue happens after s_i and is not potentially concurrent with it, then no state subsequently received could possibly form a global state with s_i. Thus, the message containing s_i can be removed from Q_i as soon as the following holds:

$$\forall j : j \neq i : V(Q_j.last)[i] > V(s_i)[i],$$

where $Q_j.last$ is the last state in Q_j.

The running time of both detection algorithms are linear in the number of global states. Unfortunately, the number of global states can be exponential in the number of processes. Even worse, the worst-case space complexity is unbounded, since the delivery of a message can be indefinitely delayed in an asynchronous system. While there are heuristics that can be used to limit the number of constructed global states, they are intrusive in that they require some kind of synchronization or limited blocking of the monitored processes. Real-time bounds on communication and the rate of change of local states can also be used, as is discussed in the next section.

3.3 Partially Synchronous Systems

In this section, we assume that each process p_i has a real-time clock C_i, and that these clocks are approximately synchronized: at any given "real" time, the difference between the clocks of two processes is no more than ϵ. We define this formally by modifying our definition of histories and linearizations slightly. Firstly, all processes (including p_0) execute "tick" events; a process's local time is the number of tick events that it has executed. If e_i is an event at p_i, then $C_i(e_i)$ is the number of tick events that p_i has executed through e_i. If H is a history with approximately synchronized clocks, then L is a linearization of H only if, in addition to the usual requirement, in all prefixes of L and every pair of processes p_i and p_j, the difference in the number of tick events executed by the two processes is at most ϵ.

In addition to approximately synchronized clocks, we assume that there are lower and upper bounds on message transmission times. This means that if process p_i executes "send m to p_j" after it has executed t_s tick events, then when p_j executes "receive m from p_i," it has executed t_r tick events, where $t_s + d_{\min} \leq t_r \leq t_s + d_{\max}$ for constants $d_{\min}$ and $d_{\max}$ (both greater than 0). These bounds will be especially important when considering messages received by the monitor p_0. Approximately synchronized clocks can be used to extend the "happens before" relation to order two events e_i and e_j even when there is no explicit communication between p_i and p_j; thus, we redefine $e_i \rightarrow e_j$:

$$e_i \rightarrow e_j \Leftrightarrow (V(e_i)[i] \leq V(e_j)[i]) \vee (C(e_i) + \epsilon < C(e_j)).^1$$

That is, e_i must happen before e_j either if e_i can causally affect e_j (as measured by vector timestamps) or if the clock times corresponding to the events show that e_i must happen first.

Our protocols will be such that each state s_i sent by a process p_i to the monitoring process p_0 will include the local time $C(s_i)$ at which the event resulting in s_i occurred, as well as the vector timestamp $V(s_i)$. The monitor can then use the vector timestamps and the clock times of these states to enumerate the levels of the lattice. The clock times can be used to further restrict the pairs of events that are potentially concurrent. With each state s_i in Q_i, the monitor can determine the latest local time at which p_i must have been in state s_i (call this $L(s_i)$). If there is a state s_i' after s_i in Q_i, then this is $C(s_i')$; if $s_i = Q_i.last$, then this is $C - d_{\max}$, where C is the monitor's current local time. If p_i had changed its local state between $C(s_i)$ and $C - d_{\max}$, then the monitor would have gotten another message from p_i by its local time C.

We can now say that two states s_i and s_j received by the monitor are potentially concurrent if both the vector time stamps and the the real-time clocks indicate this:

$$
\begin{aligned}
(V(s_i)[i] \geq V(s_j)[j]) \;\wedge\; & (V(s_j)[j] \geq V(s_i)[i]) \\
\wedge\; & ((C(s_i) - \epsilon) \leq C(s_j) \leq L(s_i) + \epsilon \\
\vee\; & (C(s_j) - \epsilon) \leq C(s_i) \leq L(s_j) + \epsilon).
\end{aligned}
\tag{5}
$$

---

[1] There is no need to take the transitive closure of the two relations because, if $d_{\min} \geq 0$, $V(e_i)[i] \leq V(e_j)[i]$ and $C(e_j) + \epsilon < C(e_k)$ then $C(e_i) + \epsilon < C(e_k)$, and if $C(e_i) + \epsilon < C(e_j)$ and $V(e_j)[j] \leq V(e_k)[j]$ then $C(e_i) + \epsilon < C(e_k)$.

Suppose now that the monitor is seeking to extend the state $\langle s_1, \ldots, s_n \rangle$ to the next level by potentially adding a new state s_i', the $(t_i + 1)^{st}$ state in Q_i. It checks to see if s_i' is potentially concurrent with the other s_j's by using Equation 5 instead of Equation 4. If s_i' is potentially concurrent with all the s_j's, then the state $\langle s_1, \ldots, s_i', \ldots, s_n \rangle$ is added to the next level of the lattice; otherwise, it is not. An exception to the last point is if $s_i' = Q_i.last$ and. s_i' was not deemed to be potentially concurrent because its latest time was too early. For example, suppose $\epsilon = 1$ and

$$C(s_i') = 3; \qquad L(s_i') = 4; \qquad C(s_j) = 6; \qquad L(s_j) = 7.$$

Because $s_i' = Q_i.last$, $L(s_i') = C - d_{max}$; as time passes on the monitor's clock, $L(s_i')$ may grow so that the two states would be judged potentially concurrent. In such cases, therefore, the decision about whether to add the state $\langle s_1, \ldots, s_i', \ldots, s_n \rangle$ to the lattice is postponed until either another message arrives from p_i or the monitor's clock advances to a point where a decision can be made. Until then, the level cannot be completely enumerated.

The conditions *possibly* Φ and *definitely* Φ can now be detected exactly as in the previous section. Each processor sends its state to the monitor whenever its vector clock changes; it includes with this message its vector time and the number of tick events it has executed. The monitor then uses this information to construct levels of the lattice, using the properties of the "potentially concurrent" states discussed above. It then reports "*possibly* Φ" or "*definitely* Φ" exactly as it would in the case of asynchronous systems.

We now argue upper bounds on detection times. Suppose that $S = \langle s_1, \ldots, s_n \rangle$ is a global state such that the last event leading to this state occurs when the monitor's local time is t. No process's local clock is higher than $t + \epsilon$ when one of the events leading to S occurs, so p_0 receives all messages necessary to construct this state by local time $t + \epsilon + d_{max}$. Local time $t + 2\epsilon$ is the latest that a process could execute an event that could be potentially concurrent with one leading to S; thus, by time $t + 2\epsilon + d_{max}$, p_0 will have completed the construction of the level containing S.

Suppose that *possibly* Φ holds of a history; this means that some consistent cut of the history supports Φ. If the last event leading to this cut occur when the monitor's local clock is t, then the monitor will finish enumerating the level of S at its local time $t + 2\epsilon + d_{max}$, detecting *possibly* Φ at that time (actually, it could detect it at time $t + \epsilon + d_{max}$ because, as noted earlier, the *possibly* protocol does not need to enumerate an entire level once it finds a state satisfying Φ).

Suppose that *definitely* Φ holds of a history; this means that there is some finite set of consistent cuts, all supporting Φ, through which all linearizations pass. If the last event leading to the last of these occurs when the monitor's local time is t, then the monitor must detect *definitely* Φ by time $t + 2\epsilon + d_{\max}$ on its local clock; this is because the last state in the level of the last cut will be enumerated by that time, and the protocol will halt.

The above discussion does not consider the amount of local computation required by the monitor. In general, this depends on the relation between ϵ and the rate at which processes can potentially change Φ. If clocks are closely synchronized, then the monitor will never have to consider more than a few state changes by any one process. If instead processes potentially change Φ very often, then the monitor may have to do significant local computation.

4 Conclusions

This paper has defined two means (*possibly* and *definitely*) by which global states in an asynchronous or loosely synchronous system can be detected. It presented algorithms by which a monitor can detect these properties in both types of systems. There are other means of detection that are also of interest. For example, we have been investigating a third type of detection, called *currently*, that occurs when the monitor learns a condition actually holds at the time of detection. One can modify our *definitely* algorithms for partially synchronous systems to detect *currently* by requiring that application processes forgo potentially rejecting the condition being detected for a well-defined amount of time. We can obtain *currently* algorithms for asynchronous systems only by forcing application processes to block.

These algorithms may be complex, both in terms of computation and storage. Although we are investigating optimizations of these algorithms, we maintain that significant complexity is required for detection to be complete. In the future, we plan to look at the kinds of information that may simplify the detection. If the property that is to be detected, Φ, has certain nice properties, then detection may be simplified. If the monitor has some knowledge of the how and when the application program potentially changes the condition to be detected, then this can also simplify detection. We have also been investigating casting the detection problem into temporal and epistomological logics. We believe that such a characterization will aid in finding sets of properties under which detection can be simplified.

Although our original application was towards distributed application management, we have also been investigating the use of these detection protcols in the scope of debugging distributed systems [3]. The constraints of a debugger are slightly different from those that arise in tool integration or distributed application management. For example, invasiveness is traditionally considered untolerable, yet in tool integration, temporarily blocking an application may be acceptable.

The work most similar in spirit to ours are the protocols developed by Spezialetti [16]. In particular, her *event holding* condition is the same specification as our protocol for detecting *currently* Φ, and the specification of her *event occurrence* condition is similar to the specification of our *possibly* Φ algorithm. However, her protocols for non–local event detection are incomplete, in that they can miss conditions that in fact held. For example, the execution in Figure 8 shows such an execution. If the messages in this figure correspond

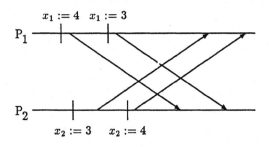

Figure 8: $\Phi \equiv (x_1 = x_2)$

to the messages generated in establishing simultaneous regions [15], then her protocol will not detect $x_1 = x_2$, yet in fact *definitely* $x_1 = x_2$ holds.

References

[1] Kenneth P. Birman and Thomas A. Joseph. Reliable communication in the presence of failures. *ACM Transactions on Computer Systems*, 5(1):47–76, February 1987.

[2] K. Mani Chandy and Leslie Lamport. Distributed snapshots: determining global states of distributed systems. *ACM Transactions on Computer Systems*, 3(1):63–75, February 1985.

[3] Robert C. B. Cooper and Keith Marzullo. Consistent detection of global predicates. In *Proceedings of the ACM/ONR Workshop on Parallel and Distributed Debugging*, pages 163–173. ACM/ONR, 1991.

[4] C. J. Fidge. Partial orders for parallel debugging. In *SIGPLAN/SIGOPS Workshop on Parallel and Distributed Debugging*, pages 183–194. ACM, 1988.

[5] David Garlan and Ehsan Ilias. Low-cost, adaptable tool integration policies for integrated environments. In *Proceedings of the Fourth Symposium on Software Development Environments*, pages 1–10. ACM SIGSOFT, 1990.

[6] David Harel and Amir Pnueli. On the development of reactive systems. In Krzysztof R. Apt, editor, *Logics and Models of Concurrent Systems*, volume 13 of *NATO ASI Series F*, pages 477–498. Springer-Verlag, New York, 1985.

[7] Leslie Lamport. Time, clocks, and the ordering of events in a distributed system. *Communications of the ACM*, 21(7):558–565, July 1978.

[8] Keith Marzullo. Tolerating failures of continuous-valued sensors. *ACM Transactions on Computer Systems*, 8(4):284–304, November 1990.

[9] Keith Marzullo, Robert C. B. Cooper, Mark Wood, and Kenneth P. Birman. Tools for distributed application management. *IEEE Computer*, 24(8):42–51, August 1991.

[10] Keith Marzullo and Laura Sabel. Using consistent subcuts for detecting stable properties. In *Proceedings of the Fifth Workshop on Distributed Algorithms and Graphs*, October 1991. To appear.

[11] Keith Marzullo and Mark Wood. Tools for distributed application management. In *Proceedings of the Spring 1991 EurOpen Conference*, pages 477–498, May 1991.

[12] Friedemann Mattern. Time and global states of distributed systems. In Michel Cosnard et. al., editor, *Proceedings of the International Workshop on Parallel and Distributed Algorithms*, pages 215–226. North-Holland, October 1989.

[13] Barton P. Miller and Jong-Deok Choi. Breakpoints and halting in distributed programs. In *Proceedings of the Eighth International Conference on Distributed Computing Systems*, pages 316–323. IEEE, 1988.

[14] Steven P. Reiss. Connecting tools using message passing in the FIELD program development environment. *IEEE Software*, 7(4), July 1990.

[15] M. Spezialetti and J. P. Kearns. Simultaneous regions: A framework for the consistent monitoring of distributed computations. In *Proceedings of the Ninth International Conference on Distributed Computing Systems*, pages 61–68. IEEE, 1989.

[16] Madelene Spezialetti. *A Generalized Approach to Monitoring Distributed Computations for Event Occurrences*. Ph.D. dissertation, University of Pittsburgh, 1989.

Using Consistent Subcuts for Detecting Stable Properties*

Keith Marzullo   Laura Sabel

Abstract

We present a general protocol for detecting whether a property holds in a distributed system, where the property is a member of a subclass of stable properties we call the *locally stable properties*. Our protocol is based on a decentralized method of constructing a maximal subset of the local states that are mutually consistent, which in turn is based on a weakened version of vector time stamps. The structure of our protocol lends itself to refinement, and we demonstrate its utility by deriving some specialized property-detection protocols, including two previously-known protocols that are known to be efficient.

1 Introduction

Chandy and Lamport [4] give a simple protocol that can be used to determine whether or not the global state of an asynchronous distributed system satisfies any given stable property. For most stable properties of interest (e.g., deadlock, termination, and lack of a token) there exist specialized protocols that are more efficient than a straightforward application of [4].

It would be useful if one could derive such special-purpose protocols by refinement of a general protocol. Unfortunately, the protocol of [4] was not developed with this in mind, and we have not found it conducive to such refinement. In order to facilitate refinement, we present a different protocol for detecting stable properties.

A naive general detection protocol is as follows: each process records in its local memory its entire local history. A separate process p_0 periodically retrieves these

*This work was supported by the Defense Advanced Research Projects Agency (DoD) under NASA Ames grant number NAG 2-593, and by grants from IBM and Siemens. The views, opinions, and findings contained in this report are those of the authors and should not be construed as an official Department of Defense position, policy, or decision. Authors' address: Cornell University Department of Computer Science, Upson Hall, Ithaca, New York   14853-7501, USA.

local histories and extracts from them the latest global state. Process p_0 then tests to see if the property holds in this global state. Unfortunately, this naive protocol is impractical since it requires unbounded local memory. This can be fixed (at a cost of generality) by having each process record only its current state and having p_0 consider some subset of these local states that could be part of a sensible global state of the system. Not all stable properties can be detected this way, but it turns out that most stable properties that have been discussed in the literature can.

In this paper, we present a protocol that can be used to detect such properties. The protocol can be easily decentralized and can be customized for different properties in order to yield efficient special-purpose protocols. We demonstrate its utility by using it to derive such protocols including two previously-known protocols that are known to be efficient.

2 Definitions

We consider an asynchronous distributed system consisting of a set of n nonfaulty processes $P = \{p_1, p_2, \ldots, p_n\}$. Between any two processes p_i and p_j there exist two unidirectional fault-free FIFO channels: $C_{i,j}$ from p_i to p_j and $C_{j,i}$ from p_j to p_i. These channels have unbounded delivery time, and processes communicate only by sending and receiving messages over these channels.

Processes execute *events*, which are partitioned into *send* events, *receive* events, and *local* events. Unless stated otherwise, e_i is an event of process p_i. The execution of an event e_i produces the local state σ_i of p_i. Thus, each local state σ_i has a corresponding event e_i that resulted in that state.

An arbitrary collection of local states may not constitute a sensible global state: the local state of a process in the collection may reflect the receipt of a message while no process' local state reflects the sending of that message. Such sets of local states are called *inconsistent*; a sensible collection of local states is called *consistent*.

A *global state* is defined to be a consistent set $\Sigma = \{\sigma_1, \sigma_2, \ldots, \sigma_n\}$ of the processes' local states. We assume that channel states are captured in the local states of the processes. There are many ways to do this, e.g., by having each process maintain a history of all messages that it sends and receives. In practice, one must ensure that the representation of the channel states uses bounded local memory.

A *consistent cut* is defined to be a set of events $C = \{e_1, e_2, \ldots, e_n\}$ such that the set of states $\{\sigma_1, \sigma_2, \ldots, \sigma_n\}$ produced by C is a global state. Thus, each consistent cut has a corresponding global state, and vice versa.

A *property* is a predicate over the global state of the system. A *stable* property is an invariant: once it becomes true, it continues to be true. The most commonly studied examples of stable properties in distributed systems are deadlock of a subset of the processes, termination of a distributed computation, and the lack of a token among the processes. There are, of course, other stable properties of interest. For example, in a token-passing system that can lose but not regenerate tokens, the predicate "there are no more than k tokens in the system" is a stable property.

An event e_i is *relevant* to a property Φ if e_i changes the value of a term in the formulation of Φ. For example, if Φ is "a subset of the processes are deadlocked" then the relevant events are those that request a resource and those that grant a resource, since Φ is formulated in terms of resource requests and grants. Note that e_i could be a local event, a send event, or a receive event, depending on how Φ is defined.

Our protocol is based on a variant of *vector clocks* [9]. In the usual definition of a vector clock V, each event e_i has an n–component vector $V(e_i)$ associated with it. $V(e_i)$ is called the *vector timestamp* of e_i. The components of $V(e_i)$ are:

- $V(e_i)[i]$ is the number of events that p_i has executed through e_i

- $V(e_i)[j], j \neq i$ is the number of events that p_i "knew" that p_j had executed when p_i executed e_i.

As an example, Figure 1 shows a space–time diagram of a two-process system with the events labeled by vector clocks.

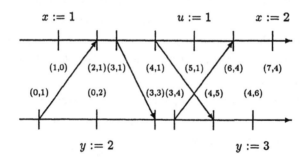

Figure 1: Execution with vector clocks.

A simple implementation of vector clocks has process p_i maintain an n–element vector V_i of counters, where p_i increments $V_i[i]$ whenever it executes an event. If e_i is a send event, then p_i includes $V_i = V(e_i)$ in the message, and if e_j is the corresponding receive event, then process p_j sets $\forall k : k \neq j : V_j[k]$ to the componentwise maximum of the previous value of $V_j[k]$ and the value of $V(e_i)[k]$.

The following three relations hold between vector clocks and global states, where $\rightarrow$ is the *happens-before* relation defined in [7]. Equation 1 defines the happens-before relation in terms of vector clocks, Equation 2 defines when two events are consistent with each other (*pairwise consistent*), and Equation 3 defines when a set of local states $\{\sigma_1, \ldots, \sigma_n\}$ produced by events $\{e_1, \ldots, e_n\}$ comprise a (consistent) global state:

$$\forall i, j : i \neq j : V(e_i)[i] \leq V(e_j)[i] \quad \equiv \quad e_i \rightarrow e_j \tag{1}$$

$$(V(e_i)[i] \geq V(e_j)[i]) \wedge$$

$$(V(e_j)[j] \geq V(e_i)[j]) \quad \equiv \quad e_i \text{ and } e_j \text{ are pairwise consistent} \tag{2}$$

$$\forall i, j : V(e_i)[i] \geq V(e_j)[i] \quad \equiv \quad \{\sigma_1, \ldots, \sigma_n\} \text{ is a global state} \tag{3}$$

Equation 2 can be derived by noting that two events e_i and e_j are inconsistent only if (without loss of generality) $e_i \rightarrow e_j \wedge \exists e_i' : e_i \rightarrow e_i' \rightarrow e_j$, or in terms of vector clocks, $V(e_i)[i] < V(e_j)[i]$. Equation 3 can be derived from Equation 2 by noting that all events in a consistent cut are pairwise consistent. Observe that $\Sigma = \{\sigma_1, \ldots, \sigma_n\}$ is a global state if and only if $C = \{e_1, \ldots, e_n\}$ is a consistent cut.

One can choose which events cause a process' vector clock to be updated. For example, some causal broadcast protocols are based on vector clocks that are updated only at broadcast events; other events (i.e., the receive and local events) do not increment the local component of V [11,1]. In our protocol, we update the vector clock only when relevant events are executed. However, if there are send events that are not relevant events, then Equation 3 need not hold. For example, suppose e_i and e_j are inconsistent relevant events: $e_i \rightarrow e_j \wedge \exists e_i' : e_i \rightarrow e_i' \rightarrow e_j$. If no such e_i' is a relevant event, then $V(e_i)[i] = V(e_j)[i]$ and $V(e_j)[j] > V(e_i)[j]$ which satisfies Equation 2. We therefore define a type of vector clock for which a weaker version of Equation 2 holds in this case.

We define the *weak vector clock* V_Φ for Φ to be the vector clock in which $V_\Phi(e_i)[i]$ counts only the events relevant to Φ that p_i has executed through e_i. Therefore, the vector timestamp associated with several events of p_i may have the same value. However, all such events result in the same local state with respect to Φ. For example, with deadlock the relevant events are the request of a resource and the granting of a resource. If a process p requests a resource and then sends an unrelated message, the send event does not change the local state of p with respect to possible deadlock and so the send event is given the same weak vector timestamp as the resource request event.

We say that two local states σ_i and σ_i' are *equivalent with respect to* Φ, written $\sigma_i \sim_\Phi \sigma_j$, if the events e_i and e_i' producing these local states have the same weak vector timestamp. Similarly, two global states $\Sigma = \{\sigma_1, \ldots, \sigma_n\}$ and $\Sigma' = \{\sigma'_1, \ldots, \sigma'_n\}$ are equivalent with respect to Φ, written $\Sigma \sim_\Phi \Sigma'$, if for all i, $\sigma_i \sim_\Phi \sigma'_i$. The following versions of Equations 1, 2, and 3 hold for both vector clocks and weak vector clocks:

$$\forall i, j: i \neq j: V_\Phi^*(e_i)[i] \leq V_\Phi(e_j)[i] \;\equiv\; \exists e'_i, e'_j: (e'_i \sim_\Phi e_i) \wedge (e'_j \sim_\Phi e_j): e'_i \to e'_j \quad (4)$$

$$(V_\Phi(e_i)[i] \geq V_\Phi(e_j)[i]) \wedge$$
$$(V_\Phi(e_j)[j] \geq V_\Phi(e_i)[j]) \;\equiv\; \exists e'_i, e'_j: (e'_i \sim_\Phi e_i) \wedge (e'_j \sim_\Phi e_j):$$
$$e'_i \text{ and } e'_j \text{ are pairwise consistent} \quad (5)$$

$$\forall i, j: V_\Phi(e_i)[i] \geq V_\Phi(e_j)[i] \;\equiv\; \exists \text{ global state } \Sigma': \Sigma' \sim_\Phi \{\sigma_1, \ldots, \sigma_n\} \quad (6)$$

Figure 2 shows weak vector clock values for the execution shown in Figure 1, where we assume that the predicate of interest references x and y, but not u nor any of the channel states. Note that although the events $x := 1$ and $y := 3$ do not form a consistent cut, their timestamps satisfy Equation 6 since there exist several cuts equivalent to this inconsistent cut (all necessarily having $\langle x = 1, y = 3\rangle$) and they are therefore consistent with respect to Φ.

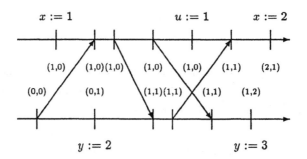

Figure 2: Execution with weak vector clocks.

Our protocol will detect a subset of the stable properties that we call *locally stable* properties. Informally, a property Φ is locally stable if no process involved in the property will change its state relative to Φ once Φ holds. For example, suppose Φ is "processes p_i and p_j are deadlocked." Φ is locally stable because once Φ becomes true, neither p_i nor p_j, the processes involved in Φ, can execute an event that could affect Φ; in particular, requesting or granting a resource.

More formally, let $\mathcal{G}$ be the set of all global states that the system can attain and let Σ be some global state in $\mathcal{G}$. Define Σ_Φ to be the set of variables of Σ that are referenced in the formulation of Φ, and given a set of processes A define

$\Sigma|A$ to be the subset of Σ that consists of the states of the processes in A. We will call Φ locally stable if it satisfies the following condition: consider any $\Sigma \in \mathcal{G}$ that satisfies Φ, and let A be the set of processes that execute no relevant events in any execution starting at Σ. Then Φ can be determined by considering only the variables in $\Sigma_\Phi|A$. Note that A is nonempty for nontrivial properties; if A is empty, then Φ can be determined without knowledge of the state of any process or channel and must therefore be constant. For this reason, we will assume in this paper that A is nonempty.

The most commonly studied stable properties—deadlock, termination, and lack of a token—are all locally stable. For example, if Σ is a deadlock state, then A includes the deadlocked processes, and so the presence of deadlock can be determined by considering only the states of the processes in A. An example of a stable property that is not locally stable is the property "there are no more than $k : k > 0$ tokens" in a system where tokens cannot be created but can be lost when passed. This is because if Σ is a state in which there are k tokens, then every process can execute a relevant event (namely, it can pass a token), and so A is empty. The property cannot be detected from the values in $\Sigma_\Phi|A$, since there are no values in this set.[1]

3 Protocol

We first assume that a process p_0 will determine whether the global state of the processes $P = \{p_1, \ldots, p_n\}$ satisfies a locally stable property Φ. Later, we will change this protocol so that any number of processes in P may concurrently assume the role of p_0.

Our protocol is based on the notion of a *consistent subcut* – a set of events whose whose timestamps satisfy Equation 6. (The state of a single process is trivially a consistent subcut.) Informally, the protocol works as follows. Whenever a process p_i executes a relevant event e_i, p_i records in a buffer B_i its local state relative to Φ and the vector time stamp $V_\Phi(e_i)$ as $B_i.\sigma$ and $B_i.V$, respectively. Process p_0 periodically collects the values of the buffers in any order. Once p_0 has received these values, p_0 determines if there exists a maximal consistent subcut among $\{B_1.V, \ldots, B_n.V\}$ such that the states associated with the timestamps in the subcut satisfy Φ. If p_0

---

[1] A need not be empty for some stable property to be not locally stable. For example, suppose Φ is again "there are no more than $k : k > 0$ tokens" and the token passing system consists of red and green processors. Furthermore, only red processors can lose tokens and a green processor never pass a token to a red processor. In this system, A = the set of green processors (green processors never execute a relevant event), yet Φ cannot be determined by examining only the states of the green processors, and so Φ is not locally stable.

- Each process $p_i \in P$ records σ_i and $V_\Phi(e_i)$ in buffer B_i upon executing a relevant event e_i.

- Periodically, p_0 collects all of the buffers B_i and extracts from them the latest subcut $\{\sigma_i : \forall j : B_j.V[i] \leq B_i.V[i]\}$.

- p_0 detects Φ if Φ holds on the latest subcut.

Figure 3: Basic Protocol

can find such a subcut, then Φ must currently hold. Note that p_0 need not examine all consistent subcuts; if $A' \subseteq A$ and $\Sigma_\Phi|A'$ supports Φ, then $\Sigma_\Phi|A$ will also support Φ, so we need examine only the maximal subcuts. Of course, Φ may be of the form $\forall p_i : \Psi(p_i)$, in which case only a full consistent cut will satisfy Φ.

Unfortunately, the number of maximal subcuts of a set of n weak vector clocks is $\Omega(2^n)$. Fortunately, it is not necessary for p_0 to examine all of these subcuts. Suppose the set of buffer values contains B_i and B_j that are inconsistent: $B_i.V[i] < B_j.V[i]$. These two states violate Equation 6, and so cannot be part of the same consistent subcut. However, B_j records the fact that p_i has executed a relevant event since B_i was recorded. Since Φ is locally stable, the state $B_i.\sigma$ of process p_i is not needed to determine whether Φ held in a global state containing $B_i.\sigma$, and so p_0 need not consider any consistent subcut containing B_i. Given the partial order $B_i \succ B_j \overset{\text{def}}{=} B_i.V[j] > B_j.V[j]$, p_0 need only find the greatest elements of $\succ$, which can be done in $\Omega(n^2)$ time.[2] We call this subcut the *latest subcut*. The latest subcut is clearly a maximal subcut, since all states that are not part of the latest subcut are inconsistent with some state in the subcut. Furthermore, all processes not in the latest subcut have executed relevant events since recording their state and so their values can be ignored. This gives us the protocol shown in Figure 3.

The soundness of this protocol is straightforward. We now argue that the protocol is complete as well; that is, if Φ holds, then our protocol will detect Φ. Let Σ be the first global state in which Φ holds. Since Φ is locally stable, there is a nonempty set of processes A none of which execute a relevant event after Σ; these

---

[2]The greatest elements of $\succ$ can be found by discarding any values B_j such that $\exists i : B_i.V[j] > B_j.V[j]$, which can be done in $O(n^2)$ time using a straightforward algorithm. And, if all values are incomparable then all the values are greatest elements of $\succ$. To determine that they are all incomparable takes n^2 comparisons, and so the problem is $\Omega(n^2)$.

processes will not change their states relative to Φ nor update their vector clocks after Σ. If p_0 initiates the protocol after Σ (i.e., when Φ holds), then p_0 will collect the states $\Sigma_\Phi | A$. From the definition of $\succ$, the state of a process p_i in A must be in any latest subcut constructed by p_0 because p_i will execute no more relevant events. Hence, p_0 will detect Φ.

3.1 Decentralization

In the above protocol, p_0's role is to collect the states, determine the latest subcut, and check if Φ holds in this subcut. We can decentralize these steps by collecting the states in a token.

Consider a token K that consists of n entries $\langle D_1, \ldots, D_n \rangle$ where each entry $D_i = (B_i.\sigma, B_i.V[i])$; that is, D_i will hold the state of p_i relevant to Φ and the local component of p_i's vector clock when it generated this state. Assume that there exists a special value $\perp$ for D_i indicating that the state has not yet been collected; so, all of the D_i in K are initially set to $\perp$.

Whenever a process wants to know whether Φ holds, it generates a token K, inserts its state and vector clock value into K, and passes the token to any other process. When a process p_j receives a token K, it takes the following steps:

1. p_j sets D_j to $(B_j.\sigma, B_j.V[j])$.

2. p_j casts out any values D_k that are not part of the latest subcut. Note that B_j must be part of the latest subcut; if it were not, then there would exist a recorded value of B_ℓ in D_ℓ such that $B_j.V[j] < B_\ell.V[j]$. This implies that p_ℓ knows of a later state of p_j than the state that is recorded in B_j, which violates the definition of B_j. Thus, only the earlier values D_k need be tested with respect to B_j. From above, the value B_k in D_k can be discarded if $B_k.V[k] < B_j.V[k]$. The value $B_k.V[k]$ is stored in $D_k.V$, so K carries enough information for p_j to make this test. If D_k is not in the latest subcut, then p_j sets D_k to $\perp$.

3. p_j determines whether the state values in K satisfy Φ. If so, then the detection is made; otherwise, p_j forwards the token to a process p_k, chosen fairly, that has $D_k = \perp$. If there is no such process, then p_j can either drop the token or, when p_j computes a new value of B_j, p_j can restart at Step 1 with this token.

The resulting protocol is summarized in Figure 4. Note that we have no *a priori* restriction on how many tokens there can be in the system at any time or on the

- Each process $p_i \in P$ records σ_i and $V_\Phi(e_i)$ in buffer B_i upon executing a relevant event e_i.

- When p_i wants to detect Φ, p_i generates a token $\langle D_1 := \perp, \ldots, D_n := \perp \rangle$, sets D_i to $(B_i.\sigma, B_i.V[i])$, and forwards the token to any other process.

When p_j receives a token:

- p_j sets D_j to $(B_j.\sigma, B_j.V[j])$.

- For each $k \neq j$, p_j sets $D_k := \perp$ if the recorded $B_k.V[k] < B_j.V[k]$.

- p_j determines if Φ holds on the state values in the token. If not, p_j forwards token to any p_k such that $D_k = \perp$.

Figure 4: Optimized Protocol

order in which the token is passed, other than that it is passed in a fair manner. These decisions can be made when the protocol is applied to a particular problem.

4 Termination Detection

We now instantiate the general protocol given above to obtain a protocol that detects *termination* in a distributed system. There are many variations of this property; the earliest that we know of is due to Dijkstra and Scholten [6]. The following definition is the same as that given in [10].

All processes are either *active* or *idle*. Only active processes can send messages. An active process may become idle at any time, and an idle process can become active only upon receipt of a message. The system is *terminated* when all processes in the system are idle and there are no messages in transit.

The events that are relevant to termination are sending a message, receiving a message, becoming idle, and becoming active. Therefore, each process will update its (weak) vector clock upon executing any of these events. The local state of a process relative to termination consists of whether the process is active or idle and whether there is a message on an incoming channel. Note that for this problem, we do not need to keep track of the contents of the messages exchanged between processes; only the number of messages is important. To capture the channel states, we have each process keep track of how many messages it has sent and received on each adjacent channel. The combined information of all of the processes will then yield the number of messages in transit on each channel: if p_i has sent more messages to p_j than p_j has received from p_i, then there is at least one message on channel $C_{i,j}$. In this way, we can represent the relevant channel states using bounded local memory.

We instantiate the general protocol given in Section 3.1 as follows.

Each process p_i maintains the following local state variables:

- $active_i$: Boolean = true if and only if p_i is active.

- $send_i[1..n]$: Integer array. $send_i[j]$ = the number of messages that p_i has sent to p_j. All are initially 0.

- $recv_i[1..n]$: Integer array. $recv_i[j]$ = the number of messages that p_i has received from p_j. All are initially 0.

When p_i sends a message to p_j, $send_i[j]$ is incremented. When p_i receives a message from p_j, $recv_i[j]$ is incremented. When p_i becomes active or idle, $active_i$ is set appropriately.

At some point, an idle process will start the detection protocol by circulating a token as described in Section 3.1. The termination condition can only be evaluated over a total global state (as opposed to a consistent proper subset of the process states), so a positive determination can be made only by the process p_f that is the last to add its state to the token.

Process p_f detects termination if and only if the following three conditions hold:

1. The timestamps in the token form a consistent cut;

2. All processes are idle: $\forall i : active_i = $ false;

3. There are no messages in transit: $\forall i, j : send_i[j] = recv_j[i]$.

We claim that item 1 is redundant: item 3 implies that the cut is a global state. Suppose by way of contradiction that when all of the states and timestamps have been collected, item 3 holds but the timestamps form an inconsistent cut. That the cut is inconsistent implies from Equation 6 that for some i, j, $B_i.V[i] < B_j.V[i]$. $B_j.V[i]$ is advanced only when p_j receives a message, and events local to p_j affect only $B_j.V[j]$. Therefore, in order for $B_j.V[i]$ to advance beyond the recorded $B_i.V[i]$, there must have been a chain of messages between p_i and p_j between the time that B_i was collected and the time that B_j was collected. This implies that there is some

k such that the recorded $send_i[k] < recv_k[i]$. This contradicts the assumption that item 3 holds.

Therefore, p_f need check only the last two items. In fact, these checks can be done incrementally. For example, we can assign a total order to the processes and have the token passed along that total order. When process p_k receives the token, it tests to see if

$$\neg active_k \wedge (\forall \ell : 1 \leq \ell < k : (send_k[l] = recv_l[k]) \wedge (send_l[k] = recv_k[l])).$$

If this condition does not hold, then p_k can drop the token. If the condition holds and $k = n$, then termination is detected; otherwise, p_k fills in D_k and passes the token to p_{k+1}.

This yields the protocol given in [8] as the *channel counting* protocol, which requires only n messages to detect termination once it holds, and which can be further refined into a protocol that is space-efficient. This is a good example of how our general protocol, which constructs consistent (sub)cuts explicitly, can be used to derive a much simpler protocol that constructs consistent cuts implicitly.

5 Deadlock Detection

We now instantiate the general protocol given in Section 3 to obtain a protocol that detects *k-out-of-m deadlock* in a distributed system. This problem was first formulated and solved in [2]. In this formulation, a process can request k resources from a pool of m resources.

A process is either active or blocked. An active process is one that is not waiting for any other process. Active processes may issue k-out-of-m requests in the following way. When an active process p_i requires k processes to carry out some request, it sends **request** messages to each of the m processes that can perform this action. Process p_i then becomes blocked, and waits until the action requested is carried out by at least k of the m processes. A process can not send any further requests while blocked.

Only active processes can carry out a requested action. If a process p_j receives a request while active, it will either become blocked or carry out p_i's requested action within finite time. In the latter case, p_j will send a **grant** message to p_i. When p_i receives k **grant** messages, it becomes active again. It then relinquishes the requests made to the rest of the processes to which it sent **request** messages by sending them **relinquish** messages. We assume that the recipient of a **relinquish** message acknowledges the message and that the sender of a **relinquish** message waits for all acknowledgements before sending another **request** message. By doing so, we guarantee that p_i can discard any **grant** messages received after k becomes zero.

The state of a process p_i relative to k-out-of-m deadlock consists of the number of grants needed for p_i to become active and the current set of processes that p_i is waiting for. We capture this state by having each process keep track of the processes on which it is blocked and the number of **grant** messages that it has sent and received on each adjacent channel.

We instantiate the general protocol given in Section 3.1 as follows. Each process p_i maintains the following local state variables:

- k_i: Integer = the number of **grant** messages required for p_i to become active (initially 0).

- $g\_send_i[1..n]$: Integer array. $g\_send_i[j]$ is the number of **grant** messages that p_i has sent to p_j (all are initially 0).

- $g\_recv_i[1..n]$: Integer array. $g\_recv_i[j]$ is the number of **grant** messages that p_i has received from p_j (all are initially 0).

- wf_i: Integer set. These are the processes that p_i is waiting for. When p_i sends a **request** message to p_j, $wf_i := wf_i \cup \{j\}$; when p_i receives a **grant** message from p_j or sends a **relinquish** message to p_j, $wf_i := wf_i - \{j\}$.

The relevant events are therefore requesting a resource, sending a **grant** message, receiving a **grant** message and sending a **relinquish** message. Deadlock is determined by constructing and reducing the system wait-for graph. This graph is constructed as follows:

- a *waits-for* edge is drawn from p_i to p_j if $wf_i \ni j \wedge (g\_send_j[i] = g\_recv_i[j]))$. That is, p_i is waiting for a resource from p_j and no grant message is in transit from p_j to p_i.

- the number of grants κ_i needed for p_i to be unblocked is $k_i - |\forall j : g\_send_j[i] - g\_recv_i[j]|$. That is, κ_i is the number of grants that p_i is waiting for less the number of grants in transit to p_i.

Deadlock is tested by reducing this graph as follows: if an edge points from p_i to p_j and p_j is active, then the edge can be erased and κ_i can be reduced by one; and if a process has $\kappa_i = 0$, then all of its outgoing edges can be erased. The system is deadlocked if and only if there are edges that cannot be removed by following these two rules.

when p_i receives token $\langle D_1, \ldots, D_n \rangle$:

    begin

        $D_i.\sigma := k_i, g\_send_i, g\_recv_i, wf_i;$

        $D_i.V := B_i.V[i];$

        for all $D_j : D_j.V < B_i.V[j] : D_j := \perp;$

        if there exists $p_j : (D_j = \perp) \wedge (\exists p_i : j \in wf_i)$

        then forward token to one of these p_j

        else begin

            construct waits–for graph;

            reduce waits–for graph;

            if graph is not fully reduced **then** signal deadlock

            else drop the token

        end

    end

Figure 5: Protocol for Detecting *k-out-of-m Deadlock*.

When a process p_i wishes to test for deadlock, p_i generates a token, fills D_i with $((wf_i, g\_send_i, g\_recv_i, k_i), B_i.V[i])$, and forwards the token to some $p_j \neq p_i$. Upon receiving a token, a process p_j sets D_j to $((wf_j, g\_send_j, g\_recv_j, k_j), B_j.V[j])$ and discards all values D_k that are inconsistent with B_j by setting D_k to $\perp$. p_j then checks to see if deadlock holds on the remaining values by constructing the waits-for graph and reducing it. If deadlock does not hold, then p_j forwards the token to any process p_k such that $D_k = \perp$.

We can improve this protocol by choosing the process to which the token is passed more carefully. Since we would like to detect deadlock as quickly as possible, the forwarding process should choose a process that is likely to add information leading to the detection of a deadlock. A reasonable choice is a process p_j such that $D_j = \perp$ and such that p_j is in wf_i for some $D_i \neq \perp$.

The full protocol is presented in Figure 5. We assume that the process p_i that generates the token does so because it suspects that it is involved in a deadlock; that is, wf_i is not empty.

This protocol can be optimized further. For example, if we restrict ourselves to RPC deadlock (1-out-of-1 requests), then $k_i = 1$ and need not be represented in the wait-for graph, and the wait-for graph is reducible if and only if it does not contain a cycle. Since any process p_i can only be waiting for one other process at a time (i.e., $|wf_i| \leq 1$), if the originating process is involved in a deadlock then the token will be passed along a cycle of blocked processes. Hence, when process p_i receives a token K it can test for a cycle in the wait-for graph very simply: if $wf_i = \emptyset$, then drop the token; otherwise, if $wf_i = \{p_j\}$, test to see if $D_j \neq \perp$ and B_i is consistent with D_j. If so, then p_i detects deadlock; otherwise, p_i forwards the token to p_j. Furthermore, if a blocked process delays receiving any **request** messages while blocked, then it is easy to show that the vector clocks are not necessary, because all states in the token are consistent at any time. Recall that when p_j receives a token, p_j adds its state and forwards the token to the process in wf_j if wf_j is nonempty and drops the token if wf_j is empty. Suppose by way of contradiction that D_i and D_j are two entries in the token such that B_i and B_j are inconsistent: $B_i.V[i] < B_j.V[i]$. Then p_i must have sent a **request** or **grant** message since its state was added to the token. Furthermore, B_j must have been added to the token after B_i, which implies that there is a path in the wait-for graph from p_i to p_j. But this means that p_i cannot become active until p_j sends a **grant** message to the process it is blocking, contradicting that p_i sent a **request** or **grant** message since its state was added to the token. Therefore, all states in the token at any time are consistent. This yields the resource-deadlock detection protocol for n-out-of-n requests presented in [3] for $n = 1$.

A similar argument can be made to show that this protocol will detect and–deadlock (m-out-of-m requests), but the argument is more complex. The resulting protocol is the resource-deadlock detection protocol for n-out-of-n requests presented in [3].

6   Conclusion

This paper presents a general protocol for detecting a class of stable properties (the locally stable properties) by constructing consistent subcuts. The protocol collects the consistent subcuts in a decentralized manner and is message efficient. We have demonstrated its use by refining it to a known protocol for termination detection, a new protocol for k-out-of-m deadlock detection, and a known protocol for 1-out-of-1 deadlock detection. In the reduction to the two known protocols, the vector clocks proved redundant since these protocols implicitly construct a consistent subcut.

The class of locally stable properties was defined in proving the protocol correct. We are interested in whether the protocol can be extended to detect a wider set of stable properties; for example, it appears easy to detect stable properties in which a process executes a bounded number of relevant events once involved in a state establishing Φ. We would like to determine what kind of properties are locally stable—for example, the locally stable properties of distributed garbage detection, termination, and global deadlock can all be expressed as detecting no token in a generalized token passing system, yet nonglobal deadlock does not seem to be so expressible.

We would also like to better understand the notion of relevant events and weak vector clocks. We have attempted to refine our protocol to other known protocols. In some cases (for example, the termination detection protocol of [5]) we were unable to do so, while in others (for example, the and deadlock protocol of [3]) we found that small changes in the definitions of relevant events and propagation of vector time stamps can greatly ease the process of refinement.

Acknowledgements We would like to thank Özalp Babaoğlu, Gil Neiger, Fred Schneider, and Sam Toueg for their contributions to the ideas in this paper. We would also like to thank Ken Birman, Navin Budhiraja, Tushar Chandra, and Mark Wood for their valuable comments on earlier drafts of this paper.

References

[1] K. Birman, A. Schiper, and P. Stephenson. Fast causal multicast. Technical Report TR-90-1105, Cornell University, April 1990. Submitted for publication.

[2] G. Bracha and Sam Toueg. A distributed algorithm for generalized deadlock detection. In *Proceedings of the Symposium on Principles of Distributed Computing*, pages 285–301. ACM SIGPLAN/SIGOPS, August 1984.

[3] K. Mani Chandy, Laura M. Haas, and Jayadev Misra. Distributed deadlock detection. *ACM Transactions on Computer Systems*, 1(2):144–156, May 1983.

[4] K. Mani Chandy and Leslie Lamport. Distributed snapshots: determining global states of distributed systems. *ACM Transactions on Computer Systems*, 3(1):63–75, February 1985.

[5] K. Mani Chandy and Jayadev Misra. An example of stepwise refinement of distributed programs: Queiscence detection. *ACM Transactions on Programming Languages and Systems*, 8(3):326–343, July 1986.

[6] Edsger W. Dijkstra and E. Scholten. Termination detection for diffusing computations. *Information Processing Letters*, 11(1):1–4, 1980.

[7] Leslie Lamport. Time, clocks, and the ordering of events in a distributed system. *Communications of the ACM*, 21(7):558–565, July 1978.

[8] Friedemann Mattern. Algorithms for distributed termination detection. *Distributed Computing*, 2(3):161–175, 1987.

[9] Friedemann Mattern. Time and global states of distributed systems. In Michel Cosnard et. al., editor, *Proceedings of the International Workshop on Parallel and Distributed Algorithms*, pages 215–226. North-Holland, October 1989.

[10] Jayadev Misra. Detecting termination of distributed computations using markers. In *Proceedings of the Symposium on Principles of Distributed Computing*, pages 290–294. ACM SIGPLAN/SIGOPS, August 1983.

[11] Larry L. Peterson. Preserving context information in an IPC abstraction. In *Proceedings of the 6th Symposium on Reliability in Distributed Software and Database Systems*, pages 22–31, March 1987.

Atomic m-register operations

(EXTENDED ABSTRACT)

Michael Merritt*   Gadi Taubenfeld*

Abstract

We investigate systems where it is possible to access several shared registers in one atomic step. Extending proofs in [Her91, LA87], we characterize those systems in which the consensus problem can be solved in the presence of faults and give bounds on the space required. We also describe a fast solution to the mutual exclusion problem using atomic m-register operations.

1 Introduction

Shared memory systems which support only atomic read and write operations have been extensively investigated, e.g. [AAD+90, Blo87, Lam86, Lam87, LA87, PB87, TM89, VA86]. This paper focuses attention on somewhat stronger systems in which it is possible to read or write several shared registers in one atomic step. We say that a system supports atomic m-register operations, if it is possible for a process to read or write m registers in one atomic step. It is known that in some contexts, such systems are more powerful than single-register primitives[Her91]. We explore this question in more detail, examining complexity bounds as well as computability. This section contains a brief overview of the results, which have been inspired by proofs in Fischer, Lynch and Paterson[FLP85], Herlihy [Her91] and Loui and Abu-Amara [LA87].

We first show that the consensus problem can be solved using n processes in systems which support atomic m-register operations if and only if $t \leq max(2m - 3, 0)$, where t is the number of possible crash failures. The theorem holds also when a process can atomically read only a single register (but can atomically write m registers).

Two known results are special case of this. Loui and Abu-Amara [LA87] proved the result for the case $m = 1$ (for any n and t), and Herlihy [Her91] proved it for the case $t = n-1$ (for any m). (Herlihy assumes that a process may atomically read only a single register.) The case where $m = 1$ and $t = n - 1$, which itself is a special case of the Loui and Abu-Amara result, is proved also in [CIL87, Her91].

Next, we prove a lower bound on the number of shared registers necessary to solve the consensus problem. We show that for any n, m and $t < n$, in a system that supports

---

*AT&T Bell Laboratories, 600 Mountain Avenue, Murray Hill, NJ 07974.

atomic m-register operations, every t-resilient consensus protocol for n processes must use at least $(t+3)\min(t+1, m) - \min(t+1, m)^2 - 1$ shared registers. That is, if $t+1 \leq m$, at least $2t + 1$ registers are needed, and when $t + 1 \geq m$, $(t+3)m - m^2 - 1$ are required. We also design a consensus protocol that uses $(t^2 + 5t + 4)/2$ shared registers.

Finally, following [PF77], we use atomic m-register operations to design a fast solution to the mutual exclusion problem. Assuming a fault-free environment, we show that there is a starvation-free solution to the mutual exclusion problem with time complexity $3\lceil\log_m(n)\rceil$. The time complexity is defined as the number of accesses to the shared memory in order to enter the critical section in the absence of contention [Lam87].

2 Asynchronous Shared Memory Systems

This section characterizes asynchronous shared memory systems which support atomic access to m registers, by stating axioms that any protocol operating in such systems satisfies. The axioms do not give a complete characterization of these systems; Only those axioms are stated that are needed to prove the lower bounds. We start with a formal description of the notion of a protocol.

An n-process protocol $P = (C, N, R)$ consists of a nonempty set C of runs, an n-tuple $N = (p_1, \ldots, p_n)$ of processes and an n-tuple $R = (R_1, \ldots, R_n)$ of sets of registers. We may think of R_i as the set of all the registers that process p_i can access. A run is a pair (f, S) where f is a function which assigns initial values to the registers and S is a finite or infinite sequence of events. When S is finite, we also say that the run is finite.

An $event$ corresponds to atomic step performed by a process. We consider here only the following types of events. (1) $read_p(r_1, \ldots, r_m, v_1, \ldots, v_m)$ – process p reads the values $v_1, \ldots, v_m$ from registers $r_1, \ldots, r_m$ respectively; (2) $write_p(r_1, \ldots, r_m, v_1, \ldots, v_m)$ – process p writes the values $v_1, \ldots, v_m$ into registers $r_1, \ldots, r_m$ respectively.

The $value$ of a register at a finite run is the last value that was written into that register, or its initial value (determined by f) if no process wrote into the register. We use $value(r, x)$ to denote the value of register r at a finite run x. A register r is said to be $local$ to process p_i if $r \in R_i$ and for any $j \neq i, r \notin R_j$. A register is $shared$ if it is not local to any process.

Let $x = (f, S)$ and $x' = (f', S')$ be runs. Run x' is a prefix of x (and x is an extension of x'), denoted $x' \leq x$, if S' is a prefix of S and $f = f'$. Let $\langle S; S' \rangle$ be the sequence obtained by concatenating the finite sequences S and the sequence S'. Then $\langle x; S' \rangle$ is an abbreviation for $(f, \langle S; S' \rangle)$. When $x \geq x'$, $(x - x')$ denotes the suffix of S obtained by removing S' from S.

For any sequence S, let S_p be the subsequence of S containing all events in S which involve p. Run (f, S) $includes$ (f', S') iff $f = f'$ and S'_p is a prefix of S_p for all $p \in N$. Runs (f, S) and (f', S') are $indistinguishable$ for process p, denoted by $(f, S)[p](f', S')$, iff $S_p = S'_p$, and $f(r) = f'(r)$ for every local register r of p. We assume throughout this paper that x is a run of a protocol if and only if all finite prefixes of x are runs. Notice that, by this assumption, if (f, S) is a run then also $(f, null)$ is a run, where $null$ is the empty sequence.

DEFINITION 1 *An asynchronous protocol that supports atomic m-register operations is a protocol whose runs satisfy axioms A1 – A3.*

AXIOMS FOR ATOMIC m-REGISTER OPERATIONS

A1 Let $\langle x; write_p(r_1, ..., r_m, v_1, ..., v_m)\rangle$ and y be finite runs where $x[p]y$.
Then $\langle y; write_p(r_1, ..., r_m, v_1, ..., v_m)\rangle$ is a run.

A2 Let $\langle x; read_p(r_1, ..., r_m, v_1, ..., v_m)\rangle$ and y be finite runs where $x[p]y$.
Then $\langle y; read_p(r_1, ..., r_m, u_1, ..., u_m)\rangle$ is a run for some value $u_1, ..., u_m$.

A3 Let $\langle x; read_p(r_1, ..., r_m, v_1, ..., v_m)\rangle$ be a run.
Then $v_i = value(r_i, x)$ for all $1 \le i \le m$.

A1 means that if a write event which involves p can happen at a run, then the same event can happen at any run that is indistinguishable to p from it. *A2* means that if a process is "ready to read" some values from some registers then an event on some other process cannot prevent it from reading, although it may prevent this process from reading some specific values which it could read previously. *A3* means that it is possible to read only the last value that is written into a register.

In order to discuss important properties of asynchronous protocols, we need the concept of a fair run. We say that process p is *enabled* at run x if there exists an event e_p such that $\langle x; e_p \rangle$ is a run. Notice that in asynchronous shared memory systems enabled process cannot become disabled as a result of an event which involves some other process. Process p is *correct* in a run y if for each run $x \le y$, if p is enabled at x then some event in $(y - x)$ involves p. A run is ℓ-fair if at least ℓ processes are correct in it.

3 Consensus

A *t-resilient consensus protocol* is a protocol for n processes, where each process has a local read-only input register and a local write-once output register. For any $(n - t)$-fair run there exists a finite prefix in which all the correct processes decide on either 0 or 1 (i.e., each correct process writes 0 or 1 into its local output register), the values written by all processes are the same, and the decision value is equal to the input value of some process.

Theorem 1 *For any n, $1 \le m$ and $t < n$, in a system that supports atomic m-register operations, there is a t-resilient consensus protocol for n processes if and only if $t \le max(2m - 3, 0)$.*

The theorem holds also when a process can atomically read only a single register (but can atomically write to m registers), or when a process can "mix" reads and writes, accessing up to m registers atomically, and choosing some to write and others to read. The same bound holds for other problems such as leader election. As mentioned in the introduction, known results are special cases of Theorem 1 [CIL87, Her91, LA87].

The proof follows [Her91], where $t = n - 1$, and [LA87], where $m = 1$ and $t = 1$. Herlihy shows that there must exist a run in which the next step by any process determines the decision value, and argues that every process p_i must write to a unique register r_i and, for every process p_j of opposite decision value, to a register r_{ij} it shares only with p_j. The bound $n - 1 = t \leq 2m - 3$ follows.

We note that for arbitrary t, a similar run must exist in which the next step by any of a group of at least $t + 1$ processes determines the decision value. There are two cases. The first assumes that there are two processes in the group that would choose different values, and argues as before that every process p_i in this group must write to a unique register r_i and, for every process p_j of opposite decision value in the group, to a register r_{ij} it shares only with p_j. The second case assumes that all the processes in the group would choose the same decision value, and a somewhat more intricate construction leads to a similar counting argument. The theorem follows.

Next, we prove a lower bound on the number of shared registers necessary to solve the consensus problem.

Theorem 2 *For any n, m and $t < n$, in a system that supports atomic m-register operations, every t-resilient consensus protocol for n processes must use at least $(t + 3) \min(t + 1, m) - \min(t + 1, m)^2 - 1$ shared registers.*

The proof of Theorem 2 uses only the requirement that the decision value is the input of some process, to show that initially there are two possible decisions. The proof is a generalization of Herlihy's construction as in Theorem 1. By slightly modifying Herlihy's solution [Her91] we can design a consensus protocol that uses $(t^2 + 5t + 4)/2$ shared registers.

Remark: Define $C_{n,t}(m)$ to be the class of problems that can be solved for n processes in the presence of up to t failures using atomic m-register operations. It is not difficult to show that $C_{n,t'}(m')$ is *strictly* contained in $C_{n,t}(m)$ when $t' > t$ and $m' \leq m$. It follows from Theorem 1 and a few other observations that: *For every n, every $1 < m \leq \lceil n/2 \rceil + 1$, and every $t > 2m - 3$, $C_{n,t}(m - 1)$ is strictly contained in $C_{n,t}(m)$.* This means that, when $t > 2m - 3$, it not possible to implement atomic m-register operations using atomic $(m - 1)$-register operations even if we assume a fixed number of processes and failures.

4 Mutual Exclusion

The mutual exclusion problem, which is one of the most studied problem in distributed computing, is to design a protocol that guarantees mutually exclusive access to a critical section among a number of competing processes. Such a protocol is starvation-free if any process that is trying to enter its critical section, eventually is in its critical section. In this section we use the atomic m-register operations to design a fast solution to the mutual exclusion problem. The time complexity is defined as the number of accesses to the shared memory in order to enter the critical section in the absence of contention [Lam87].

Lamport notes that all the published starvation-free mutual exclusion algorithms (for $m = 1$) require a process to execute at least $O(n)$ operations to shared memory in the absence of contention [Lam87]. By slightly modifying the solution in [PF77], it is possible to design, for $m = 1$, a starvation-free solutions to the mutual exclusion problem with time complexity $O(\log_2(n))$. By designing a solution for m processes, replacing the binary tree in [PF77] with an m-ary tree, and assuming a fault-free environment, we obtain the following theorem.

Theorem 3 *For any n and $m \geq 1$, in a system with n processes that supports atomic m-register operations, there is a starvation-free solution to the mutual exclusion problem with time complexity $3\lceil \log_m(n) \rceil$.*

For systems that support only atomic read and write operations (i.e., $m = 1$) Burns and Lynch proved that any deadlock-free solution to the mutual exclusion problem must use at least as many shared registers as processes [BL80]. They also showed that this lower bound is tight. Their result holds also for every $m > 1$.

5 Discussion

We have proved several results for the consensus and the mutual exclusion problems in systems where it is possible to atomically access m shared registers. For mutual exclusion, we have obtained a speed-up of only $\log_2(m)$, in a model which groups m register operations together into a single step. Is it possible to find a speed-up linear in m? It would also be interesting to solve other problems in this model.

In [Lam86], Lamport has defined three general classes of shared read/write registers— safe, regular, and atomic—depending on their properties when several reads and/or writes are executed concurrently. The weakest possibility is *safe* register, in which it is assumed only that a read not concurrent with any writes obtains the correct value. A *regular* register is a safe register in which a read that overlaps a write obtains either the old or new value. An *atomic* register, is a safe register in which the reads and writes behave as if they occur in some definite order. In this paper we have considered only the case where it is possible to atomically access m *atomic* registers. An interesting question is what happens when the m registers are only regular or safe.

References

[AAD+90] Y. Afek, H. Attiya, D. Dolev, M. Gafni, M. Merritt, and N. Shavit. Atomic snapshots. In *Proceedings of the 9th Annual ACM Symposium on Principles of Distributed Computing*, pages 1–13, August 1990. Submitted for publication.

[ABD+90] H. Attiya, A. Bar-Noy, D. Dolev, D. Peleg, and R. Reischuk. Renaming in an asynchronous environment. *Journal of the ACM*, 37(3):524–548, July 1990.

[BL80] J. E. Burns and A. N. Lynch. Mutual exclusion using indivisible reads and writes. In *18th annual allerton conference on communication, control and computing*, pages 833–842, 1980.

[Blo87]   B. Bloom. Constructing two-writer atomic registers. In *Proc. 6th ACM Symp. on Principles of Distributed Computing*, pages 249–259, 1987.

[CIL87]   B. Chor, A. Israeli, and M. Li. On processor coordination using asynchronous hardware. In *Proc. 6th ACM Symp. on Principles of Distributed Computing*, pages 86–97, 1987.

[FLP85]   M. J. Fischer, N. A. Lynch, and M. S. Paterson. Impossibility of distributed consensus with one faulty process. *Journal of the ACM*, 32(2):374–382, April 1985.

[Her91]   M. Herlihy. Wait-free synchronization. *ACM Trans. on Programming Languages and Systems*, 11(1):124–149, January 1991.

[LA87]   C. M. Loui and H. Abu-Amara. Memory requirements for agreement among unreliable asynchronous processes. *Advances in Computing Research*, 4:163–183, 1987.

[Lam86]   L. Lamport. On interprocess communication, parts I and II. *Journal of Distributed Computing*, 1(2):77–101, 1986.

[Lam87]   L. Lamport. A fast mutual exclusion algorithm. *ACM Trans. on Computer Systems*, 5(1):1–11, 1987.

[PB87]   G. L. Peterson and J. E. Burns. Concurrent reading while writing ii: the multiwriter case. In *Proc. 28th IEEE Symp. on Foundations of Computer Science*, pages 383–392, 1987.

[PF77]   G. L. Peterson and M. J. Fischer. Economical solutions for the critical section problem in a distributed system. In *Proc. 9th ACM Symp. on Theory of Computing*, pages 91–97, 1977.

[TM89]   G. Taubenfeld and S. Moran. Possibility and impossibility results in a shared memory environment. In *3rd International Workshop on Distributed Algorithms*, 1989. Lecture Notes in Computer Science, vol. 392 (eds.: J.C. Bermond and M. Raynal), Springer-Verlag 1989, pages 254–267.

[VA86]   P. M. B. Vitanyi and B. Awerbuch. Atomic shared register access by asynchronous hardware. In *Proc. 27th IEEE Symp. on Foundations of Computer Science*, pages 223–243, 1986. Errata, Ibid., 1987.

A Robust Distributed Mutual Exclusion Algorithm

Sampath Rangarajan*   Satish K. Tripathi[†]

Institute for Advanced Computer Studies
University of Maryland
College Park, MD 20742

Abstract

Correct operation of a distributed system with replicated resources requires that mutual exclusion be maintained among independent requests to these resources at different sites in the system. In this paper, we propose "asymptotically high resiliency" as a performance measure for fault-tolerant mutual exclusion algorithms for systems where sites can fail. We then present an efficient highly fault-tolerant algorithm for mutual exclusion. The algorithm is totally distributed in nature and is shown to have a message overhead of $O(\sqrt{N}\log N)$ where N is the number of sites. The algorithm provides mutual exclusion with a *resiliency* (that is, probability that permission for a mutual exclusion request is obtained in spite of site failures) approaching 1 asymptotically with an increase in N.

1   Introduction

Reliability of a distributed system can be increased by replicating a resource at different sites. For example, a file can be replicated at different sites to improve its availability in case of site failures. Correct operation of a distributed system with such replicated data requires that mutual exclusion be maintained among independent requests to copies of the same data in different sites. There are a number of other applications in distributed

---

*E-mail: sampath@umiacs.umd.edu

[†]Also with the Department of Computer Science, University of Maryland at College Park, College Park, MD 20742. E-mail: tripathi@cs.umd.edu

systems with the requirement that exclusive access to a resource be given to a single process in the system. These include atomic commitment and other synchronization problems.

All mutual exclusion algorithms are instances of what are called *quorum consensus protocols*. In a quorum consensus protocol, an operation (or a process performing that operation) can proceed only if it gets permission from a group of other processes. Such a group is called a quorum group. Collection of all quorum groups will constitute a quorum set. Permission from all processes from any one of the groups in the quorum set is sufficient for an operation to proceed, and to ensure mutual exclusion, every pair of groups in the quorum set should have a non-empty intersection [4]. In the rest of the paper, without loss of generality, we assume that each process runs on a different processor or *site*. Henceforth, we will use the terms process and site interchangeably. A *distributed algorithm* for mutual exclusion is one where each site bears equal responsibility in controlling mutual exclusion. Each site in a distributed algorithm serves as the arbitrator for the same number of sites. In a non distributed algorithm, some sites will bear more responsibility than some others in controlling mutual exclusion.

There are two main issues in the design of fault-tolerant quorum consensus algorithms for mutual exclusion. The first issue is the message overhead involved per mutual exclusion request and the second issue is the *resiliency* of the algorithm. Given the possibility of site failures, one measure of resiliency is the probability that if a mutual exclusion request is initiated by a site, then a quorum group can be found by the algorithm (which of course, includes the initiating site) such that all sites in the quorum group are available (a site is available, if it has not failed) to control mutual exclusion. We will call this measure *site resiliency* and denote it by R_{site}. A weaker measure of resiliency is the probability that *at least* one quorum group (in the whole system) can be found such that all sites in that group are available. We will call this measure *system resiliency* and denote it by R_{sys}. System resiliency in some sense gives us the probability that mutual exclusion requests initiated through at least one site in the system will succeed (in spite of site failures). Note that resiliency is dependent not only on how the quorum groups are formed, but also how the quorum groups are used by an algorithm. Hence, the term *resiliency of an algorithm* (rather than resiliency of a quorum grouping procedure). In the rest of the paper, we will use the term resiliency when we want to refer collectively to both the above measures. It is to be noted that we are concerned only with the fault tolerance of mutual exclusion algorithms. So, as far as we are concerned, a mutual exclusion request may not be satisfied *only* due to site failures. We *do not* consider the situation where a mutual exclusion request is not granted because some other site has already been granted permission and is using the replicated resource.

The algorithm presented in [13] requires $O(N)$ messages per mutual exclusion request and is not very resilient. In this approach the quorum set contains one quorum group with all sites being members of this group. In the *primary site* approach [3], the quorum set contains only one quorum group with the primary site as the only member of that group. Such an approach is not distributed and is very inexpensive in terms of the number of messages required for mutual exclusion, but its resiliency is entirely dependent on the failure of the primary site. In the majority consensus (or majority voting) method [16] each site is assigned a single vote and any group of majority of sites will form a quorum group. In this case, it is clear that any two quorum groups will have a non-empty intersection. Such an approach leads to high resiliency, but requires at least $\lceil \frac{N+1}{2} \rceil$ messages per mutual exclusion request. An extension of the above method is *weighted voting* [8] [14] where each site is assigned a number of votes (not necessarily one), and any group of nodes with the majority of votes will form a quorum group. Again, its is easy to see that every pair of quorum groups will have a non-empty intersection. A weighted voting scheme where one node is assigned all the votes will be equivalent to the primary site approach. The majority voting scheme is a distributed algorithm whereas weighted voting is not.

Quorum sets can be constructed explicitly such that any two groups in the set have a non-empty intersection. Such a concept of *coteries* was proposed in [7]. Unlike the majority voting and the weighted voting schemes, a logical structure is introduced among the sites to form quorum groups and such a construction requires that a list of quorum groups be maintained explicitly (there could be exponential number of such quorum groups). In this case, the message complexity depends on the size of the quorum groups. Maekawa [11] showed that quorum groups each of size $\sqrt{N}$ can be (explicitly) constructed such that every pair of such groups will have an intersection of exactly one (site). There are N such quorum groups each associated with one site. To perform an operation, a site requires permission from all the sites in its associated group. It was shown that for a *distributed algorithm* (where each site bears equal responsibility in controlling mutual exclusion), the minimum size of quorum groups is $\sqrt{N}$. Such an algorithm has a low message complexity $O(\sqrt{N})$, but is not very resilient (as will be shown later). Other algorithms where quorum groups are formed by imposing a logical structure include a token based approach [15], an extension to this approach based on forming spanning trees among sites [12] and another tree structured quorum protocol [1]. The token based approaches [15] [12] have poor resiliency and are not distributed algorithms. The tree structured protocol in [1] is also not a distributed algorithm and in the best case requires $O(\log N)$ messages and in the worst case requires $\lceil \frac{N+1}{2} \rceil$ messages. The algorithm is resilient to some failure patterns, but there are situations where mutual exclusion is unobtainable even with just $\log N$ failures.

There is a trade-off between resiliency and message complexity of mutual exclusion algorithms and it is very difficult to compare these algorithms without a uniform measure which relates these two parameters. In this paper, we propose *asymptotically high resiliency* as such a measure. The resiliency (be it site or system) of an algorithm is asymptotically high if it *does not* decrease when the system size increases. Based on such a measure, we design a totally distributed algorithm with asymptotically high resiliency where the message overhead per mutual exclusion request is $O(\sqrt{N \log N})$. Our algorithm is based on grouping processors into subgroups and we could control the resiliency by controlling the group size.

In Section 2, we present asymptotically high resiliency as a performance measure for fault-tolerant mutual exclusion algorithms and discuss the resiliency of existing algorithms. In Section 3, we discuss the algorithm due to Maekawa. Section 4 presents the design of our algorithm. In Section 5, the resiliency and message complexity of our algorithm are analyzed. Conclusion are presented in Section 6.

2 Resiliency of Fault Tolerant Algorithms

Algorithms with a higher message overhead for mutual exclusion seem to have higher resiliency. As we discussed in the last section, a common measure of resiliency is needed to compare message overhead of different algorithms. We assume that sending data from one site to any other site in the distributed system (connected by an underlying reliable network) constitutes a message. We also assume that a message can be sent directly from any site to any other site through the network.

2.1 Asymptotically high resiliency

Let us use the term *availability* to denote the probability that a site is non-faulty and is available to reply to mutual exclusion requests. We denote the availability of a site by p. A mutual exclusion algorithm with N sites, is defined to have asymptotically high site (system) resiliency if R_{site} (R_{sys}) *does not* decrease when $N \rightarrow \infty$. That is, R_{site} (R_{sys}) stays a constant or increases towards 1 when $N \rightarrow \infty$. It should be noted that such a measure of resiliency is not just for measuring the resiliency of algorithms for large systems. In fact, such a measure gives us an understanding about the behavior of an algorithm for a system of any size. For example, an algorithm with asymptotically high resiliency can guarantee that its resiliency when run on a (say) 11 site system will be at least as good as its resiliency when run on a 10 site system.

2.2 Resiliency of existing mutual exclusion algorithms

Let us first look at some of the distributed algorithms that we discussed in the last section. The algorithm in [13] does not have asymptotically high resiliency because it requires all sites to be available for a mutual exclusion request to succeed. When the system size grows, the probability of this event goes to zero. The majority consensus and hierarchical consensus approaches both have asymptotically high system and site resiliency. For a majority consensus approach it can be shown that the both $1 - R_{site}$ and $1 - R_{sys}$ goes to zero exponentially with an increasing system size (N). That is, the resiliency goes to 1. But both these algorithms may require up to N messages per mutual exclusion request. It can be shown that the distributed algorithm in [11] does not have asymptotically high resiliency (both site and system) although it requires only $\sqrt{N}$ messages. We will consider this algorithm and its resiliency in more detail in the next section.

For the non distributed algorithms like the ones in [8] [15] [12] and [1] resiliency will depend on which sites bear more responsibility in the mutual exclusion decisions. The token based algorithms in [15] and [12] will not have asymptotically high resiliency (because the resiliency is determined by the site with the token). The algorithms in [8] and [1] may or may not have asymptotically high resiliency.

3 Maekawa's Algorithm

In [11], it is shown how N sites (processors) in a distributed system can be divided into groups with $\sqrt{N}$ sites per group, such that any two groups have a non-null intersection. If S_i and S_j are two groups of $\sqrt{N}$ sites each, then $S_i \cap S_j \neq 0$. There are N such groups, where each group is *associated* with a particular site. It can be shown then that each site belongs to $\sqrt{N}$ groups. The situation, where N is not a square leads to some dummy sites being added. The groups are formed using finite projective planes [11]. An example is shown in Figure 1(from [11]) where there are 21 sites, and 21 groups with 5 sites per group (the number closest to 21 which is a square is 25). Also, each site belongs to 5 groups and any two groups have a non-null intersection. S_i refers to the group *associated* with site i. For the mutual exclusion algorithm using this grouping strategy in [11], the 21 groups would be the quorum groups. If a site i requires mutual exclusion, it gets permission from all the sites in its associated group S_i. For example, if site 1 requests mutual exclusion, it needs to obtain permission from sites 2, 3, 4 and 5. So, each site has its own quorum group (which is its associated group).

As each group is of size $O(\sqrt{N})$, and each site requires permission from all the sites

$$S_{20} = \{5, 9, 10, 15, 20\}$$
$$S_{21} = \{5, 8, 11, 14, 21\}$$
$$S_5 = \{5, 7, 12, 17, 18\}$$
$$S_{19} = \{5, 6, 13, 16, 19\}$$
$$S_{16} = \{4, 9, 11, 16, 18\}$$
$$S_{17} = \{4, 8, 10, 17, 19\}$$
$$S_4 = \{4, 7, 13, 14, 20\}$$
$$S_{15} = \{4, 6, 12, 15, 21\}$$
$$S_{12} = \{3, 9, 12, 14, 19\}$$
$$S_{13} = \{3, 8, 13, 15, 18\}$$
$$S_3 = \{3, 7, 10, 16, 21\}$$
$$S_{11} = \{3, 6, 11, 17, 20\}$$
$$S_9 = \{2, 9, 13, 17, 21\}$$
$$S_8 = \{2, 8, 12, 16, 20\}$$
$$S_7 = \{2, 7, 11, 15, 19\}$$
$$S_2 = \{2, 6, 10, 14, 18\}$$
$$S_{18} = \{1, 18, 19, 20, 21\}$$
$$S_{14} = \{1, 14, 15, 16, 17\}$$
$$S_{10} = \{1, 10, 11, 12, 13\}$$
$$S_6 = \{1, 6, 7, 8, 9\}$$
$$S_1 = \{1, 2, 3, 4, 5\}$$

Figure 1: Grouping processors with sqrt(N) processors per group

in its associated group for mutual exclusion, the total number of messages per mutual exclusion request is $O(\sqrt{N})$. Let us consider the resiliency of this algorithm. If p is the availability of a site, then $R_{site} = p^{\sqrt{N}-1}$. When N increases, it can be seen that site resiliency goes to zero. There are N quorum groups, but they are overlapping. For *upper bound* calculations on system resiliency, we will assume that there are N independent quorum groups. Then, the algorithm is system resilient if at least one group has all the sites available. So, $R_{sys} = 1 - (1 - p^{\sqrt{N}-1})^N$. Again, when N increases, it can be shown that R_{sys} goes to zero. So, this algorithm does not have asymptotically high system or site resiliency.

Although the algorithm is not resilient to site failures, it has many attractive characteristics. It is a totally distributed algorithm where every site has equal responsibility in controlling mutual exclusion. Also, it is shown in [11] that for a totally distributed mutual exclusion algorithm, a lower bound on the size of quorum groups is $\Omega(\sqrt{N})$ which implies that $O(\sqrt{N})$ messages are necessary. Other distributed algorithms like the majority voting, hierarchical consensus etc. obtain higher resiliency by increasing the size of quorum groups and hence the number of messages. The question we would like to answer is: can we increase the number of messages by a small number (from Maekawa's bound) and achieve asymptotically high resiliecny? The underlying problem is that in

$$S_{1,2,3} = \{\{1,2,3\},\{4,5,6\},\{7,8,9\}\}$$
$$S_{10,11,12} = \{\{1,2,3\},\{10,11,12\},\{13,14,15\}\}$$
$$S_{16,17,18} = \{\{1,2,3\},\{16,17,18\},\{19,20,21\}\}$$
$$S_{4,5,6} = \{\{4,5,6\},\{10,11,12\},\{16,17,18\}\}$$
$$S_{13,14,15} = \{\{4,5,6\},\{16,17,18\},\{19,20,21\}\}$$
$$S_{19,20,21} = \{\{7,8,9\},\{10,11,12\},\{19,20,21\}\}$$
$$S_{7,8,9} = \{\{7,8,9\},\{13,14,15\},\{16,17,18\}\}$$

Figure 2: Grouping processors using subgroups

Maekawa's algorithm even if one site in unavailable in a group, the associated site will not be able to get mutual exclusion.

A way to get around this problem is to define the concept of a subgroup permission (or vote). Instead of a site asking for permission from a group of sites, what if it is made to ask for permission from a group of *subgroups* of sites? In this case, depending on how the algorithm is designed, subgroup permission may be obtained even if not all sites inside the subgroup are available. In the next section, we propose such a grouping strategy and show how an algorithm can be designed such that with a very small increase in the size of a quorum group (and hence the number of messages) asymptotically high resiliency can be achieved.

4 An Algorithm Based on Subgroups

4.1 Grouping processors using subgroups

Consider a system with N sites. We group these sites into $\frac{N}{G}$ groups (not to be confused with a quorum group) of G sites each. We will call each such group a *subgroup*. We then construct $\frac{N}{G}$ quorum groups such that each quorum group is made up of $K = \sqrt{\frac{N}{G}}$ subgroups with each subgroup containing G sites. Now, each such quorum group will be associated with G sites. An example of such a grouping strategy is shown in Figure 2. Again, we consider a 21 site system. Each subgroup contains 3 sites ($G = 3$) and there are 7 quorum groups ($\frac{N}{G} = 7$). Each quorum group consists of $K = 3$ subgroups ($\frac{N}{G} = 7$, but the closest square to 7 is 9, and the number of subgroups is $\sqrt{\frac{N}{G}}$). Each quorum group is associated with 3 sites. For example, quorum group $S_{1,2,3}$ is associated with sites 1, 2 and 3. The intersection between every pair of quorum groups is exactly one subgroup. For example, quorum groups $S_{1,2,3}$ and $S_{10,11,12}$ intersect at subgroup $\{1,2,3\}$. Compare such a grouping strategy to Maekawa's grouping strategy for a system of the same size (21 sites) shown in Figure 1.

4.2   The Algorithm

When a site initiates a mutual exclusion request (only an available site can initiate such a request), it asks for permission from all the subgroups in its associated group. If permission is granted by all subgroups, then the request is successful. Permission from a subgroup is obtained if a *majority* of the sites in the subgroup grant their permission.

The protocol will be as follows. Let A_i be the associated group for site i. Let $A_i = \{SG_1, SG_2, ..., SG_K\}$ where SG denotes a subgroup. Note that i belongs to exactly one of these subgroups. Let us refer to this as the *local subgroup*. When site i initiates a mutual exclusion request, it sends messages to all the sites in each of the subgroups (including its local subgroup) $SG_1...SG_K$. Let $Maj(SG_j) = 1$ denote the event that a majority of sites in SG_j have granted their permission and $Maj(SG_j) = 0$ denote the event that a majority of sites in SG_j *did not* grant their permission. Then, the request is successful if

$$\forall_j : (1 \leq j \leq K) \ \{Maj(SG_j) = 1\}$$

For example, in Figure 2, if site 1 initiates a mutual exclusion request, it will send request messages to sites $2, 3, 4, 5, 6, 7, 8$ and 9. It then receives reply messages from these sites. Suppose it receives reply messages granting permission from (say) $2, 4, 5, 7, 8$ and 1 (it always has its own permission), then it has received permission from a majority of the sites in *each* of the subgroups and so the request is deemed successful.

Let us consider the overhead for such an algorithm. Each site is associated with only one quorum group. So, it has to maintain a list of the sites in its quorum group and also information about how these sites are grouped into subgroups. Further, a request involves sending messages to G sites in each of the $\sqrt{\frac{N}{G}}$ subgroups for a total of $\sqrt{N * G}$ sites. Exactly the same protocol as in [11] can be followed for message exchange. To avoid deadlocks, time stamps have to be used to determine priority. We omit the details here. The reader is referred to [11] and [4]. In [11], it is shown that each request to a site involves exchanging 3 to 5 messages. The algorithm in [11] may deadlock as shown in [4]. It is also shown in [4] that by including additional messages such deadlocks in Maekawa's algorithm could be removed and the message overhead in this case would involve exchanging 3 to 7 messages per request to a site. So, in our case, a mutual exclusion request would involve exchanging $c\sqrt{N * G}$ messages where c is between 3 and 7.

4.3 Correctness

The fact that the algorithm is *deadlock free* and *starvation free* directly follows from the proof of the protocol in [11] and its modification in [4], as we propose the use of exactly the same protocol for message exchange. We need to show mutual exclusion. No two sites can have their mutual exclusion requests satisfied at the same time as shown below. First, assume the contrary. Then,

1. The two sites which have obtained mutual exclusion must have obtained permission from all the subgroups in their respective quorum groups.

2. If the above two sites have the same associated quorum group, then all the subgroups in that quorum group must have given their permission to *both* the sites.

3. If the above sites have different quorum groups, still there should be one subgroup which has given permission to *both* the sites, because every pair of quorum groups intersect at one subgroup.

4. According to the specification of the algorithm, since a site grants permission to only one request at a time, and since permission from a *majority* of the sites in a subgroup are required for that subgroup's permission, at most one request is granted permission by a subgroup. This contradicts (2) and (3).

5. Hence only one site can have its mutual exclusion request granted at any one time.

In the next section we consider both site and system resiliencies for our algorithm and show that asymptotically high resiliency can be achieved by proper selection of the group size G. The group size also determines the message overhead per mutual exclusion request, because the message overhead is given by $O(\sqrt{N * G})$.

5 Analysis of Resiliency and Message Overhead

From the definition of site and system resiliencies it is clear that if the system is site resilient, then it is system resilient. Let us first calculate the site resiliency of our algorithm and show that it is asymptotically high with the proper choice of size of subgroups. We will assume that individual availabilities of sites are greater than 0.5. That is, $p > 0.5$. Also, unless otherwise stated, all logarithm is to the base e. In order to do so, we need the following corollary from [5] [6].

5.1 Analysis of Resiliency

Corollary 5.1 *Let q and p be positive numbers such that $q + p = 1$. Then,*

$$\sum_{t=k}^{m} \binom{m}{t} q^t p^{m-t} \leq exp\left[(m-k)\log(\frac{mp}{(m-k)}) + k\log(\frac{mq}{k})\right]$$

where $k \geq mq$.

The above corollary can be used to derive the following result.

Corollary 5.2 *Let Z be a binomial random variable with parameters n and q such that $0 \leq q \leq 1$. Then*

$$P(Z \geq \epsilon n q) \leq \exp\left(-\left(n q \epsilon \log(\epsilon)\right.\right.$$
$$\left.\left. + n(1-q)\left[\frac{1-q\epsilon}{1-q}\right]\log\left[\frac{1-q\epsilon}{1-q}\right]\right)\right), \quad 1 \leq \epsilon$$

From the above corollary, by substituting $\epsilon = \frac{1}{2q}$, we can derive the following lemma.

Lemma 5.1 *Let Z be a binomial random variable with parameters n and q such that $0 \leq q < \frac{1}{2}$. If $\epsilon = \frac{1}{2q}$, th en*

$$P(Z \geq \frac{n}{2}) \leq \exp\left(-\frac{n}{2}\left(\log\left(\frac{1}{2q}\right) - \log(2(1-q))\right)\right)$$

Let us now compute the probability that a majority of the sites in a subgroup of size G are *not* available, given that individual site availability is p. Let $q = 1 - p$. The following lemma gives that result.

Lemma 5.2 *The probability that a majority of sites in a subgroup of size G are not available is given by $P[NO_{maj}] \leq exp(-c_1 G)$, where c_1 is a constant equal to $\frac{(\log(\frac{1}{2q}) - \log(2(1-q)))}{2}$*

Proof: Let Z be the binomial random variable which gives the number of sites that are *not* available in a group. This random variable has parameters G (the size of the group) and q (the probability that an individual site is not available). Then, from Lemma 5.1,

$$P[NO_{maj}] = P(Z \geq \frac{G}{2}) \leq exp(-c_1 G)$$

where $c_1 = \frac{(\log(\frac{1}{2q}) - \log(2(1-q)))}{2}$

$\square$

Let us consider the site resiliency R_{site} of our algorithm. A site requesting mutual exclusion will get it (in the absence) of any other requests, if it gets permission from each of the subgroups. A permission from a subgroup needs a majority of the sites in that subgroup to be available. There are $\sqrt{\frac{N}{G}}$ subgroups in a quorum group and so $R_{site} = (1 - P[NO_{maj}])^{\sqrt{\frac{N}{G}}}$. That is, from Lemma 5.2,

$$R_{site} \geq (1 - exp(-c_1 G))^{\sqrt{\frac{N}{G}}} \qquad (1)$$

We should choose G properly such that asymptotically high site resiliency is achieved. The following theorem relates group size to resiliency.

Theorem 5.1 If $G \geq c_2 \log N$ where c_2 is a properly chosen constant, then $R_{site} \to 1$, when $N \to \infty$.

Proof: From Equation 1

$$R_{site} \geq (1 - exp(-c_1 G))^{\sqrt{\frac{N}{G}}}$$

Let $G = c_2 \log N$. Then,

$$R_{site} \geq (1 - N^{-c_1 c_2})^{\sqrt{\frac{N}{c_2 \log N}}}$$

Let $c_2 = \frac{1}{2c_1} + \epsilon$, where $\epsilon \geq 0$. Then ,

$$R_{site} \geq (1 - N^{-(\frac{1}{2} + c_1 \epsilon)})^{\sqrt{\frac{N}{(\frac{1}{2c_1} + \epsilon) \log N}}}$$

We know that for any positive b, $(1 - \frac{1}{b})^b \to \frac{1}{e}$ when $b \to \infty$. Using this fact in the above equation and after si mplification, for large N,

$$R_{site} \geq exp\left(-\left(\frac{1}{N^{c_1 \epsilon} \sqrt{(\frac{1}{2c_1} + \epsilon) \log N}}\right)\right)$$

Now, using the above equation and the fact that $e^{-x} = 1 - x + \frac{x^2}{2!} - \dots$ and ignoring higher order terms,

$$1 - R_{site} \leq \frac{1}{N^{c_1 \epsilon} \sqrt{(\frac{1}{2c_1} + \epsilon) \log N}}$$

Then,

$$R_{site} \geq 1 - \frac{1}{N^{c_1 \epsilon} \sqrt{(\frac{1}{2c_1} + \epsilon) \log N}}$$

It is easy to see that when $N \to \infty$, $R_{site} \to 1$. Note that c_2 was selected to be $\frac{1}{2c_1} + \epsilon$, where $\epsilon \geq 0$. That is c_2 should be at least as big as $\frac{1}{2c_1}$.

□

Given that each subgroup is of size $c_2 \log N$, the constant c_2 determines how fast the site resiliency goes towards 1. For example, from the above theorem, if $\epsilon = \frac{1}{c_1}$, which means $c_2 = \frac{3}{2c_1}$, then $R_{site} \geq 1 - O(\frac{1}{N\sqrt{\log N}})$.

Let us consider an example. For asymptotically high site resiliency, from the above theorem, it is sufficient for c_2 to be $\geq \frac{1}{2c_1}$. $c_1 = \frac{(\log(\frac{1}{2q}) - \log(2(1-q)))}{2}$. Let the individual site availability be 0.9. Then, $c_1 = 0.51$. Then, $c_2 \geq 0.9789$. That means, if a subgroup size is $G = 0.9789 \log N$ we are assured of asymptotically high resiliency. For example, if $N = 100$, then the subgroup size turns out to be $G = 5$. So, we will form $\frac{N}{G} = 20$ subgroups of 5 sites each and form quorum groups as explained in Section 4. There will be 20 such quorum groups. Each quorum group will have 5 subgroups (The nearest larger square to 20 is 25. We have to add a few dummy sites to make the construction work as explained before).

The site resiliency given by Thoerem 5.1 is a lower bound on the system resiliency R_{sys}. So, if an algorithm has asymptotically high site resiliency, it has asymptotically high system resiliency. Tighter bounds for R_{sys} can be developed.

5.2 Message Complexity

Consider the message complexity of our algorithm. Given that the algorithm has asymptotically high site resiliency, a site issuing a request has to send messages to only the sites in all the subgroups in its quorum group (for asymptotically high probability of success). There are $\sqrt{\frac{N}{G}}$ subgroups with G sites per subgroup. We have $G = O(\log N)$. So, the message complexity is $O(\sqrt{N \log N})$. Compare this to the majority voting scheme where at least $\frac{N}{2}$ messages are required if all sites are available and an increasing number of messages with lesser availability. The hierarchical voting scheme with a message overhead of $N^{0.63}$ messages with all sites available and may be up to N messages with decreasing availability. Note that we assume that the site availability $p > 0.5$ and our message complexity does not increase with decreasing availability. For our example, with $N = 100$ and $p = 0.9$, $\sqrt{0.9789 N \log N} < 22$. As explained before, a mutual exclusion request would involve exchanging $c\sqrt{N * G}$ messages where c is between 3 and 7. If $c = 3$, then in our example, 66 messages are exchanged.

Once asymptotic resiliency is reached, the resiliency of a majority voting scheme goes towards 1 at a faster rate than our algorithm. So, there is a trade-off between message complexity and resiliency. The point we would like to stress is as follows. With a message complexity $\sqrt{N \log N}$, we can achieve asymptotically high resiliency in our algorithm. But, if higher resiliency is required, we can achieve that by increasing the subgroup size G. For example, an instance of our algorithm with $G = N$ (that is, there is only one subgroup in the system with all N processors in it) is nothing but a majority voting scheme. So, we have the flexibility of controlling the resiliency by controlling the subgroup size.

6   Conclusions

We proposed asymptotically high resiliency as a performance measure for fault-tolerant mutual exclusion algorithms. We also presented a distributed fault-tolerant algorithm for mutual exclusion which has an asymptotically high resiliency to site failures. The message overhead of the algorithm is $O(\sqrt{N \log N})$. Although some of the existing algorithm have asymptotically high resiliency, they have a much larger message overhead. We further showed that the resiliency of our algorithm can be increased by (sub)grouping sites appropriately which leads to an increase in message overhead. Thus, there is a trade-off between message overhead and resiliency. Majority voting scheme is an instance of our algorithm where all the sites belong to a single (sub) group.

References

[1] D. Agarawal and A. El Abbadi, "An Efficient Solution to the Mutual Exclusion Problem," *ACM Symposium on Principles of Distributed Computing*, pp. 193-200, 1989.

[2] M. Ahamad, M. H. Ammar and S. Y. Cheung, "Multi-Dimensional Voting," Technical Report, College of Computing, Georgia Institute of Technology, 1991.

[3] P. A. Alsberg and J. D. Day, "A Principle for Resilient Sharing of Distributed Resources," *Proceedings of the Second International Conference on Software Engineering*, pp. 562-570, October 1976.

[4] B. A. Sanders, "The Information Structure of Distributed Mutual Exclusion Algorithms," *ACM Trans. Computer Systems*, vol. 5, no. 3, pp. 284-299, August 1987.

[5] H. Chernoff, "A Measure of Asymptotic Efficiency for Tests of a Hypothesis Based on the Sum of Observations," Annals of Mathematical Statistics, vol.23, pp. 493–507, 1952.

[6] P. Erdos and J. Spencer, "Probabilistic Methods in Combinatorics," Academic Press, 1974.

[7] H. Garcia-Molina and D. Barbara, "How to Assign Votes in a Distributed System," *JACM*, vol. 32, no. 4, pp. 841-860, October 1985.

[8] D. K. Gifford, "Weighted Voting for Replicated Data," *Proceedings of the Seventh Symposium on Operating System Principles*, pp. 150-159, 1979.

[9] L. Lamport, "The Implementation of Reliable Distributed Multiprocess Systems," *Computer Networks*, no. 2, 1978, pp. 95-114.

[10] L. Lamport, "Time, Clocks and the Ordering of Events in a Distributed System," *Comm. of ACM*, vol. 21, no. 7, pp. 558-565, 1978.

[11] M. Maekawa, "A $\sqrt{N}$ Algorithm for Mutual Exclusion in Decentralized Systems," *ACM. Trans. Computer Systems*, pp. 145-159, May 1985.

[12] K. Raymond, "A Tree Based Algorithm for Mutual Exclusion," *ACM Transactions on Computer Systems*, vol. 7, no. 1, pp. 61-77, February 1989.

[13] G. Ricart and A. K. Agarwala, "An Optimal Algorithm for Mutual Exclusion in Computer Networks," *Communication of the ACM*, vol. 24, no. 1, pp. 9-17, January, 1981.

[14] D. Skeen, "A Quorum Based Commit Protocol," *Proceedings of the 6th Berkeley Workshop on Distributed Data Management and Computer Networks*, pp. 69-80, February, 1982.

[15] I. Suzuki and T. Kasami, "A Distributed Mutual Exclusion Algorithm," *ACM Transactions on Computer Systems*, vol. 3, no. 4, pp. 344-349, November 1985.

[16] R. H. Thomas, "A Majority Consensus Approach to Concurrency Control for Multiple Copy Databases," *ACM Transactions on Database Systems*, vol. 4, no. 2, pp. 180-209, June 1979.

MESSAGE DELAYING SYNCHRONIZERS

Lior Shabtay and Adrian Segall
Dept. of Computer Science
Technion, Israel Institute of Technology
Haifa, Israel 32000
email : liors@cs.technion.ac.il, segall@cs.technion.ac.il

Abstract

Delaying messages is one of the techniques that can be employed in order to create correct asynchronous protocols using the synchronizer mechanism. In this paper, we show that message delaying cannot be implemented with certain synchronizers and that the synchronizers must be altered before message delaying can be applied. We present three different techniques that solve the problem and work for most synchronizers.

1   Introduction

In this paper we are dealing with distributed protocols in two network models : the *synchronous* model and the *asynchronous* model. In the asynchronous model, messages sent by a node to one of its neighbors are received by that neighbor in FIFO order within a finite undetermined time and code can be performed by nodes only upon receiving a message. In the synchronous model, all link delays are bounded by some quantity T. The network contains a global clock that ticks synchronously at all network nodes at time intervals T. Code can be performed upon receipt of a message and/or at clock ticks. Nodes are allowed to send messages only at clock ticks.

Synchronizers are tools for transforming protocols written for a *synchronous model* to protocols that run on an *asynchronous model*. The synchronous protocol will be referred here as the *original* protocol, its messages as the *original-protocol messages* and its code as the *original-protocol code*. The original-protocol code is composed of the *original-protocol pulse code* which is the part performed at the synchronous model pulses and the *original-protocol message code* which is the part performed in the synchronous protocol upon receipt of a message.

The asynchronous protocol created by the synchronizer generates a sequence of 'clock-pulses' at each node of the network, that occur asynchronously at different nodes. At each pulse, the node performs the original-protocol pulse code and sends original-protocol messages.

The methodology of synchronizers was introduced in [Awer 1], where three synchronizers were presented : the α synchronizer, which possesses an overhead of $O(|E|)$ in communication complexity and $O(1)$ in time complexity per each pulse, the β synchronizer with overhead of $O(|V|)$ in communication and $O(D)$ in time complexity per pulse (where D is the diameter of the network), and the γ synchronizer, that enables trade-off between these values.

All synchronizers in [Awer 1], [PU 1] and [AP 1] work in a very similar way : a node may perform a new pulse as soon as it knows that it has received all the original-protocol messages sent to it by its neighbors at the former pulse. However, it is shown in [LT 1] that those synchronizers do not ensure proper execution, in the sense that the created asynchronous-protocol is not necessarily an exact simulation of the original synchronous-protocol. The problem is that the synchronizers allow original-protocol messages sent by a node at a given pulse to arrive at a neighbor node *before* the time when the latter has performed that same pulse. Obviously, such a situation cannot occur in a synchronous model.

One way to solve this problem is suggested in [LT 1], [FLS 1] and [ER 1]. The idea is that when a message is received too early at a node, it is not treated immediately. Instead, such messages are saved and treated only after the pulse is performed at the node. In Sec. 2 and 3 of this paper, we show that this remedy works with synchronizer α, but cannot be applied to synchronizers β and γ. The problem is that in the latter, when a node receives an original-protocol message, it may not be able to distinguish whether the message is a timely message and should be treated immediately or is an early one and should be saved to be treated at a later time.

In the rest of the paper, we present several other solutions for this problem.

2 Synchronizer α^*

A *synchronizer* is a transformation that transforms a synchronous protocol to an asynchronous one by generating a sequence of 'clock-pulses' at each node of the network, where the pulses occur asynchronously at different nodes. A new phase of the protocol starts at each node after every clock pulse. In the created asynchronous protocol, we denote the series of pulses at node i by $t_i(0)$, $t_i(1)$, ..., and original-protocol messages are allowed to be sent by i only at times $\{t_i(n), n \geq 0\}$.

Definition 2.1 *A synchronous protocol combined with a synchronizer, or in short a combined protocol is the asynchronous protocol created by the synchronizer when given the original (synchronous) protocol as input.*

Definition 2.2 *In a combined protocol, $M^k(n)$ is defined as the ordered set of all messages sent to k by all its neighbors i, at their own time $t_i(n)$, ordered according to the timing of their arrival at node k. Members of $M^k(n)$ contain the information about the neighbor by which the message was sent.*

In synchronizer α [Awer 1], a node i is said to be 'safe' with respect to $t_i(n)$, if all messages of the original protocol sent by i at $t_i(n)$ have already been received by the respective neighbors. When a node learns that it is safe, it sends a SAFE message to all its neighbors. Node i generates $t_i(n+1)$ when it learns that all its neighbors are safe.

Synchronizers β and γ [Awer 1] work somewhat differently, but still a node performs a pulse when it learns that all its neighbors are safe with respect to the former pulse.

Lemma 2.1 *The (combined) protocol created by combining any synchronous protocol with one of the synchronizers α, β or γ satisfies the following property :*
the original-protocol message sent by i to neighbor k at time $t_i(n)$ arrives at node k before $t_k(n+1)$ and after $t_k(n-1)$.

Proof: When node k performs $t_k(n+1)$, its neighbor i is safe with respect to $t_i(n)$, which means that the message sent by i to k at $t_i(n)$ had already been received. On the other hand, when node i performs $t_i(n)$ and sends a message to k, node k is safe with respect to $t_k(n-1)$, which means that $t_k(n-1) < t_i(n)$. $\qquad\Box$

If a combined protocol satisfies the property in lemma 2.1, a message sent by a node i to a neighbor k and received during $(t_k(n), t_k(n+1))$, was sent at time $t_i(n)$ or $t_i(n+1)$. In other words, the message belongs to $M^k(n)$ or $M^k(n+1)$. However, if indeed a message sent at $t_i(n+1)$ is received by k during $(t_k(n), t_k(n+1))$ and is processed, the combined protocol does not behave as the original protocol since in the latter, all messages sent at $t_i(n+1)$ arrive during $(t_k(n+1), t_k(n+2))$. It was shown in [LT 1] that such misbehavior can occur with any of the synchronizers α, β, γ.

In order to implement α^*, which is the message delaying version of α that ensures proper execution, we need a *fifo queue of messages* at each node.

When an original-protocol message is received too early at some node, the message is pushed into the fifo. The fifo is emptied by the node processor immediately after performing the next pulse and each message that is poped out of the fifo is treated as if it had just arrived.

From Lemma 2.1, the original-protocol messages received by a node i during $(t_i(n), t_i(n+1))$ belong either to $M^i(n)$ or to $M^i(n+1)$. Since the messages of $M^i(n)$ received during that interval are treated immediately, the messages that are in the fifo at time $t_i(n+1)$ belong to $M^i(n+1)$. These messages are treated only after $t_i(n+1)$, as if they were received in the correct interval.

In synchronizer α^*, it is easy to recognize whether a message belongs to $M^i(n)$ or to $M^i(n+1)$, since in each interval, nodes send SAFE messages to all neighbors when they are safe. Thus, a node i can decide whether a message belongs to $M^i(n)$ or to $M^i(n+1)$ by simply counting the number of SAFE messages received from that neighbor. Since there are only two possibilities, a one bit counter is sufficient for this purpose.

The code of α^* is presented in Table 1.

3 Synchronizer β_1^*

In synchronizer β, an initialization phase creates a directed tree with a root node s. When recognizing that itself as well as all nodes in its sub-tree are safe, each node sends a SAFE

Messages

CONFIRM – Delivered by the data link when all original-protocol messages sent are confirmed to have been received by the other end of the link.

SAFE – Sent to all neighbors to signal that the node is safe with respect to the current pulse.

Variables

G_i : The set of neighbors of i.

Z_i : A counter, contains the number of the latest pulse performed by node i (a one bit counter is sufficient).

$N_i(l)$: A counter that counts the number of SAFE messages received from neighbor l (a one bit counter is sufficient).

st_i : 1 when all original-protocol messages sent at the last pulse are confirmed, 0 otherwise.

Procedures

msgCode : The original-protocol message code for treating messages of type msg.

PulseCode : The original-protocol pulse code.

Initialization

$$Z_i = -1. \qquad N_i(l) = 0, \forall l \in G_i. \qquad st_i = 1.$$

Algorithm for node i

    When receiving an original-protocol message named msg from neighbor l

        if $Z_i = N_i(l)$ *then*   ; (timely message)

             call msgCode (msg, l)

        else push (msg, l) ; (early message)

    When receiving START, CONFIRM or SAFE from neighbor l *do*

        if SAFE *then* $N_i(l) \leftarrow N_i(l) + 1$

        if CONFIRM *then*

            $st_i \leftarrow 1$

            send SAFE to all $k \in G_i$

        if $(st_i = 1)$ and $(\forall j \in G_i, \ Z_i < N_i(j))$ *then*

            $Z_i \leftarrow Z_i + 1; \ st_i \leftarrow 0$

            call PulseCode

            while not fifo_empty *do*

                pop $(message, neighbor)$

                call msgCode $(message, neighbor)$; $(msg$ = the type of message

Table 1: The α^* synchronizer

message to its father. When the root node s finds out that all nodes in the network are safe, it broadcasts an AWAKE message along the tree. This message informs the nodes that they should perform the next pulse.

As shown below, a modification similar to the one of Sec. 2 does not work on β. The reason is that a node may not be able to distinguish between timely and early messages. Therefore, the node cannot decide whether to treat this message immediately or to push it into the fifo.

We shall prove this claim by introducing two execution examples, of two different protocols combined with synchronizer β, on the same network. In both executions, node i receives identical messages in identical order. In both executions, node i receives during $(t_i(n), t_i(n+1))$ one original-protocol message from neighbor k. Yet, in one execution, this message belongs to $M^i(n)$ and in the other, it belongs to $M^i(n+1)$. The two execution examples take place in the network shown in Fig. 1a) with the spanning tree of Fig. 1b).

In both examples, node s performs $t_s(0)$ and sends an AWAKE message to nodes i and k. No original-protocol messages are sent by s at $t_s(0)$. Nodes i and k receive the message at approximately the same time and perform $t_i(0)$ and $t_j(0)$ respectively. In the first example, node k sends at time $t_j(0)$ an original-protocol message to node i. Node i has no original-protocol messages to send at time $t_i(0)$. Thus, node i sends immediately a SAFE message, and after a while it receives the original-protocol message sent by node k.

In the second example, both i and k have no original-protocol messages to send at times $t_i(0)$ and $t_j(0)$ respectively. Thus, they send SAFE messages to s immediately after performing pulse 0. When the two SAFE messages are received by s, the latter sends an AWAKE message to i and k, while performing $t_s(1)$. The AWAKE message to i is delayed. When the AWAKE message to k arrives, the latter performs $t_j(1)$ and sends an original-protocol message to node i. This message arrives at node i before the arrival of the AWAKE message from node s and thus before $t_i(1)$.

In both executions, node i receives the same messages, in the same order : an AWAKE message from node s, and then an original-protocol message from node k. But, in the first example, the message belongs to $M^i(0)$ and should be processed immediately, while in the second, it belongs to $M^i(1)$ and should be delayed.

The same problem occurs in synchronizer γ and in the synchronizer for hypercube topology presented in [PU 1]. In fact, it will occur in any synchronizer that does not require nodes to send synchronizer messages to all neighbors, unless some other mechanism to identify early messages is employed. Synchronizer β_1, to be introduced next, constructs such a mechanism by requiring two waves in each direction on the tree, as opposed to one wave in β. In this way, each node will be given a time $t_i^{READY}(n)$ between $t_i(n)$ and $t_i(n+1)$, such that all messages that arrive before $t_i^{READY}(n)$ are timely messages, i.e. belong to $M^i(n)$, and those arriving after $t_i^{READY}(n)$ are early messages, i.e. belong to $M^i(n+1)$.

Synchronizer β_1 needs the same initialization as β, i.e. a creation of a spanning tree, with root s say, for the network graph. The messages used by synchronizer β_1 are SAFE, READY, and AWAKE. A SAFE message is sent by each node to its father when all nodes

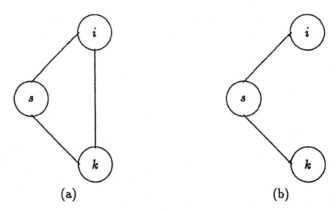

Figure 1: Three nodes network and spanning tree example

in its sub-tree, including itself, are safe. When node s finds out that all nodes in the network are safe, it starts a PIF of READY messages along the tree [Seg 1]: a node sends a READY message to all its neighbors in the tree except its father when it receives a READY message from its father, and sends a READY message to its father when it has received READY messages from all its neighbors in the tree. When node s receives a READY message from all neighbors in the tree, it knows that all nodes in the network have received the READY message, and it broadcasts an AWAKE message on the tree. This message tells the nodes to perform the next pulse.

Lemma 3.1 *When performing β_1, a node i that receives an original-protocol message during $(t_i(n), t_i(n+1))$ can decide whether this message belongs to $M^i(n)$ or to $M^i(n+1)$.*

Proof: Let $t_i^{READY}(n)$ be the first time after $t_i(n)$ when node $i \neq s$ receives a READY message. This message was sent by the father of node i in the tree. We shall show that all original-protocol messages received by node i during $(t_i(n), t_i^{READY}(n))$ belong to $M^i(n)$ and those received during $(t_i^{READY}(n), t_i(n+1))$ belong to $M^i(n+1)$.

To prove this, assume that the original-protocol message is received by node i before $t_i^{READY}(n)$. This means that node s has not performed $t_s(n+1)$ yet. Since s is the first to perform every pulse, no message belonging to $M^i(n+1)$ had been sent yet, and thus the message belongs to $M^i(n)$. If the original-protocol message is received after $t_i^{READY}(n)$, then all nodes in the network are safe with respect to the n-th pulse. Thus, the message does not belong to $M^i(n)$.

When node s receives an original-protocol message during $(t_s(n), t_s(n+1))$, the message belongs to $M^s(n)$, because no node performs pulse $(n+1)$ before node s does. Therefore, the statement of the Lemma is ensured at node s in a trivial way, and no message-delaying mechanism is needed at node s. □

The message-delaying version of β_1 is referred to as β_1^*. In the latter, an original-protocol message received after $t_i^{READY}(n)$ and before $t_i(n+1)$ is placed in a fifo and the messages in the fifo are processed immediately after $t_i(n+1)$. The communication complexity of β_1^* is $4|V|$ and the time complexity is $2D$ — both double the complexity of β.

4 A generalization

The method described in Sec. 3 can be generalized to other synchronizers that have the problem of distinguishing between timely and early messages. The method works for synchronizers that use SAFE messages to inform nodes that all their neighbors are safe, but do not necessarily send SAFE messages to all neighbors.

Before discussing this method, we present a new definition. This definition has been used first in [AP 1].

Definition 4.1 *Node i is said to be ready for pulse$(n+1)$, if all the messages sent to i from its neighbors at pulse(n) have already arrived.*

If all the neighbors of node i are safe with respect to pulse(n), then node i is ready for pulse$(n+1)$. The idea of synchronizer β_1^* is that a node i may receive messages belonging to $M^i(n+1)$ only after receiving a READY message that informs the node that it is ready for pulse$(n+1)$.

As indicated in Sec. 3, with a synchronizer in which node i performs $t_i(n+1)$ when it is informed that all its neighbors are safe with respect to $t_i(n)$, a node cannot distinguish between timely and early messages unless SAFE messages are sent to all neighbors. However, similarly to Sec. 3, a new synchronizer can be constructed, where a node changes its state to *ready* when it realizes that all its neighbors are safe with respect to some pulse(n). The ready information is propagated in the network in exactly the same way the safe information was formerly propagated. Node i performs $t_i(n+1)$ when it realizes that it and all its neighbors are ready for pulse$(n+1)$.

This technique solves the problem, since a node that receives an original-protocol message before changing its state to *ready* knows that this message was sent at the current pulse (no node performs the next pulse before all its neighbors know that they are ready). If the message is received when the node is in *ready* state, it knows that the message was sent at the later pulse.

For example, in synchronizer β, SAFE messages are forwarded downtree to node s, and at a finite time, node s knows that all nodes in the network are safe. At that time, node s broadcasts this knowledge along the tree by sending messages that cause the nodes to perform the new pulse. In synchronizer β_1, these messages only cause the nodes to realize that they are ready. The fact that the nodes know they are ready is forwarded downtree to node s, and at the end of this stage, node s knows that all nodes in the network know that they are ready. At that time, node s broadcasts this knowledge along the tree by sending messages that cause the nodes to perform the new pulse.

The general method proposed in this section will be used to design synchronizer γ_1^* in Sec. 6.

5 Synchronizers β_2^*, β_3^*

A different method for solving the problem of distinguishing between timely and early messages has been suggested for the SOS synchronizer presented in [CCGZ 1] and for dealing with synchronizers in dynamic networks in [AS 1]. The idea is that a suffix is added to each original-protocol message, containing the number of the pulse at which the message is sent. In [FLS 1], the assumption that the number of the pulse is contained in each original-protocol message is included in the synchronous model.

When implementing this scheme with a synchronizer that has the property in lemma 2.1, a one-bit suffix is sufficient, since the messages received by node i during $(t_i(n), t_i(n+1))$ belong either to $M^i(n)$ or to $M^i(n+1)$. The implementation of this scheme for β will be referred to as β_2^*. Note that synchronizer β_2^* can be implemented only if alteration of the original-protocol messages is allowed.

Synchronizer β_3^* is another implementation, that does not require changes in the format of the original-protocol messages. Yet, the communication complexity of combined-protocols created with β_3^* may be up to twice the complexity achieved with β_1^* or β_2^*, depending on the complexity of the original protocol. The idea of β_3^* is that each node sends two messages for each original-protocol message. The first is a synchronizer message named PULSE_NUM, that contains the number of the pulse at which the message is sent. Immediately afterwards, it sends the original-protocol message. Each node maintains an array. The array at node i is $pulse\_num_i(l)$, $\forall l \in G_i$. When a PULSE_NUM message is received from neighbor k, node i sets $pulse\_num_i(k)$ as the pulse number contained in this PULSE_NUM message. When the original-protocol message is received from neighbor k during $(t_i(n), t_i(n+1))$, the message is processed immediately if $n = pulse\_num_i(k)$ and is delayed otherwise.

6 Synchronizers γ_1^*, γ_2^*, γ_3^*

In synchronizer γ [Awer 1], an initialization phase creates a partition of the network into clusters. The partition is defined by a spanning forest of the network. Each tree in the forest defines a cluster of nodes. Between each two neighboring clusters, one preferred link is selected. Inside each cluster there is a *leader node*. We say that a cluster is safe if all nodes in the cluster are safe.

SAFE messages are forwarded down along each tree in the forest. When a cluster leader knows that all nodes in the cluster are safe, it broadcasts this knowledge along the tree by using CLUSTER_SAFE message. Each node forwards this message to all its sons and along all adjacent preferred links.

The next phase determines the time at which all neighboring clusters are known to be safe. In this phase, each leaf node in the cluster sends a NEIGHBORS_SAFE message to its father after receiving a CLUSTER_SAFE message on all adjacent preferred links. Each intermediate node sends a NEIGHBORS_SAFE message to its father after receiving a NEIGHBORS_SAFE message from all its sons and CLUSTER_SAFE from all adjacent preferred links. At the end of this process, the leader node knows that all neighboring clusters are also safe.

In synchronizer γ, the leader node knows at this stage that all the nodes in the cluster are ready (all the nodes in the cluster and in the neighboring clusters are safe), so it broadcasts an AWAKE message over the cluster tree, causing all nodes in the cluster to perform the next phase. As with β, message delaying cannot be performed on γ, since a node cannot distinguish early messages from timely ones.

Synchronizer γ_1 is a version of γ created by using the method described in Sec. 4. In synchronizer γ_1, when the leader node knows that all nodes in the cluster are ready, it spreads this knowledge to all nodes in the cluster and to the neighboring clusters. This is done by broadcasting a CLUSTER_READY message along the tree. Each node forwards the message to all its sons and to all adjacent preferred links.

The next phase determines the time at which all the neighboring clusters are known to be ready. In this phase, each leaf node in the cluster sends a NEIGHBORS_READY message to its father after receiving a CLUSTER_READY message on all adjacent preferred links. Each intermediate node sends a NEIGHBORS_READY message to its father after receiving a NEIGHBORS_READY message from all its sons and a CLUSTER_READY message from all adjacent preferred links. At the end of this process, the leader node knows that all the nodes in the neighboring clusters are also ready.

Now, the leader node can perform the next pulse and also can broadcast an AWAKE message over the cluster tree, causing all nodes in the cluster to perform the next pulse.

Synchronizers γ_2^* and γ_3^* are derived from synchronizer γ in exactly the same way that β_2^* and β_3^* are derived from synchronizer β.

The same three techniques can be used for solving the problem for the hypercube-topology synchronizer presented in [PU 1].

References

[Awer 1] B. Awerbuch, *Complexity Of Network Synchronization*, Journal of the Association for Computing Machinery, Vol. 32, No. 4, October 1985, pp. 804-823.

[Awer 2] B. Awerbuch, *Reducing Complexities of the Distributed Max-Flow and Breadth-First-Search Algorithms by Means of Network Synchronization*, NETWORKS, Vol. 15 (1985) 425-437.

[FLS 1] A. Fekete, N. Lynch and L. Shrira, *A Modular Proof of Correctness for a Network Synchronizer*, 2nd International Workshop on DISTRIBUTED ALGORITHMS, Amsterdam, July 1987.

[LT 1] K.B. Lakshmanan and K. Thulasiraman, *On The Use Of Synchronizers For Asynchronous Communication Networks*, 2nd International Workshop on DISTRIBUTED ALGORITHMS, Amsterdam, July 1987.

[PU 1] D. Peleg and J.D. Ullman, *An Optimal Synchronizer for the Hypercube*, SIAM Journal of COMPUT. Vol 18, No. 4, pp. 740-747, August 1989.

[CCGZ 1] C.T. Chou, I. Cidon, I.S. Gopal and S. Zaks, *Synchronizing Asynchronous Bounded Delay Networks*, IEEE Transactions on communications, Vol. 38, No. 2, February 1990, Pages 144-147.

[AP 1] B. Awerbuch and D. Peleg, *Network Synchronization with Polilogarithmic Overhead*, Technical Report CS90-19, August 1990, The Weizmann Institute of Science, Department of Applied Mathematics and Computer Science, Israel.

[ER 1] S. Even and S. Rajsbaum, *Lack of Global Clock Does Not Slow Down the Computation in Distributed Networks*, Technical Report, October 1988, Computer Science Department, Technion — Israel Institute of Technology.

[AS 1] B. Awerbuch and M. Sisper, *Dynamic Networks are as fast as Static Networks*, Proc. 29-th IEEE symp. on Foundations of Computer Science 1988, pages 206-220.

[LTC 1] K.B. Lakshmanan and K. Thulasiraman and M.A. Comeau, *An Efficient Distributed Protocol for Finding Sortest Paths in Networks with Negative Weights*, IEEE Transactions on Software Engineering, Vol. 15, No. 5, MAY 1989.

[SM 1] B. Schieber and S. Moran, *Slowing Sequential Algorithms for Obtaining fast Distributed and Parallel Algorithms : Maximum Matchings*, Proc. 5-th ACM Symp. on Pronciples of Distributed Computing, pages 282-292, ACM August 1986.

[Seg 1] A. Segall, *Distributed Network Protocols*, IEEE Transactions on Information Theory, vol. IT-29, no. 1, January 1983.

AUTHOR INDEX

Lecture Notes in Computer Science

For information about Vols. 1–504
please contact your bookseller or Springer-Verlag